# Microtubule Protocols

# METHODS IN MOLECULAR MEDICINE™

## John M. Walker, SERIES EDITOR

METHODS IN MOLECULAR MEDICINE™

# Microtubule Protocols

Edited by

## Jun Zhou

*Nankai University, Tianjin, China*

HUMANA PRESS ✳ TOTOWA, NEW JERSEY

Production Editor: Christina Thomas
Cover design by: Nancy K. Fallat

Cover illustration: Immunofluorescence images of microtubules (green) and DNA (blue) in Cos-7 cells. The image was taken in Dr. Harish C. Joshi Laboratory, Department of Cell Biology, Emory University School of Medicine, Atlanta, GA.

Printed in the United States of America. 10 9 8 7 6 5 4 3 2 1

eISBN: 978-1-59745-442-1

Library of Congress Cataloging in Publication Data

Microtubule protocols / edited by Jun Zhou.
    p. ; cm. -- (Methods in molecular medicine ; 137)
  Includes bibliographical references and index.
  ISBN: 978-1-58829-642-9 (alk. paper)
  1. Microtubules--Research--Methodology. I. Zhou, Jun. II. Series.
  [DNLM: 1. Microtubules. 2. Drug Delivery Systems. 3.
Microtubules--drug effects. 4. Neoplasms--drug therapy. W1 ME9616JM
v.137 2007 / QU 350 M626 2007]
  QH603.M44M518 2007
  616.99'4061--dc22
                          2006027756

# Preface

Microtubules are essential components of the cytoskeleton consisting of alpha- and beta-tubulin subunits. They play critical roles in a variety of cellular processes such as cell shaping, intracellular trafficking, cell division, and cell migration. The critical roles that microtubules play make them a very suitable target for the development of chemotherapeutic drugs against human cancer. The effectiveness of microtubule-interacting drugs has been validated by the successful use of several vinca alkaloids and taxanes in cancer treatment. Our knowledge of microtubules and microtubule-interacting drugs has greatly evolved over the past decades, unarguably in part due to the advance of experimental techniques. This book contains a comprehensive collection of essential and up-to-date methods used to study the biology of microtubules and the mechanisms of action of microtubule-interacting drugs. These convenient and reproducible methods range from the purification and characterization of microtubule proteins, analysis of post-translational modifications of tubulin, and determination of microtubule structure, to the visualization of microtubule and spindle behavior, measurement of microtubule dynamics, and examination of microtubule-mediated cellular processes. This book also includes techniques to examine drug-microtubule/tubulin interactions and methods for the development and discovery of novel agents that target microtubules. The book intends to offer a collection of state-of-the-art techniques for both basic and clinician researchers who are interested in microtubules and microtubule-interacting drugs.

*Jun Zhou*

# Contents

# Contributors

LINDA A. AMOS, PhD • *MRC Laboratory of Molecular Biology, Cambridge, UK*

JOSÉ MANUEL ANDREU, PhD • *Centro de Investigaciones Biológicas, CSIC, Madrid, Spain*

SUSAN L. BANE, PhD • *Department of Chemistry, State University of New York at Binghampton, Binghamton, NY*

RUBÉN MARTÍNEZ BUEY, PhD • *Centro de Investigaciones Biológicas, CSIC, Madrid, Spain*

DANIEL W. BUSTER, PhD • *Albert Einstein College of Medicine, Bronx, NY*

MARIE-FRANCE CARLIER, PhD • *Laboratoire d'Enzymologie et Biochimie Structurales, CNRS, Gif-sur-Yvette, France*

JOHN J. CORREIA, PhD • *University of Mississippi Medical Center, Jackson, MS*

SALVATORE DEBONIS, • *Institut de Biologie Structurale, CEA-CNRS-UJF, Grenoble, France*

JOSÉ FERNANDO DÍAZ, PhD • *Centro de Investigaciones Biológicas, CSIC, Madrid, Spain*

AUDREY DORLEANS, MS • *Laboratoire d'Enzymologie et Biochimie Structurales, CNRS, Gif-sur-Yvette, France*

VINCENT GACHE, PhD • *INSERM, Unitt 366, DRDC/CS, CEA-Grenoble, Grenoble, France*

BENOÎT GIGANT, PhD • *Laboratoire d'Enzymologie et Biochimie Structurales, CNRS, Gif-sur-Yvette, France*

KEIKO HIROSE, PhD • *National Institute of Advanced Industrial Science and Technology, Tsukuba, Japan*

MARCEL KNOSSOW, DSc • *Laboratoire d'Enzymologie et Biochimie Structurales, CNRS, Gif-sur-Yvette, France*

FRANK KOZIELSKI, PhD • *Institut de Biologie Structurale, CEA-CNRS-UJF, Grenoble, France*

CHUNYONG LIU, PhD • *Nankai University, Tianjin, China*

MIN LIU, MS • *Nankai University, Tianjin, China*

SHARON LOBERT, PhD • *University of Mississippi Medical Center, Jackson, MS*

SYLVIE LUCHE, BS • *DRDC/CS, CEA-Grenoble, Grenoble, France*

THOMAS H. MACRAE, PhD • *Dalhousie University, Halifax, Canada*

CRAIG A. MANDATO, PhD • *McGill University, Montreal, Canada*

ADAM I. MARCUS, PhD • *Emory University, Atlanta, GA*

SUSAN L. MOOBERRY, PhD • *Southwest Foundation for Biomedical Research,*

PAUL A. O'CONNELL, MSc • *Dalhousie University, Halifax, Canada*
RUDOLF OLDENBOURG, PhD • *Marine Biological Laboratory, Woods Hole, MA*
ANDREI V. POPOV, PhD • *INSERM, Unit 366, INSERM, Unit 366, DRDC/CS, CEA-Grenoble, Grenoble, France*
SERGEY POPOV, PhD • *University of Illinois at Chicago, Chicago, IL*
RUDRAVAJHALA RAVINDRA, PhD • *State University of New York, Binghamton, NY*
VLADIMIR RODIONOV, PhD • *University of Connecticut Health Center, Farmington, CT*
IRINA SEMENOVA, PhD • *University of Connecticut Health Center, Farmington, CT*
DAVID J. SHARP, PhD • *Albert Einstein College of Medicine, Bronx, NY*
ANDREJ SHEVCHENKO, PhD • *Max Planck Institute of Molecular Cell Biology and Genetics, Dresden, Germany*
DIMITRIOS SKOUFIAS, PhD • *Institut de Biologie Structurale, CEA-CNRS-UJF, Grenoble, France*
ANDREY TSVETKOV, PhD • *University of Illinois at Chicago, Chicago, IL*
RICHARD H. WADE, PhD • *Institut de Biologie Structurale, Grenoble, France*
PATRICE WARIDEL, PhD • *Max Planck Institute of Molecular Cell Biology and Genetics, Dresden, Germany*
ANNA A. ZAYDMAN, BS • *State University of New York, Binghamton, NY*
AMELIA B. ZELNAK, MD • *Emory University School of Medicine, Atlanta, GA*
TONG ZHANG, BS • *McGill University, Montreal, Canada*
JUN ZHOU, PhD • *Nankai University, Tianjin, China*

# 1

## Microtubules

*An Overview*

### Richard H. Wade

#### Summary

Microtubules are found in all eukaryotes and are built from $\alpha\beta$-tubulin heterodimers. The $\alpha$-tubulins and $\beta$-tubulins are among the most highly conserved eukaryotic proteins. Other members of the tubulin family have come to light recently and, like $\gamma$-tubulin, appear to play roles in microtubule nucleation and assembly. Microtubule assembly is accompanied by hydrolysis of GTP associated with $\beta$-tubulin so that microtubules consist principally of "GDP-tubulin" stabilized by a short "GTP cap." Microtubules are polar, cylindrical structures some 25 nm in diameter. Protofilaments made from tubulin heterodimers run lengthwise along the microtubule wall with the $\beta$-tubulin subunit at the microtubule plus end. The crystallographic structures of tubulins are essential to understand in detail microtubule architecture and interactions with stabilizing and destabilizing drugs and proteins.

**Key Words:** Tubulin family; tubulin structures; microtubule assembly; microtubule structure.

## 1. Introduction

Microtubules are key actors in the cytoskeleton of eukaryotic cells where they play important roles in organizing the spatial distribution of organelles throughout interphase and of chromosomes during cell division. They can be extremely stable as in cilia and flagella or very dynamic as in the mitotic spindle. Microtubules are approx 25-nm diameter hollow tubes with walls made from tubulin heterodimers ($\alpha\beta$-tubulin) stacked head-to-tail to form "protofilaments" that run lengthwise along the wall of the tube conferring the structural and a dynamic polarity that is an essential feature of microtubules. When microtubules assemble, one end (called the plus end) grows more quickly than the other. In the cell, microtubules usually grow outwards from the microtubule-organizing center (MTOC) towards the cell membrane with the plus

From: *Methods in Molecular Medicine, Vol. 137: Microtubule Protocols*
Edited by: Jun Zhou © Humana Press Inc., Totowa, NJ

end leading. They often remain attached to the MTOC by their minus end and the resulting microtubule networks engage in intracellular transport, in cell division and, in the form of rings of doublets, in ciliary and flagellar motility. They interact with many ligands including the dynein and kinesin motor proteins for which they provide directional pathways. Molecules isolated from plants can strongly influence microtubule assembly and stability. Such drugs have important fundamental and medical applications.

## 2. Tubulin Family

### 2.1. α-Tubulin and β-Tubulin Go Together

The α-tubulin and β-tubulin monomers have similar masses of about 55 kDa as measured on sodium dodecyl sulfate-polyacrylamide gels and they both have about 450 amino acid residues. The monomers interact noncovalently to form the very stable tubulin heterodimer, which is the functional form of the protein. Tubulin, actin, and elongation factor (EF)-1α are among the most highly conserved eukaryotic proteins and their amino acid sequences have been used to construct phylogenetic trees of eukaryotic organisms that can be compared to trees obtained by analysis of small ribosomal subunit RNA *(1)*. Because of the expression of several tubulin genes and topf posttranslational modifications, several isoforms of α- and β-tubulin often coexist in eukaryotic cells. These evolutionarily conserved tubulin isotypes are distinguished by a characteristic stretch of some 15 or more amino acid residues, rich in acidic amino acids, at their COOH-terminus and also to some extent by their $NH_2$-terminal region. Different α- and β-tubulin isotypes with highly conserved sequences are often expressed independently, and each isotype tends to have similar gene expression patterns within and between species. Different organisms possess various numbers of evolutionarily conserved isotypes, some unicellular organisms have a unique αβ-tubulin, others, like the vertebrates, can have a large number of tubulin genes and many pseudogenes *(2,3)*. The degree of sequence conservation within the α- and β-tubulin subfamilies is around 60% and there is typically around 40% similarity between the two types of tubulin. There are distinct sequence differences in a few specific regions of α- and β-tubulin.

Tubulin undergoes a number of posttranslational modifications, including for example, acetylation of a specific lysine near the $NH_2$-terminus of α-tubulin by tubulin acetyltransferase, and removal of the C-terminal tyrosine residue of α-tubulin by a tubulin specific carboxypeptidase (tubulin detyrosinase) that acts preferentially on polymerized tubulin. Another enzyme, tubulin-tyrosine ligase can reattach tyrosine to the C-terminus of α-tubulins. Up to seven glutamate residues (polyglutamylation) can be added to a glutamate residue near the C-terminus of both α- and β-tubulin. Other posttranslational modifications include phosphorylation and polyglycylation.

There are at least three alternative explanations for the existence of multiple tubulin isotypes. Each has a measure of experimental support, implying that the isotypes probably have multiple biological functions *(2,4,5)*.

Explanation 1: the isotypes result from a slow sequence drift of duplicated genes and have no functional significance. In this case, isotypes should be functionally interchangeable and there are several reports appearing to support this.

Explanation 2: the existence of isotypes could improve an organism's capacity to respond to environmental challenges. For example, the temperature dependent transcription of some of the nine β-tubulin isotypes in the plant *Arabidopsis thaliana* could favor adaptation to extreme conditions.

Explanation 3: individual tubulin genes code for proteins participating in a set of distinct αβ heterodimers each with a specific microtubule function. This is known as the multitubulin hypothesis. Among several clear examples of different αβ-tubulins influencing microtubule structure and function, the nematode *Caenorhabditis elegans* is particularly striking. Somatic cells in this nematode have microtubules with 11 protofilaments but, in the six touch cells (mechanosensory neurons), the highly expressed tubulins mec-12 (α-tubulin) and mec-7 (β-tubulin) participate, as a cell specific tubulin heterodimer, in the formation of 15 protofilament microtubules. These have been shown by mutational studies to be essential for touch sensitivity *(5–7)*.

## 2.2. Other Tubulins Like Microtubule Minus Ends

It was a surprise when, in 1989, a new tubulin with only about 30% identity to α- and β-tubulins was discovered in the fungus *Aspergillus nidulans (8)*. This protein forms a distinct subfamily, γ-tubulin, that appears to exist in all eukaryotes where it is involved in microtubule nucleation and stabilization. It is located in MTOC *(3,9)*. In many cells, γ-tubulin is found associated with a number of other proteins in a 2.2-MDa tubulin ring complex (g-TuRC). A smaller tubulin complex (g-TuSC) also exists. A number of in vitro experiments have confirmed the role of γ-TuRCs in microtubule nucleation from centrosomes and spindle pole bodies. Both γ-tubulin and γ-TuRCs have been shown to bind to microtubule minus ends. Although γ-tubulin in complex with pericentrin *(10,11)* is involved in microtubule nucleation within MTOCs a detailed picture of microtubule nucleation and early growth remains a challenge.

Four other tubulins have now been identified *(3,5)* and they all appear to localize to the centriole or basal body regions or to affect the functions of these "organelles." A δ-tubulin mutant leads to defective flagella assembly from basal bodies in the green alga *Chlamydomonas*. ε-Tubulin locates to the region around the centriole in a cell cycle dependent manner but is not directly involved in microtubule nucleation. η-Tubulin may tether γ-tubulin to basal bodies and a mutation of this gene in *Paramecium* leads to basal body duplication defects.

Finally, ζ-tubulin is found in the basal body region in trypanosomes and in the centriolar region in some animal cells. Significantly, these tubulins all appear to be involved in the duplication of centrioles and basal bodies (*3*).

## 2.3. Prokaryotes Have Tubulin "Ancestors"

FtsZ (filamenting temperature sensitive strain) is a 40-kDa protein that forms filamentary structures in bacteria. It is involved in cell division in many bacteria by forming, together with other proteins, a contractile ring at the site of septation (*12*). FtsZ is ubiquitous in eubacteria, and is also found in some archeabacteria, in chloroplasts and in some mitochondria. It binds and hydrolyses GTP. It has limited sequence similarity to tubulin (typically around 33% identical/strongly similar residues in amino acid sequence alignments) with a higher degree of similarity in the approx 250 N-terminal residues. The nucleotide binding motif GGGTGTG is very similar to the tubulin signature motif [SAG]GGTG[SA]G (*13*). Typical FtsZ sequences are around 360 residues in length. The crystal structures of FtsZ and tubulin are very similar and these proteins appear to form a distinct GTPase family (*14–16*).

Two genes coding for tubulin-like proteins, BtubA and BtubB (bacterial tubulin a and b, respectively) have been reported in the bacterial genus *Prosthecobacter*. These prokaryotic tubulins have about 35% sequence identity to their eukaryotic counterparts and only around 10% identity to bacterial FtsZ (*17,18*). BtubA and BtubB have been expressed in *Escherichia coli*, allowing their assembly to be studied in vitro. When mixed together in the presence of GTP, they form, in a 1:1 ratio, long protofilament bundles, four to seven protofilaments wide. BtubA and BtubB genes, but not FtsZ, are present in four species of the *Prosthecobacter* genus. The very strong structural similarity to eukaryotic tubulins (*19*) supports horizontal gene transfer from a eukaryote.

## 3. Microtubule Assembly and Disassembly

The purification of tubulin in large quantities starting from pig or bovine brain where it makes up some 20% of the protein content was essential to allow widespread in vitro studies of many aspects of microtubule behavior and structure, for an example *see* **ref. 20**. Both α-tubulin and β-tubulin bind guanosine triphosphate but GTP is attached permanently at the N-site (nonexchangeable) in α-tubulin and only the β-subunit is active as a GTPase with exchangeable nucleotide attached to the E-site. Tubulin assembly requires GTP, magnesium ions and a temperature higher than approx 30°C. GTP hydrolysis occurs as microtubules assemble. The overall mass of assembling microtubules can be followed by measuring the solution turbidity in a spectrophotometer. Microtubules disassemble at low temperatures and cycling between soluble tubulin at 4°C and microtubules at 37°C is an important part of tubulin purification protocols.

Because GTP hydrolysis accompanies their growth, microtubules are mostly made up of "GDP-tubulin" with the growing end capped by GTP tubulin, probably limited to the endmost tubulin dimer *(21,22)*. Microtubules are also formed in the presence of the slowly hydrolyzable GTP analogue GMPCPP, showing that GTP hydrolysis is not in itself essential for microtubule growth *(23)*. "GMPCPP" microtubules are more stable than "GDP" microtubules, suggesting that GDP-tubulin is required for microtubule disassembly. Although constrained to be straight within the microtubule wall, GDP-tubulin has long been thought to have a natural curved conformation as witnessed by the ring-like structures produced when microtubules lose their GTP-cap and disassemble. Electron cryomicroscopy shows how, during disassembly, protofilaments peel off microtubule extremities to form circular oligomers some 34 nm in diameter *(24)*. These GDP-tubulin oligomers can then recycle into assembly competent tubulin dimers by exchanging GTP for GDP *(25)*.

The growth of individual microtubules can be followed using various light microscopy techniques (digital interference contrast microscopy [DIC], fluorescence microscopy, dark field microscopy). The first direct observations of in vitro microtubule assembly in 1986 revealed that individual microtubules both grow and shrink actively *(26)*. Plus ends grow about 10 times more quickly than minus ends. Microtubule growth at both ends is interrupted at random times by "catastrophic" shrinkage and then by "rescue" initiating a new phase of rapid growth. This behavior is called dynamic instability and was previously proposed by Mitchison and Kirchner to explain the behavior of microtubules during mitosis *(27)*. Dynamic instability is now accepted as an important aspect of microtubule behavior both in interphase and in the establishment of interactions between the microtubule spindle and chromosomes during mitosis.

In interphase cells, microtubules are usually nucleated in the region surrounding the centrosome (MTOC) that is rich in ring-like structures containing γ-tubulin. During cell division, microtubules nucleate at the two spindle poles, where γ-tubulin complexes are also located, then grow and shrink dynamically. They ultimately attach to kinetochores on chromatid pairs and align chromosomes to the midplane at metaphase. Interestingly, kinetochores bind preferentially to GTP- rather than GDP-microtubules in vitro, and attach to the plus-end in preference to other positions on the sides or at the minus end of microtubules *(28)*.

Microtubule networks, and the polar distribution of microtubules in the mitotic spindle, provide pathways for dynein and kinesin allowing these ATP dependent microtubule motors to participate in cell division and to engage in intracellular transport and organization. Both motor proteins can detect the underlying structural polarity of microtubules and move in specific directions with unerring accuracy *(29)*.

## 4. Microtubule Organization and Structure

The quest for information on microtubule structure has extended from the 1960s until the present. Initial successes were obtained by electron microscopy of (1) thin sections of glutaraldehyde/tannic acid fixed cells from various organisms and, (2) of negatively stained axonemal microtubules. These observations helped to establish the complex organization of cilia and flagella based on nine microtubule doublets encircling a central pair of microtubules. As thin section fixation and staining methods improved it became possible to directly count thirteen "protofilaments" in cross sections of the cylindrical, 25-nm diameter, microtubule wall. Images of individual microtubules obtained using negative stain showed the protofilaments are aligned lengthwise along the microtubule and associate laterally with an approx 0.9 nm shift from one protofilament to the next. As a result, tubulin subunits in neighboring protofilaments describe a 12-nm pitch, left-handed, helical pathway around the microtubule. Subsequently, microtubules with different numbers of protofilaments have been found in a variety of organisms and specialized cells. Genetic methods have shown β-tubulin is particularly important for determining microtubule architecture (*6,7,30*).

Microtubules assembled from pure tubulin in vitro have a wide range of protofilament numbers, typically in the range 10 to 16 (*31,32*). The moiré pattern contrast of well preserved microtubules observed in their fully hydrated state by electron cryomicroscopy shows that the microtubule architecture allows this wide range of protofilament numbers to be readily accommodated by a global rotation of the surface lattice. Protofilaments run parallel to the microtubule axis only for 13 protofilament microtubules, in all other cases the protofilaments describe long-pitch helices.

**Figure 1** represents a three-dimensional reconstruction of microtubules at about 14 Å resolution obtained by electron cryomicroscopy and computerized three-dimensional reconstruction methods (*32*). The outer surface shows parallel, compact protofilaments separated by deep grooves. The approx 40 Å (4 nm) periodicity visible along each protofilament corresponds to the spacing between the α- and β-tubulin subunits that are indistinguishable at the resolution of the reconstruction. The main contacts between protofilaments are at the bottom of the grooves that are pierced by holes about 15 Å by 20 Å in size giving direct access to the microtubule lumen. The outermost surface of the protofilaments appears quite smooth and the inner surface significantly rougher.

Much effort has been made to relate the tubulin dimer orientation to microtubule polarity and it is now established that β-tubulin caps the microtubule plus end and α-tubulin the minus end (*33–36*). Growth at the microtubule plus end is achieved by the interaction of the α-subunit of an incoming tubulin dimer

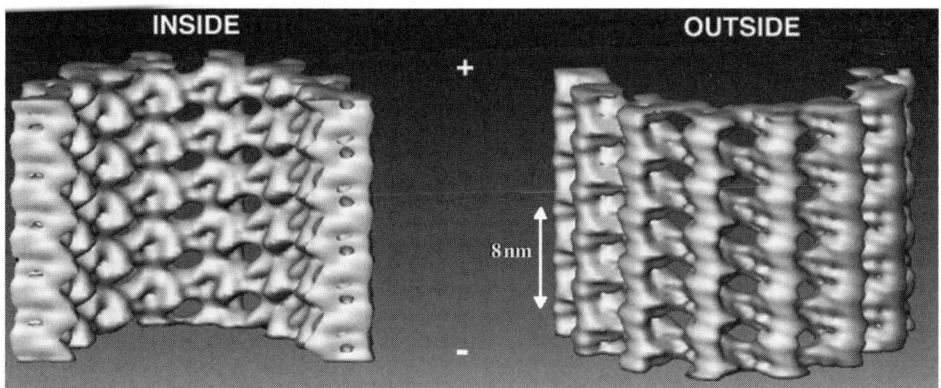

Fig. 1. Microtubule structure obtained by electron cryomicroscopy and three-dimensional image reconstruction *(32)*. The surface representation is been cut lengthwise to show the both the inner and outer surfaces. Protofilaments run vertically, with the microtubule plus end at the top. Contacts between the protofilaments are seen to establish across the deep grooves between the protofilaments. The inner microtubule surface is significantly "rougher" than the outside.

with an *in situ* β-subunit thereby activating hydrolysis of GTP at the E-site in β-tubulin. The α-tubulin capped minus end will behave differently because GTP in α-tubulin is buried within the αβ interface and is neither exchangeable nor hydrolyzable. The organization of the microtubule surface lattice is directly visible in three-dimensional reconstructions using images obtained by electron cryomicroscopy of microtubules "decorated" with kinesin motor domains. Identical tubulin subunits are in contact from protofilament to protofilament around the microtubule wall. This organization is known as the B lattice *(37,38)*. Importantly, this implies that standard 13 protofilament microtubules have a lattice mismatch, or "seam," between a single pair of protofilaments (**Fig. 2**) *(38,39)*. This is a direct consequence of the microtubule assembly mechanism involving endwise addition of tubulin dimers onto a sheet-like lattice at the microtubule extremity. This sheet rolls up into a cylinder as the microtubule grows *(40)*. Consequently, microtubule growth is not a helical process and microtubules are cylindrical, rather than helical, structures. The seam may turn out to be important for microtubule disassembly by providing a "weak link" facilitating the peeling apart of protofilaments.

## 5. Tubulin at Atomic Resolution

An atomic resolution structural model of the tubulin heterodimer was obtained in 1998 by electron crystallography of taxol stabilized tubulin sheets obtained in

**A  Front**                              **B  Back**

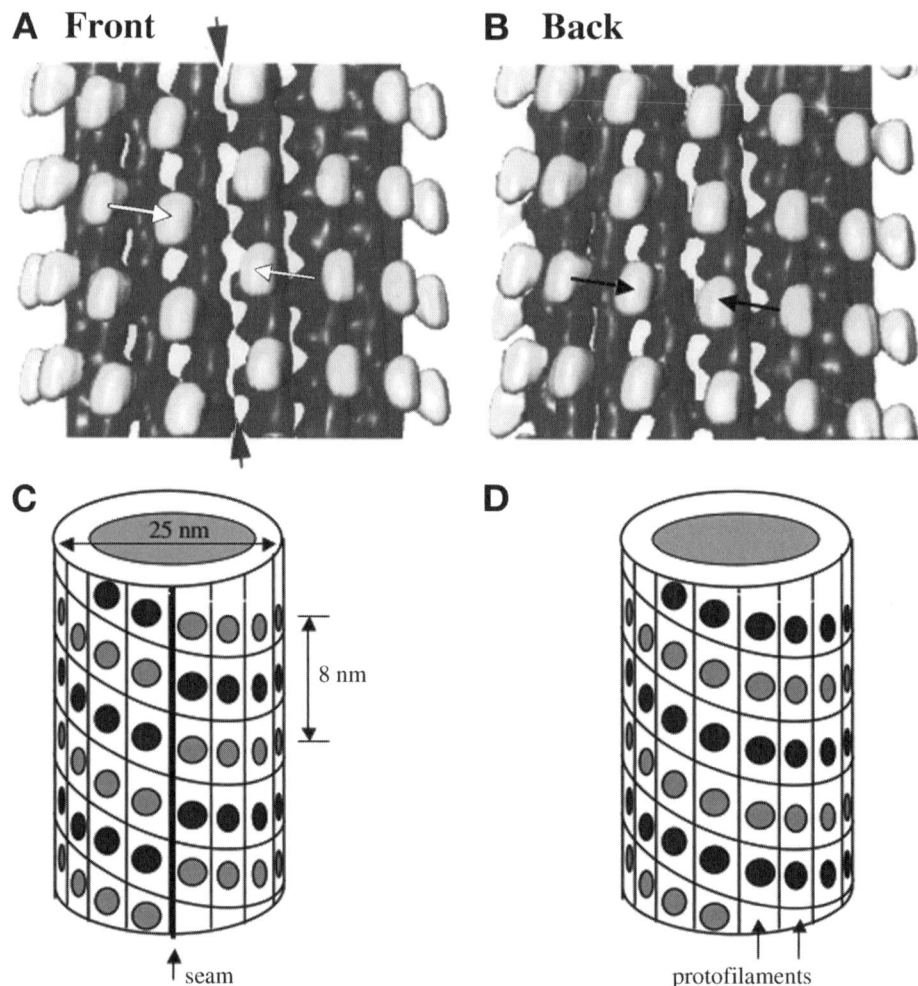

Fig. 2. Microtubule organization. The surface lattice organization can be visualized in three-dimension image reconstructions of microtubules specifically decorated with monomeric kinesin motor domains. Using images obtained by electron cryomicroscopy and tomographic reconstruction methods, similar tubulin subunits are highlighted by kinesin monomers and follow a helical path around the microtubule. This is interrupted by a seam (arrows), compare the views of both sides of the reconstruction (**A,B**). (**C,D**) Drawings of front and back views of a 13 protofilament microtubule, α-subunits black and β-subunits gray. Protofilaments run vertically as stacks of tubulin heterodimers. Similar subunits follow a helical path around the cylindrical microtubule wall and after a full turn the subunits switch from one type to the other across the so-called "seam."

the presence of zinc ions *(41)*. The protofilaments in these sheets have the same longitudinal organization as in microtubules, giving a putative model of the inter- and intradimer contacts along the protofilaments. However, lateral contacts between protofilaments are different in the two lattices because neighboring protofilaments are aligned antiparallel in "zinc" sheets and parallel in microtubules.

The crystallographic structures of the $\alpha$- and $\beta$-subunits are practically identical (**Fig. 3A**). Both have a compact structure with three functional domains. The N-terminal domain forms a Rossmann fold with six $\alpha$-helices and a six-stranded parallel $\beta$-sheet with seven loops involved in nucleotide binding. The $\alpha$-helix H7 connects the N-terminal and the intermediate domain that runs from residues 206–381 (**Fig. 3B**). This domain consists of a four-stranded, mainly parallel, $\beta$-sheet flanked by two $\alpha$-helices on one side and one on the other. In $\beta$-tubulin this domain includes the taxol-binding pocket. The C-terminal domain has two long, almost parallel, $\alpha$-helices (H11 and H12) running along the full length of each subunit. The C-terminal residues (10 in $\alpha$-tubulin and 18 in $\beta$-tubulin) are disordered and are not visible in the structures.

The $\alpha$-tubulin subunit has GTP in the nucleotide binding site (N-site) whereas the E-site in $\beta$-tubulin contains GDP. Taxol is also visible in the structure of $\beta$-tubulin. Significantly, the nucleotide is positioned at the interface between monomers along the protofilaments. The $\alpha$-tubulin N-site is hidden within the intra dimer interface both in the free heterodimer and in assembled microtubules. The $\beta$-tubulin GTP binding site is partially exposed in the heterodimer explaining how nucleotide exchange is possible in solution but impossible when this site is buried within microtubule protofilaments.

Recently the crystallographic structure of tubulin complexed with the $\alpha$-helical stathmin-like domain (SLD) of the neural protein RB3 has been determined by X-ray diffraction *(42)*. In this structure, the helical segment of RB3-SLD binds two $\alpha\beta$-tubulin heterodimers in a curved configuration and the $NH_2$-terminal region that is conserved in the stathmin family forms a $\beta$-hairpin capping the endmost $\alpha$-tubulin subunit. Interestingly, the tubulin structures have a distinct "curved" conformation compared to the "straight" conformation of the zinc sheet structures. Notably, the intermediate domains are rotated by about $10°$ in the curved conformation and there are local differences in loops involved in contacts along and between protofilaments.

Crystals of the tubulin-RB3-SLD complex are proving to be a very powerful tool for examining the interaction of tubulin with microtubule destabilizing drugs (**Fig. 4**). The colchicine binding site is found to be at the interface between the $\alpha$- and $\beta$- subunits of the same tubulin heterodimer *(43)*. Colchicine is mostly buried in the $\beta$-tubulin intermediate domain between $\beta$-strands S8, S9, loop T7

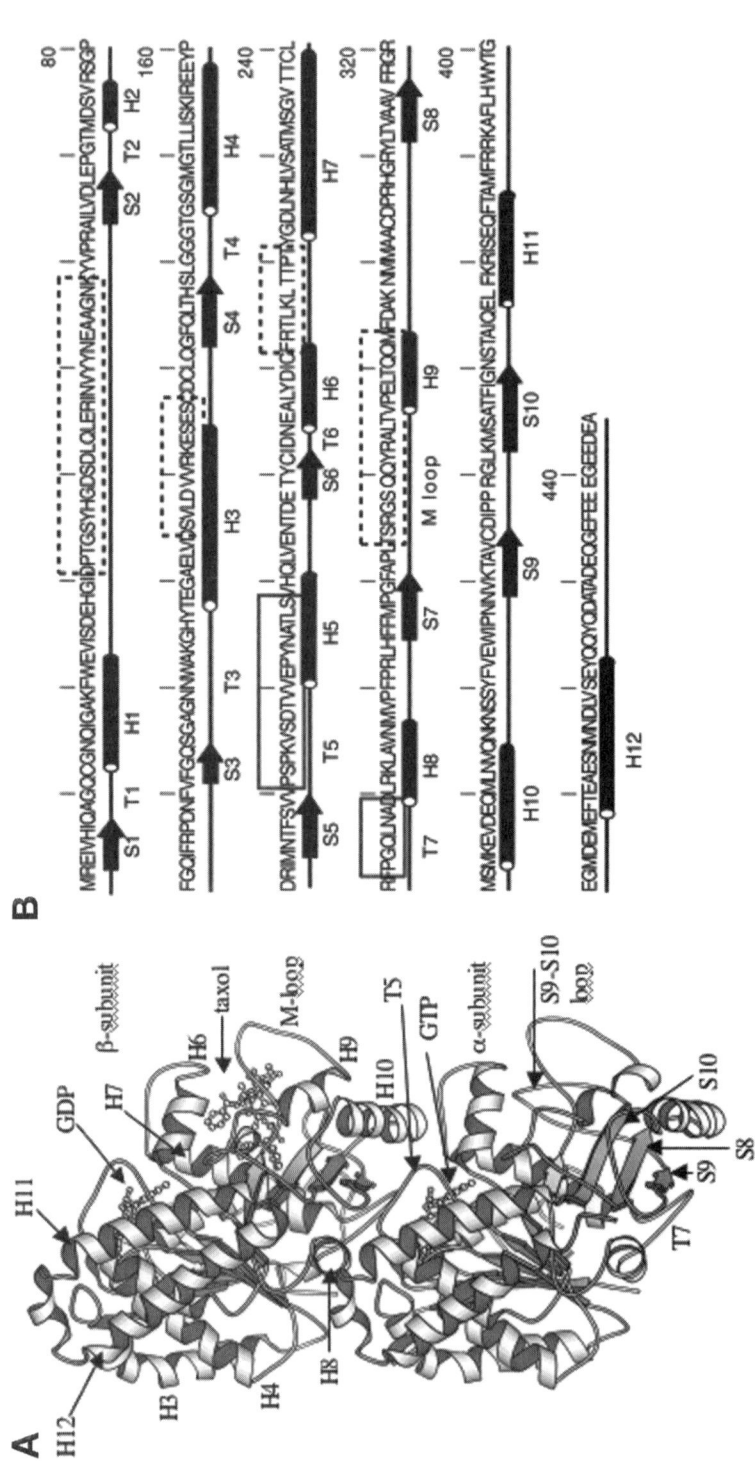

Fig. 3. (A) Structure of the tubulin heterodimer obtained by electron crystallography (*41*). Key structural features are indicated, β-strands S, α-helices H, and turns T. Note that GTP binds near the top of the subunits where it is exposed in the β-subunit whereas the α-tubulin nucleotide is enclosed between the two subunits. There is a taxol binding pocket in β-tubulin whereas this region in α-tubulin is occupied by longer loop between S9 and S10 (B) Sequence of pig brain β-tubulin with corresponding structural features. The boxed regions are important for lateral (dashed lines) and longitudinal contacts (full lines), respectively.

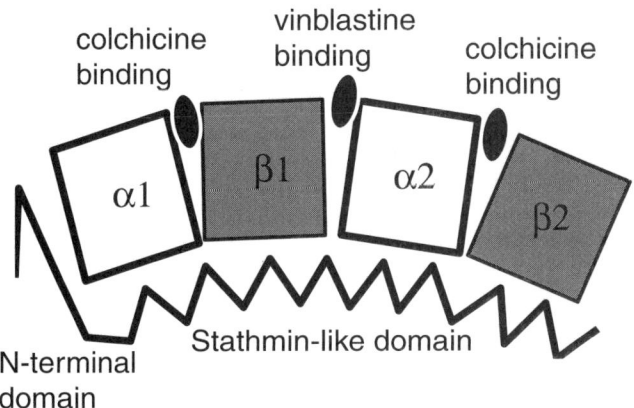

Fig. 4. Schematic representation of the curved configuration of two tubulin dimers, α1β1 and α2β2, interacting with a stathmin-like helix based on the structure of a tubulin–RB3-SDL complex determined by X-ray crystallography *(42)*. The curved tubulin conformation may prevent lateral contacts between protofilaments thus explaining tubulin sequestration by the stathmin-related proteins. Colchicine has been shown to bind between subunits of the same dimer whereas vinblastine binds between dimers *(42–44)*.

and α-helices H7, H8. It interacts with loop T5 of the α-tubulin subunit and locks-in the curved tubulin conformation. Podophyllotoxin was shown to bind to the same site. On the other hand, the vinca alkaloid, vinblastine has been known for some time to bind to a different site and this has now been localized in the tubulin-colchicine–RB3-SLD complex by soaking crystals in vinblastine solutions before flash freezing *(44)*. The binding site lies between two tubulin heterodimers and includes tubulin residues located toward the inner lumen of microtubules and involved in longitudinal contacts along microtubules. In solution, vinblastine is known to induce the formation of spiral-like tubulin aggregates. This is also the case for a tubulin–RB3-SLD complex in which 24 amino acid residues have been removed from the $NH_2$-terminus of the RB3-SLD.

The crystal structure of a human γ-tubulin GTP- γS complex was obtained recently at 2.7 Å resolution *(45)*. This is the first structure of a monomeric tubulin and has the highest resolution obtained to date. Although γ-tubulin is in the GTP state (GTP-γS is a nonhydrolysable GTP analogue) it has a curved configuration similar to that observed in the αβ-tubulin–RB3-SLD complex. One of the crystal packing interactions resembles the lateral interactions between protofilaments in microtubules showing that monomeric γ-tubulin in the curved configuration might self-associate by forming stable lateral interactions. This may allow the γ-tubulin template in the γ-TuRC to nucleate microtubules by interacting longitudinally with αβ-tubulin whilst enhancing lateral interactions between the tubulin heterodimers.

## 6. Microtubules at Pseudo-Atomic Resolution: Interactions With Drugs and Proteins

Although a microtubule structure has now been solved at 8 Å resolution *(46)*, the best way to obtain a microtubule model at pseudo-atomic resolution is to dock the tubulin dimer structure obtained by crystallography into a three-dimensional microtubule reconstruction obtained by electron cryomicroscopy *(32,47)*. The most important feature revealed by such models concerns the lateral interactions between tubulin subunits in adjacent protofilaments (**Fig. 5**). These interactions take place mainly in the regions at the bottom of the grooves between protofilaments where the three-dimensional image reconstructions show significant density. The M-loop situated between β-strand S7 and α-helix H9 in one tubulin subunit interacts with H3 and loops H1-S2, H2-S3 in the tubulin subunit in the neighboring protofilament. In β-tubulin, this loop is influenced by the antimitotic drug, taxol, that binds in a nearby pocket, occupied in α-tubulin by an eight-residue insert in the loop between S9 and S10 *(41,47)*. The stabilization of microtubules by taxol blocks cell division leading to the use of this drug for chemotherapy of certain forms of cancer *(48)*.

Microtubules assembled in the presence of taxol have fewer protofilaments than a drug-free control assembly, suggesting that taxol influences the M loop configuration leading to both microtubule stability and a reduced number of protofilaments. Note that taxol resistant human ovarian cancer cells have β-tubulins with mutations near the beginning of the M-loop and in the S9-S10 loop *(49)*.

Other microtubule stabilizing agents including the epothilones, sarcodictyin A, eleutherobin also influence microtubule architecture. Sarcodictyin A, epothilone B, and eleutherobin all modify the average numbers of protofilaments in vitro assembled microtubules so apparently, like taxol, they also effect lateral interactions between tubulin subunits in adjacent protofilaments. Taxotere increases the average protofilament number, indicating that this drug must have different interactions within the taxol-binding region of β tubulin. Interestingly, GMPCPP, a nonhydrolysable GTP analogue, is unique in producing microtubules with a well-defined number of protofilaments. Typically 96% of the microtubule population has 14 protofilaments. The implication that the nucleotide influences lateral interactions between protofilaments is, at first site, surprising because the exchange-

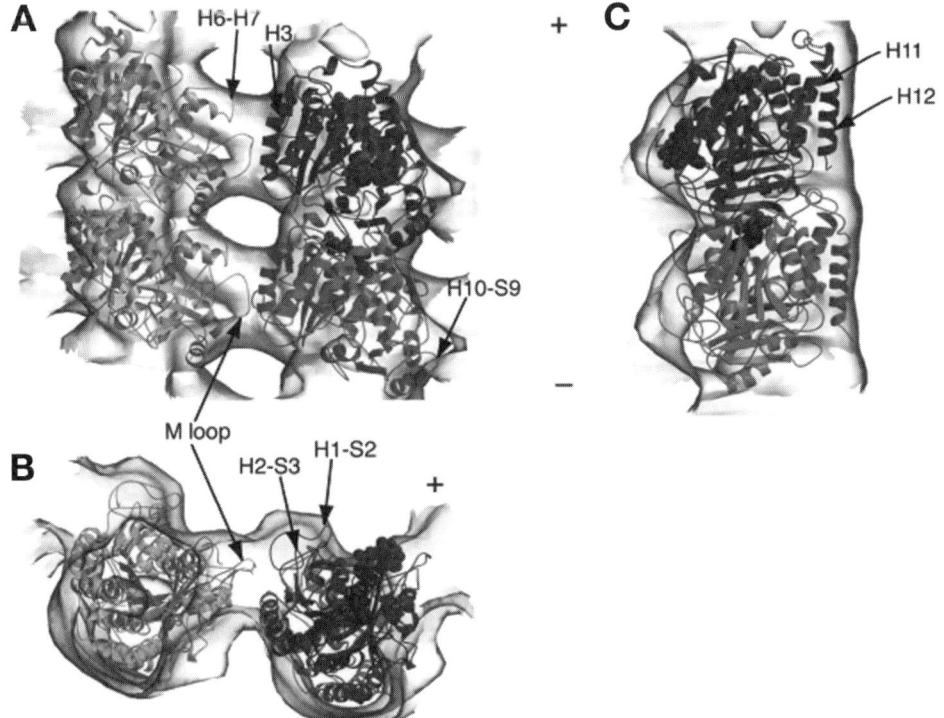

Fig. 5. The tubulin crystal structure docked into the microtubule map, gray isodensity surface. The α- and β-tubulin subunits are shown, respectively, as lighter and darker gray ribbons. (**A**) Part of two protofilament neighbors viewed from the outer surface of the microtubule, with the microtubule plus end at the top. (**B**) Protofilaments seen end-on, looking from the microtubule plus end. (**C**) Side view of a protofilament as seen across the thickness of the microtubule wall. Helices H11, H12 run along the smooth outer surface on the right.

able site on the β subunit lies at the longitudinal interface between tubulin dimers along the protofilaments. However, several regions directly involved with nucleotide binding could possibly effect the position of helices or loops involved in lateral interactions *(32,41,47)*.

In addition to the data obtained by crystallography many other results support the idea that GTP-tubulin is straight and GDP-tubulin is bent as in the curved protofilaments that peel off the plus end of microtubules as microtubules disassemble *(24,25)*. Interestingly, the existence of two tubulin conformations in microtubules was detected in an investigation showing the tubulin dimer spacing along microtubule protofilaments is slightly greater for tubulin-

GMPCPP (GTP configuration) than for tubulin-GDP *(50)*. Because microtubules normally contain GDP-tubulin in the curved configuration it is the lateral interactions within the microtubule wall that hold GDP-tubulin in place until released during disassembly. The curved conformation observed for tubulin dimers complexed with the $\alpha$-helical RB3 stathmin-like domain can be fitted to protofilaments at the microtubule plus end by computer modelling *(32,42)*. The helical stathmin-like domain is found to be offset to one side of a protofilament provoking a curvature that prevents the lateral contacts between the M loop and the neighboring protofilament perhaps accounting for tubulin sequestration by proteins in the stathmin family.

## References

1. Baldauf, S. L., Roger, A. J., Wenk-Siefert, I., and Doolittle, W. F. (2000) A kingdom level phylogeny of eukaryotes based on combined protein data. *Science* **290,** 972–977.
2. Luduena, R. F. (1998) Multiple forms of tubulin: different gene products and covalent modifications. *Int. Rev. Cytol.* **178,** 207–275.
3. Dutcher, S. K. (2003) Long-lost relatives reappear: identification of new members of the tubulin superfamily. *Curr. Opin. Microbiol.* **6,** 634–640.
4. Wilson, P. G. and Borisy, G. G. (1997) Evolution of the multi-tubulin hypothesis. *BioEssays* **19,** 451–454.
5. McKean, P. G., Vaughan, S., and Gull, K. (2001) The extended tubulin superfamily. *J. Cell Sci.* **114,** 2723–2733.
6. Savage, C. and Chalfie, M. (1991) Genetic aspects of microtubule biology in the nematode *Caenorhabditis elegans. Cell Motil. Cytoskel.* **18,** 159–163.
7. Fukushiga, T., Siddiqui, Z. K., Chou, M., et al. (1999) *Mec-12,* an $\alpha$-tubulin required for touch sensitivity in *C. elegans. J. Cell Sci.* **112,** 395–403.
8. Oakely, C. E. and Oakely, B. R. (1989) Identification of $\gamma$-tubulin, a new member of the tubulin superfamily encoded by mipA gene of *Aspergillus nidulans. Nature* **338,** 662–664.
9. Oakely, B. R. (1992) $\gamma$-tubulin: the microtubule organiser ? *Trends Cell Biol.* **2,** 1–5.
10. Dictenberg, J. B., Zimmerman, W., Sparks, C. A., et al. (1998) Pericentrin and $\gamma$-tubulin form a protein complex and are organised into a novel lattice at the centrosome. *J. Cell Biol.* **141,** 163–174.
11. Zimmerman, W. C., Sillibourne, J., Rosa, J., and Doxsey, S. J. (2004) Mitosis-specific anchoring of gamma tubulin complexes by pericentrin controls spindle organization and mitotic entry. *Mol. Biol. Cell.* **15,** 3642–3657.
12. Goehring, N. W. and Beckwith, J. (2005) Diverse paths to midcell: assembly of the bacterial cell division machinery. *Curr. Biol.* **15,** R514–R526.
13. Erickson, H. P. (1997) FtsZ, a tubulin homologue in prokaryote cell division. *Trends Cell Biol.* **7,** 362–370.

14. Löwe, J. and Amos, L. A. (1998) Crystal structure of the bacterial cell division protein FtsZ. *Nature* **391,** 203–206.
15. Nogales, E., Downing, K. H., Amos, L. A., and Löwe, J. (1998) Tubulin and FtsZ form a distinct family of GTPases. *Nature Struct. Biol.* **5,** 451–458.
16. Oliva, M. A., Cordell, S. C., and Lowe, J. (2004) Structural insights into FtsZ protofilament formation. *Nat. Struct. Mol. Biol.* **11,** 1243–1250.
17. Jenkins, C., Samudrala, R., Anderson, I., et al. (2002) Genes for the cytoskeletal protein tubulin in the bacterial genus *Prosthecobacter*. *Proc. Natl. Acad. Sci. USA* **99,** 17,049–17,054.
18. Sontag, C. A., Staley, J. T., and Erickson, H. P. (2005) In vitro assembly and GTP hydrolysis by bacterial tubulins BtubA and BtubB. *J. Cell Biol.* **169,** 233–238.
19. Schlieper, D., Oliva, M. A., Andreu, J. M., and Löwe, J. (2005) Structure of bacterial tubulin BtubA/B: evidence for horizontal gene transfer. *Proc. Natl. Acad. Sci. USA* **102,** 9170–9175.
20. Asnes, C. F. and Wilson, L. (1979) Isolation of bovine brain microtubule proteins without glycerol: polymerisation kinetics change during purification cycles. *Anal. Biochem.* **98,** 64–73.
21. Carlier M. F. (1991) Nucleotide hydrolysis in cytoskeletal assembly. *Curr. Opin. Cell Biol.* **3,** 12–17.
22. Caplow, M. (1992) Microtubule dynamics. *Curr. Opin. Cell Biol.* **4,** 58–65.
23. Hyman, A. A., Salser, S., Drechsel, D. N., Unwin, N., and Mitchison, T. J. (1992) Role of GTP hydrolysis in microtubule dynamics: information from a slowly hydrolyzable analogue, GMPCPP. *Mol. Biol. Cell* **3,** 1155–1167.
24. Mandelkow, E. M., Mandelkow, E, and Milligan, R. A. (1991) Microtubule dynamics and microtubule caps: a time-resolved cryo-electron microscopy study. *J. Cell Biol.* **114,** 977–991.
25. Melki, R., Carlier. M. -F., Pantaloni, D., and Timasheff, S. N. (1989) Cold depolymerization of microtubules to double rings: geometric stabilization of assemblies. *Biochem.* **28,** 9143–9152.
26. Horio, T. and Hotani, H. (1986) Visualization of the dynamic instabilty of individual microtubules by dark-field microscopy. *Nature* **321,** 605-607.
27. Mitchison, T. and Kirschner, M. (1984) Dynamic instability of microtubule growth. *Nature* **312,** 237–242.
28. Severin, F. F., Sorger, P. K., and Hyman, A. A. (1997) Kinetochores distinguish GTP from GDP forms of the microtubule lattice. *Nature* **388,** 888–891.
29. Bloom, G. S. and Endow, S. A. (1995) Motor proteins 1: kinesins. *Protein Profile* **12,** 1105–1171.
30. Raff, E. C., Fackenthal, J. D., Hutchens, J. A., Hoyle, H. D., and Turner, F. R. (1997) Microtubule architecture specified by a β-tubulin isoform. *Science* **275,** 70–73.
31. Chrétien, D. and Wade, R. H. (1991) New data on the microtubule surface lattice. *Biol. Cell* **71,** 161–174
32. Meurer-Grob, P., Kasparian, J., and Wade, R. H. (2001) Microtubule structure at improved resolution. *Biochem.* **40,** 8000–8008.

33. Mitchison, T. J. (1993) Localisation of an exchangeable GTP binding site at the plus end of microtubules. *Science* **261,** 1044–1047.
34. Hirose, K., Fan, J., and Amos, L. A. (1995) Re-examination of the polarity of microtubules and sheets decorated with kinesin motor domain. *J. Mol. Biol.* **251,** 329–333.
35. Fan, J., Griffith, A. D., Lockhart, A., and Cross, R. A. (1996) Microtubule minus ends can be labelled with a phage display antibody specific to alpha-tubulin. *J. Mol. Biol.* **259,** 325–330.
36. Wade, R. H. and Hyman, A. A. (1997) Microtubule structure and dynamics. *Curr. Opin. Cell Biol.* **9,** 12–17.
37. Amos, L. A. and Klug, A. (1974) Arrangement of subunits in flagellar microtubules. *J. Cell Sci.* **14,** 523–549.
38. Song, Y. -H. and Mandelkow, E. (1993) Recombinant kinesin motor domain binds to beta-tubulin and decorates microtubules with a B surface lattice. *Proc. Natl. Acad. Sci. USA* **90,** 1671–1675.
39. Metoz, F., Arnal, I., and Wade R. H. (1997) Tomography without tilt: three-dimensional imaging of microtubule-motor complexes. *J. Struct. Biol.* **118,** 159–168.
40. Chrétien, D., Fuller, S. D., and Karsenti, E. (1995) Structure of growing microtubule ends: two-dimensional sheets close into tubes at variable rates. *J. Cell Biol.* **129,** 1311–1328.
41. Nogales, E., Wolf, S. G., Downing, K. A. (1998) Structure of the $\alpha\beta$ tubulin dimer by electron crystallography. *Nature* **391,**199–203.
42. Gigant, B., Curmi, P. A., Martin-Barbey, C., et al. (2000) The 4 Å X-ray structure of a tubulin:stathmin-like domain complex, *Cell* **102,** 809–816.
43. Ravelli, R. B. G., Gigant, B., Curmi, P. A., et al. (2004) Insight into tubulin regulation from a complex with colchicine and a stathmin-like domain. *Nature* **428,** 198–202.
44. Gigant, B., Wang, C., Ravelli, R. B. G., et al. (2005) Structural basis for the regulation of tubulin by vinblastine. *Nature* **435,** 519–522.
45. Eldaz, H., Rice, L. M., Stearns, T., and Agard, D. A. (2005) Insights into microtubule nucleation from the crystal structure of human γ-tubulin. *Nature* **435,** 523–527.
46. Li, H., DeRosier, D. J., Nicholson, W. V., Nogales, E., and Downing, K. H. (2002) Microtubule structure at 8 Å resolution. *Structure* **10,** 13,417–13,428.
47. Nogales, E., Whittaker, M., Milligan, R. A., and Downing, K. H. (1999) High resolution model of the microtubule. *Cell* **96,** 79–88.
48. Wood, K. W., Cornwell, W. D., and Jackson, J. R. (2001) Past and future of the mitotic spindle as an oncology target. *Curr. Opin. Pharmaco.* **1,** 370–377.
49. Giannakakou, P., Sackett, D. L., Kang, Y. -K., et al. (1997) Paclitaxel-resistant human ovarian cancer cells have mutant β-tubulins that exhibit impaired paclitaxel-driven polymerisation, *J. Biol. Chem.* **272,** 17,118–17,125.
50. Hyman, A. A., Chrétien, D., Arnal, I., and Wade R. H. (1995) Structural changes accompagnying GTP hydrolysis in microtubules: information from a slowly hydrolyzable analogue guanylyl-($\alpha,\beta$)-methelyne-diphosphonate. *J. Cell Biol.* **128,** 117–125.

# 2

## Large Scale Purification of Brain Tubulin With the Modified Weisenberg Procedure

### José Manuel Andreu

#### Summary

This method is a modification of the initial procedure employed to purify tubulin from mammalian brain. It consists of tissue homogenization, elimination of cell membranes, ammonium sulfate fractionation, and batch anion exchange, followed by selective precipitation with magnesium chloride. Half gram of electrophoretically homogenous, active, concentrated calf brain tubulin is typically purified in 9 h, dialyzed overnight, and stored under liquid nitrogen. Prior to use the protein is equilibrated in the experimental buffer and its concentration measured. This tubulin preparation has been very extensively characterized. Frozen aliquots have been found to retain microtubule assembly activity after 10 yr of storage.

**Key Words:** Tubulin; W-tubulin; purification; assembly; microtubules; brain.

## 1. Introduction

Tubulin was extracted from sperm tail (*1*), and it was first purified from mammalian brain employing biochemical procedures by Weisenberg, Borisy, and Taylor (*2*) as the colchicine-binding protein proposed to be the subunit of microtubules, found in most eukaryotic cells. Tubulin constituting microtubules is a αβ-heterodimer of homologous GTP-binding subunits; the GTP bound to β-tubulin is hydrolyzed upon microtubule assembly and GDP-tubulin tends to disassemble, which confers their dynamic properties to microtubules. The single sites of colchicine and taxol binding are at β-tubulin (for a review *see* **ref. 3**).

Conditions for the in vitro assembly of microtubules were also first discovered by Weisenberg (*4*), and subsequently applied for the preparation of microtubule proteins (tubulin, microtubule associated and motor proteins) by cycles of assembly and disassembly (*5,6*), from which tubulin can be purified from microtubule bound proteins by phosphocellulose ion-exchange chromatogra-

From: *Methods in Molecular Medicine, Vol. 137: Microtubule Protocols*
Edited by: Jun Zhou © Humana Press Inc., Totowa, NJ

phy *(7,8)* (known as PC-tubulin) and cycled again to remove the inactive protein fraction *(9)*. Numerous modified assembly cycle procedures have appeared in the literature over the years, including the use of high concentrations of organic acids, such as glutamate, MES, and PIPES buffers to prepare MAP-depleted tubulin *(10–12)*.

The original Weisenberg method of tubulin purification *(2)* was improved by Weisenberg and Timasheff *(13)*, and subsequently modified to give a very well-characterized preparation of stabilized tubulin *(8,14–17)* employing the so called modified Weisenberg procedure (W-tubulin). This preparation was employed to first demonstrate microtubule assembly from purified tubulin without the need of microtubule-associated proteins *(18)*. Both W-tubulin and PC-tubulin assemble under appropriate conditions.

Although we use the cycle procedure *(6)* with modifications *(19)* to prepare microtubule proteins, we prefer the modified Weisenberg procedure to directly purify active tubulin (instead of PC-tubulin or commercial sources) owing to the relatively large quantities of highly purified W-tubulin, which can be prepared with relatively small manpower (two day, including preparation of materials) and equipment. This method *(16,17)* is currently in use in the author's laboratory for the purification of bovine brain tubulin *(20,21)*, from which microtubules are prepared for the investigation of microtubule stabilizing drugs binding at the Taxol and other binding sites *(22)*, employing fluorescent competition methods (*see* Chapter 17), which are susceptible of high-throughput application. Bovine W-tubulin has been subjected to in-depth biochemical and biophysical studies. Very similar tubulin has been successfully prepared in other laboratories from pig, dog, lamb *(8,23)*, or chicken brain *(24)* employing the modified Weisenberg procedure.

Brain remains the main source to prepare large quantities of tubulin for microtubule research, although tubulin has been purified from a variety of other sources for specific purposes, employing diverse methods. Yeast tubulin can be purified in milligram quantities *(25)*. Mammalian tubulin has been expressed as inclusion bodies in bacteria, folded in tens of micrograms quantities and assembled into microtubules *(26)*. Recently, a bacterial tubulin has been discovered; its structure is very similar to eukaryotic tubulin, however, it has not assembled into microtubules *(27)*.

## 2. Materials

### 2.1. Centrifuges and Reservations

1. Book four preparative centrifuges, four medium size rotors (1.5-L capacity, six-hole). This is reduced to two centrifuges and rotors after the first 2–3 h (a full preparation can also be started with two centrifuges and rotors and, if necessary

half a preparation can be made with one centrifuge). Prepare 24 centrifuge bottles and a few spares (large mouth, complete with screw caps and sealing inner caps with their O-rings) (*see* **Note 1**). Prepare two smaller rotors with 16 tubes (*see* **Note 2**). Carefully inspect the centrifuge bottles, keep separated for tubulin preparation, and do not use any bottles with cracks or other signs of aging; do not oven-dry after washing, because this may damage polycarbonate. Clean rotors and centrifuges after use.

2. Phone the appointed abattoir for the approximate time of slaughter; be there slightly in advance.
3. Make bench space (about 2 m) in the cold room and cover with disposable absorbent paper.
4. Reserve space in a liquid nitrogen tank for about 30 cryogenic vials (1-mL size; not necessarily sterile for tubulin storage).

## 2.2. Chromatography Gels and Dialysis Membrane

These and the following materials are prepared the day before purification.

1. Swell 15 g DEAE Sephadex A50 in 1.5 L of 10 m$M$ sodium phosphate buffer pH 7.0, using a 2-L beaker or an open bottle, in a boiling water bath during 5 h. Store in the fridge. Equilibrate and wash the day of the preparation (*see* **Subheading 3.**). Discard after use and wash the containers separate from other materials.
2. Take 450 mL of previously swollen, gravity sedimented Sephadex G-25 medium in water, pack a 2.5 × 80-cm column, wash with 1 L of 10 m$M$ sodium phosphate buffer pH 7.0 and store in the cold room overnight. Equilibrate with PMG buffer (*see* **Subheading 2.4.**, **item 3**) the day of the preparation. After use, empty the column and wash the G-25 gel three times with distilled water by sedimentation and aspiration; add a bit of sodium azide and store at 4°C.
3. Preboiled and washed dialysis membrane (6-mm diameter) is stored in 10% ethanol in the cold. Rinse well before use and discard afterward.

## 2.3. Solid Chemicals

1. 800 g very pure ammonium sulfate; 125 g biochemical or analytical grade ammonium sulfate.
2. 300 mg GTP, and 0.5 mL of neutralized 0.1 $M$ GTP stock solution (*see* **Note 3**). Leave at −20°C the preweighed aliquots required to add to the buffers the following day.

## 2.4. Buffers and Stock Solutions

Prepared with deionized and freshly purified water (Milli-Q or equivalent).

1. Stock phosphate buffer: (A) 0.2 $M$ NaH$_2$PO$_4$ (27.6 g NaH$_2$PO$_4$ H$_2$O to 1 L). (B) 0.2 $M$ Na$_2$HPO$_4$ (28.39 g Na$_2$HPO$_4$ to 1 L). (C) 0.2 $M$ Sodium phosphate buffer stock (0.2 $M$ NaPi): mix equal volumes of A and B. After dilution to 10 m$M$, its pH is 7.00 +/− 0.05 at 20–25°C. Prepare 1 L.

2. 1 *M* MgCl$_2$ (20.33 g of analytical grade MgCl$_2$ 6H$_2$O to 100 mL).
3. PMS buffer: 10 m*M* sodium phosphate, 0.5 m*M* MgCl$_2$, 0.24 *M* sucrose, pH 7.0. (Prepare 8 L: 657.2 g sucrose [biochemical grade], 400 mL 0.2 *M* NaPi, 4 mL 1 *M* MgCl$_2$; check that pH is about 6.95 without adjustment and store in cold room.)
4. PMG buffer: 10 m*M* sodium phosphate, 0.5 m*M* MgCl$_2$, 0.1 m*M* GTP, pH 7.0. Add the GTP before use. (Prepare 4.5 L buffer: 225 mL 0.2 *M* NaPi 0.2 *M*, 2.25 mL 1 *M* MgCl$_2$, confirm pH 7.0; store in the cold room; separate 0.5 L of this buffer; to the 4 L buffer left, add 0.214 g GTP Li$_2$ before use.)
5. 0.8 *M* KCl-PMG buffer: 10 m*M* sodium phosphate, 0.8 *M* KCl, 0.5 m*M* MgCl$_2$, 0.1 m*M* GTP, pH 7.0. Add the GTP before use (prepare 1 L: 59.65 g KCl, 50 mL 0.2 *M* NaPi, 0.5 mL 1 *M* MgCl$_2$ 1 *M*, adjust to pH 7.0 with NaOH and store in the fridge; separate into two portions of 0.5 L; before use, add 0.027 g GTP Li$_2$ to 0.5 L).
6. 0.4 *M* KCl-PMG buffer: 10 m*M* sodium phosphate, 0.4 *M* KCl, 0.5 m*M* MgCl$_2$, 0.1 m*M* GTP, pH 7.0. Add the GTP before use. (Prepare 1 L: mix 0.5 L 0.8 *M* KCl-PMG without GTP with 0.5 L PMG buffer without GTP, adjust to pH 7.0 with NaOH and store in the fridge; add 0.054 g GTP Li$_2$ before use.)
7. 1 *M* Sucrose-PMG buffer: 10 m*M* sodium phosphate, 1 *M* sucrose, 0.5 m*M* MgCl$_2$, 0.1 m*M* GTP, pH 7.0. Add the GTP before use. Prepare 0.5 L, or 0.25 L if the preparations were coming out small. For 0.5 L: 171.15 g sucrose (biochemical grade), 25 mL 0.2 *M* NaPi, 0.250 mL 1 *M* MgCl$_2$, adjust to pH 7.0 and store in the cold room; before use add 0.5 mL of neutralized 0.1 *M* GTP.

## *2.5. Other Items*

1. Homogenizer: a 1- to 1.5-L domestic blender (300 W) with a glass jar with bottom blades (Osterizer) or a Waring blender, which is employed only for tubulin preparation (keep a spare jar and blade set).
2. The filtering system for batch ion-exchange consists of a 1-L funnel with a medium-coarse porous plate (keep a spare one), two suction flasks with rubber adaptors for the funnel, and a vacuum trap. After use, the porous glass plate must be cleaned with chromic mixture and thoroughly rinsed with distilled water.
3. After the centrifuge rotors and tubes, the blender, buffers, and the G-25 column, take to the cold room a fraction collector, a powerful magnetic stirrer and assorted magnetic bars, cheese cloth (a coarse gauze), a simple balance to equilibrate centrifuge bottles, forceps, scissors, trays (one covered with aluminum foil for brain dissection), assorted spatula, magnetic bars, Pasteur pipets and rubber bulbs, plastic beakers (two of 5 L, two of 3 L, one of 2 L), glass beakers (250 mL, two of 100 mL, 50 mL, 10 mL), graduated cylinders (two of 2 L, one 500 mL), a washing flask, and wiping paper (put all these in a small cart). Do not forget a copy of the **Subheading 3.**
4. Take a portable refrigerator with plenty of ice, gloves, and plastic bags to the car; drain out the water before use.

## 3. Methods

The approximate timing at several steps is given in parenthesis. *See* **Fig. 1** for a scheme. All operations are carried out in the cold room except otherwise noted (it is better to avoid summer for this preparation). Keep rotors in the cold room the night before use and have the centrifuges precooled (*see* **Note 4**).

### 3.1. Collection of Bovine Brains (t = −1 to −1.5 h)

Eight bovine brains from 15- to 18-mo-old animals (approx 3 kg; better 12 brains from younger calf if they were available) are collected at the local abattoir within 30 min of slaughter (still warm), immediately put into ice within plastic bags in a portable fridge, and transported to the laboratory cold room (*see* **Note 5**)

### 3.2. Tissue Homogenization (t = 0)

1. Remove meninges, clots, and superficial blood vessels with forceps. Cut and discard any distinctly dark brain portions. Take the whole brains (including the brain stem) and wash them with PMS.
2. Rapidly mince tissue to small pieces with scissors. Wash twice with one volume of PMS, employing two layer of cheese cloth to separate the liquid.
3. Divide tissue into two equal parts of approx 1.4 L each in 5-L beakers, add slightly more than one volume (1.6 L) of PMS, stir and homogenize it (maintaining the tissue to buffer ratio in the several portions) with the domestic blade homogenizer at maximal speed (during 45 s in a Osterizer or 30 s in a Waring blender).

### 3.3. Removal of Cell Membranes and Debris (t = 1 h)

1. Fill 24 centrifuge bottles, to fit into four rotors (excess homogenate not fitting into the bottles will be discarded), equilibrate and carefully close them with sealing caps. (**Caution:** do not fill the bottles to the top, but fill them as if they were uncapped, in order to make sure that liquid will not be spilled at the rotor angle; watch centrifuges accelerate and never over-speed derated rotors.) Centrifuge using two or four precooled medium size six-hole rotors at 19,000$g$ (at average rotor radius, $R_{ave}$) 30 min at 4°C (*see* **Note 1**). This step can be done with four centrifuges at a time, or with two centrifuges consecutively while all the tissue is being homogenized; only two centrifuges are required after this step.
2. While the homogenate is centrifuging, add GTP to the buffers, start equilibrating the Sephadex G25 column with 800 mL PMG, aspirate the excess liquid from the settled DEAE-Sephadex, add one volume of PMG and adjust to pH 6.8.

### 3.4. Fractionation With (NH₄)₂SO₄

1. Take the supernatant S1, measure its volume (about 2.3–2.5 L) and slowly add 177 g/L of solid $(NH_4)_2SO_4$ (32% saturation) with continuous magnetic stirring

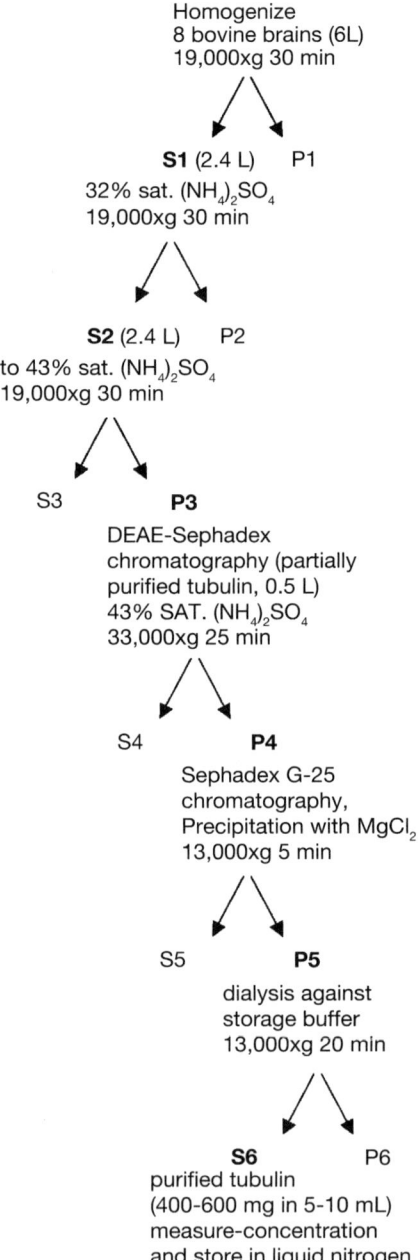

Fig. 1. Scheme summarizing the preparation of tubulin with the modified Weisenberg procedure. For sodium dodecyl sulfate-polyacrylamide gel electrophoresis of the fractions during purification (*see* **refs. 8** and **15**).

($[NH_4]_2SO_4$ can be preweighted for 2.2 L and adjusted during addition). When the salt is completely dissolved wait 10 min and centrifuge at 19,000$g$ 30 min at 4°C (this will take two rotors).

2. Wash the DEAE-Sephadex for a second time and adjust pH to 6.8.
3. Take the supernatant S2 (approximately same volume as S1) and slowly add 71 g/L solid $(NH_4)_2SO_4$ (this brings the solution to 43% saturation) with continuous stirring, wait 10 min and centrifuge at 19,000$g$ 30 min at 4°C.
4. Wash the DEAE-Sephadex for a third time and check pH (6.8).
5. Discard the supernatant S3, take the tubulin-containing pellet P3 and resuspend it with 50 mL PMG, using a spatula and a magnetic stirrer, avoiding foam formation; use another 50 mL PMG to wash the bottles and mix well to resuspend all precipitate.

### 3.5. DEAE-Sephadex A50 Batch-Chromatography (t = 4.5 h)

1. Most of this step can be performed on ice, out from the cold room. Take the resuspended pellet P3 and mix well with the preequilibrated DEAE Sephadex A50 (about 500 mL settled gel from which the excess liquid has been taken out). Keep 5 min on ice and remove the nonadsorbed material by filtration on a sintered glass funnel on a suction flask, until the wet gel cake cracks.
2. Resuspend the gel with 400 mL of 0.4 $M$ KCl-PMG, keep 5 min on ice and filter as above to remove nontubulin eluting materials. Repeat the resuspension and filtration for a second time.
3. Resuspend the gel with 250 mL of 0.8 $M$ KCl-PMG, keep 5 min on ice and filter with the sintered glass funnel, collecting the eluted tubulin on a clean suction flask. Repeat the resuspension and filtration for a second time with the 250 mL left of 0.8 $M$ KCl. After this ion exchange step the yield is about 1 g of 90% pure tubulin, which is collected by $(NH_4)_2SO_4$ precipitation and further purified by selective $MgCl_2$ precipitation.
4. Slowly add 124 g $(NH_4)_2SO_4$ (analytical grade) per 500 mL filtrate, with constant stirring, back in the cold room, wait 10 min after the salt is dissolved and centrifuge at 33,000$g$ ($R_{ave}$; *see* **Note 2**) 25 min at 4°C in two small eight-tube rotors. This will give pellets P4, from which the supernatant should separate well by inverting the tubes.
5. Meanwhile prepare the Sephadex G-25 column and fraction collector for the next step.

### 3.6. Sephadex G25 Column Chromatography (t = 6.5 h)

This is made to remove the $(NH_4)_2SO_4$, which would interfere with the $MgCl_2$ precipitation. Carefully resuspend the precipitate P4 with PMG to a final volume smaller than 80 mL to obtain a turbid solution, through which one can still see. Load the solution into the equilibrated Sephadex G-25 column (2.5 × about 80 cm) in the cold room and run it to practically maximal gravity flow, collecting 6 mL fractions.

### 3.7. Tubulin Precipitation With $MgCl_2$ (t = 8.5 h)

1. Wait until the protein is coming out from the G-25 column (this can be noticed between tubes 20 and 40, in 10–15 fractions, which tend to make foam). Add 40 m$M$ $MgCl_2$ to each protein-containing fraction (0.24 mL of 1 $M$ $MgCl_2$), cap with Parafilm and mix well by inverting the tubes.
2. Immediately combine the fractions in which a milky precipitate has rapidly (seconds) formed and collect the protein by sedimentation at 13,000$g$ (*see* **Note 2**) 5 min at 4°C (pellet P5). Tail fractions in which a precipitate is slowly (minutes) formed are not pooled, because they may decrease the overall yield. The supernatant of this $MgCl_2$ precipitation may turn turbid upon standing in the cold room, but it is discarded because it has been observed to contain inactive tubulin.

## 3.8. Dialysis and Storage

1. Dissolve pellet P5 with a minimal volume (5–10 mL) of 1 $M$ sucrose-PMG (storage buffer) in a small glass beaker, carefully using a magnetic stirrer or a 5-mL syringe with a thick needle, until a cloudy viscous solution is obtained (it will turn more clear upon dialysis).
2. Dialyze overnight at 4°C against 500 mL of 1 $M$ sucrose-PMG, employing prewashed thin dialysis bag, in a Parafilm-capped graduate cylinder, with good stirring.
3. Next morning, centrifuge the dialyzed tubulin and some dialysis buffer at 13,000$g$ ($R_{ave}$) 20 min at 4°C in Pyrex centrifuge tubes (*see* **Note 6**). Save the clear viscous protein supernatant S6 and discard aggregates in the bottom of the tube; keep some dialysis buffer for reference.
4. Measure the tubulin concentration. Dilute 5 µL S6 into 1 mL of 6 $M$ neutral guanidinium chloride (in duplicate), take the ultraviolet spectrum of the diluted samples, employing as reference similar dilutions of buffer, and measure the absorbance at 275 nm. The extinction coefficient of tubulin in this solvent is $E_{275} = 1.09$ g/(L/cm) *(17)*. Tubulin at this final stage of purification is more than 98% homogeneous in polyacrylamide gel electrophoresis with sodium dodecyl sulfate. The tubulin concentration is usually between 50 and 100 g/L. Store in 20- to 30-mg aliquots in cryotubes under liquid nitrogen (with the concentration, operator initials, and preparation number noted). Record the total volume. A typical yield is 400–600 mg tubulin after some training (preparations ranging between 200 and 1200 mg have been obtained).

## 3.9. Preparation for Use and Activity

1. Rapidly melt the tubulin aliquot(s) and keep the tube(s) in ice. Remove the sucrose and the excess GTP, and equilibrate the protein in the experimental buffer by chromatography in appropriate cooled Sephadex G25 columns. Clarify the protein by centrifugation. Measure the protein concentration as previously listed, or after dilution in neutral 1% sodium dodecyl sulfate ($E_{276} = 1.07$ g/(L/cm); *28)*, or directly in neutral buffer after scattering correction, employing the extinc-

tion coefficient of native tubulin, $E_{276} = 1.16$ g/(L/cm) *(17)*. Alternately, the very concentrated tubulin stock solution can be directly diluted >100-fold in experimental buffer, only for experiments in which the residual components of the storage buffer do not interfere, or for less rigorous tests. Unassembled tubulin ages rapidly in most buffers without stabilizers *(29)*, requiring activity controls by the end of experiments; it is normally discarded after 4–6 h on ice. Tubulin without stabilizer should not be frozen again, because upon melting it will precipitate in many buffers.

2. Tubulin polymerization can be easily monitored by turbidity and negative stain electron microscopy, and quantified by pelleting the polymers formed *(30)*. W-tubulin cooperatively assembles into microtubules from a critical concentration (Cr) above which essentially all excess protein goes into the polymers *(18,31,32)*. For an example, in 10 m$M$ sodium phosphate, 3.4 $M$ glycerol, 6 m$M$ MgCl$_2$, 1 m$M$ GTP, pH 6.5 buffer at 37°C, the Cr value is 4 $\mu M$ tubulin. W-tubulin is also fully active in ligand binding. Unassembled W-tubulin $\alpha\beta$-dimers contain $1.88 \pm 0.16$ guanine nucleotide *(32)* and bind $0.97 \pm 0.05$ colchicine molecules per dimer *(17)*, whereas assembled W-tubulin binds $0.99 \pm 0.05$ taxol molecules per dimer *(33)*.

## 4. Notes

1. We use Sorvall SLA 1500 and GSA rotors at 12,000 rpm in Sorvall RC-5C centrifuges (giving 22,000$g$ at R$_{max}$, 19,000$g$ at R$_{ave}$, and 15,000$g$ at R$_{min}$), and 250-mL polycarbonate bottles with sealing caps (Sorvall, cat. nos. 03939 and 03278) filled with less than 240 mL each.
2. For example, Sorvall SS34 rotors and unsealed tubes. These give at 19,000 rpm, 43,000$g$ at R$_{max}$, 33,000$g$ at R$_{ave}$, and 23,000$g$ at R$_{min}$. At 12,000 rpm, they give 17,000$g$ at R$_{max}$, 13,000$g$ at R$_{ave}$, and 9000$g$ at R$_{min}$.
3. For tubulin work, we normally use GTP Lithium salt (Roche or Sigma), which has a better stability than the sodium salt, and store it dry at –20°C.
4. Tubulin is unstable during purification and the longer the time employed the smaller the yield. Everything should be ready in advance, and the preparation requires a good coordination. Performing several consecutive preparations usually increases the yield. It is recommended (although not necessary) that one or two trained people help the operator with the large volumes of material in the initial **steps 3.2** and **3.3.** If not enough brains or centrifuges are available, or should the operator find difficult to rapidly handle the 6 L volume of homogenate, the preparation may be downscaled to half, obtaining half or even more tubulin.
5. We have used only bovine brains from veterinary inspected young animals, which comply with human health regulations, and were not subject to any restrictions owing to bovine spongiform encephalopathy. All homogenization and centrifugation procedures are performed wearing disposable gloves, and human contact with the bovine tissue is avoided. Should any local restrictions apply to bovine brain, it is better to switch to pig or other vertebrate brain for tubulin preparation.
6. For example, 12-mL thick wall Sorvall Pyrex tubes with adaptors in a SS34 rotor at 12,000 rpm.

## References

1. Shelanski, M. L. and Taylor, E. W. (1968) Properties of the protein subunit of central-pair and outer-doublet microtubules of sea urchin flagella. *J. Cell. Biol.* **38,** 304–315.
2. Weisenberg, R. C., Borisy, G. G., and Taylor, E. W. (1968) The colchicine binding protein of mammalian brain and its relation to microtubules. *Biochemistry* **7,** 4466–4478.
3. Nogales, E. (2000) Structural insights into microtubule function. *Annu. Rev. Biochem.* **69,** 277–302.
4. Weisenberg, R. C. (1972) Microtubule formation in vitro in solutions containing low calcium concentrations. *Science* **177,** 1104–1105.
5. Shelanski, M. L., Gaskin, F., and Cantor, C. R. (1973) Microtubule assembly in the absence of added nucleotides. *Proc. Natl. Acad. Sci. USA* **70,** 765–768.
6. Karr, T. L., White H. D., Coughlin, B. A., and Purich D. L. (1982) A brain microtubule protein preparation depleted of mitochondrial and synaptosomal components. *Methods Cell Biol.* **24,** 51–60.
7. Weingarten, M. D., Lockwood, A. H., Hwo, S., and Kirschner, M. W. (1975) A protein factor essential for microtubule assembly. *Proc. Natl. Acad. Sci. USA* **72,** 1858–1862.
8. Williams, R. C. and Lee, J. C. (1982) Preparation of tubulin from brain. *Methods Enzymol.* **85,** 376–385.
9. Ashford, A. J., Andersen, S. S. L., and Hyman, A. (1998) Purification of tubulin from bovine brain, in *Cell Biology: A Laboratory Handbook,* (Celis, J., ed.), Academic Press, San Diego, CA, pp. 205–212.
10. Hamel, E. and Lin, C. M. (1981) Glutamate-induced polymerization of tubulin: characteristics of the reaction and application to the large-scale purification of tubulin. *Arch. Biochem. Biophys.* **209,** 29–40.
11. Hamel, E., del Campo, A. A., Lowe, M. C., Waxman, P. G., and Lin, C. (1982) Effects of organic acids on tubulin plymerfization and associayted guanosine 5'-triphosphate hydrolysis. *Biochemistry* **21,** 503–509.
12. Castoldi, M. and Popov, A. V. (2003) Purification of brain tubulin through cycles of polymerization-depolymerization in a high-molarity buffer. *Prot. Expr. Purif.* **32,** 83–88.
13. Weisenberg, R. C. and Timasheff, S. N. (1970) Aggregation of microtubule subunit protein. Effects of divalent cations, colchicine and vinblastine. *Biochemistry* **9,** 4110–4116.
14. Lee, J. C., Frigon, R. P., and Timasheff, S. N. (1973) The chemical characterization of calf brain microtubule protein subunits. *J. Biol. Chem.* **248,** 7253–7262.
15. Lee, J. C. (1982) Purification and chemical properties of brain tubulin. *Methods Cell Biol.* **24,** 9–30.
16. Andreu, J. M. and Timasheff, S. N. (1982) Interaction of tubulin with single ring analogues of colchicine. *Biochemistry* **21,** 534–543.

17. Andreu, J. M. and Timasheff, S. N. (1982) Conformational states of tubulin liganded to colchicine, tropolone methyl ether and podophyllotoxin. *Biochemistry* **21,** 6465–6476.

18. Lee, J. C. and Timasheff, S. N. (1975) The reconstitution of microtubules from purified calf brain tubulin. *Biochemistry* **14,** 5183–5187.

19. De Pereda, J. M., Wallin, M., Billger, M., and Andreu, J. M. (1995) Comparative study of the colchicine binding site and the assembly of fish and mammalian microtubule proteins. *Cell Motil. Cytoskel.* **30,** 153–163.

20. Andreu, J. M. (1982) Interaction of tubulin with non-denaturing amphiphiles. *EMBO J.* **1,** 1105–1110.

21. Diaz J. F., Barasoain, I., Souto, A. A., Amat-Guerri, F., and Andreu, J. M. (2005) Macromolecular accessibility of fluorescent taxoids bound at a paclitaxel binding site in the microtubule surface. *J. Biol. Chem.* **280,** 3928–3937.

22. Buey, R. M., Barasoain, I., Jackson, E., et al. (2005) Microtubule interactions with chemically diverse stabilizing agents: thermodynamics of binding to the paclitaxel site predicts cytotoxicity. *Chem. Biol.* **12,** 1269–1279.

23. Devred, F., Barbier, P., Douillard, S., Monasterio, O., Andreu, J. M., and Peyrot V. (2004) Tau induces ring and microtubule formation from alpha-beta tubulin dimers under non-assembly conditions. *Biochemistry* **43,** 10,520–10,531.

24. Sanchez, S. A., Brunet, J. E., Jameson, D. M., Lagos, R., and Monasterio, O. (2004) Tubulin equilibrium unfolding followed by time-resolved fluorescence and fluorescence correlation spectroscopy. *Protein Sci.* **13,** 81–88.

25. Davis, A., Sage, C. R., Wilson, L., and Farrel, K. W. (1993) Purification and characterization of tubulin from the budding yeast Saccharomyces cerevisiae. *Biochemistry* **32,** 8823–8835.

26. Shah, C., Xu, C. Z. Q., Vickers, J., and Williams, R. C. (2001) Properties of microtubules assembled from mammalian tubulin synthesized in *Escherichia coli*. *Biochemistry* **40,** 4844–4852.

27. Schlieper, D., Oliva, M. A., Andreu, J. M., and Löwe, J. (2005) Structure of bacterial tubulin BtubA/B: evidence for horizontal gene transfer. *Proc. Natl. Acad. Sci. USA* **102,** 9170–9175.

28. Andreu, J. M., Gorbunoff, M. J., Lee, J. C., and Timasheff, S. N. (1984) Interaction of tubulin with bifunctional colchicine analogues: an equilibrium study. *Biochemistry* **23,** 1742–1752.

29. Prakash, V. and Timasheff, S. N. (1982) Aging of tubulin at neutral pH. *J. Mol. Biol.* **160,** 499–515.

30. Andreu, J. M. and Timasheff, S. N. (1986) The measurement of large protein assemblies by turbidity and other techniques. *Meth. Enzymol.* **130,** 47–59.

31. Lee, J. C. and Timasheff, S. N. (1977) In vitro reconstitution of calf brain microtubules: effects of solution variables. *Biochemistry* **16,** 1754–1764.

32. Diaz, J. F., Menendez, M., and Andreu, J. M. (1993) Thermodynamics of ligand-induced assembly of tubulin. *Biochemistry* **32,** 10,067–10,077.

33. Diaz, J. F. and Andreu, J. M. (1993) Assembly of purified GDP-tubulin into microtubules induced by taxol and taxotere: reversibility, ligand stoichiometry and competition. *Biochemistry* **32,** 2747–2755.

# 3

## Purification and Mass Spectrometry Identification of Microtubule-Binding Proteins From Xenopus Egg Extracts

**Vincent Gache, Patrice Waridel, Sylvie Luche, Andrej Shevchenko, and Andrei V. Popov**

### Summary

Microtubule-binding proteins are conveniently divided into two large groups: MAPs (microtubule-associated proteins), which can stabilize, anchor, and/or nucleate microtubules, and motors, which use the energy of ATP hydrolysis for a variety of functions, including microtubule network organization and cargo transportation along microtubules. Here, we describe the use of Taxol-stabilized microtubules for purification of MAPs, motors, and their complexes from Xenopus egg extracts. Isolated proteins are analysed using sodium dodecyl sulfate gel electrophoresis and identified by various mass spectrometry and database mining technologies. Found proteins can be grouped into three classes: (1) known MAPs and motors; (2) proteins previously reported as associated with the microtubule cytoskeleton, but without a clearly defined cytoskeletal function; (3) proteins not yet described as having microtubule localization. Sequence-similarity methods employed for protein identification allow efficient identification of MAPs and motors from species with yet unsequenced genomes.

**Key Words:** Tubulin; microtubule; microtubule-associated protein; MAP; motor; Xenopus; egg extracts; mass spectrometry; proteomics.

## 1. Introduction

Microtubule cytoskeleton plays multiple roles both in interphase and in mitosis. Microtubules polymerize from $\alpha\beta$ tubulin heterodimers *(1,2)* and are organized in the cell by a number of accessory proteins, called motor proteins and MAPs (microtubule-associated proteins) *(3,4)*. Motor proteins, which are represented by the cytoplasmic dynein and the members of kinesin superfamily, use the energy of ATP hydrolysis for a variety of functions including generating force to move along microtubules *(5)*. The minimal definition of a

From: *Methods in Molecular Medicine, Vol. 137: Microtubule Protocols*
Edited by: Jun Zhou © Humana Press Inc., Totowa, NJ

MAP is a protein, which can bind in vitro to microtubules, but more often by MAPs we understand proteins, which also colocalize with microtubules in the cell *(6)*, coprecipitate with microtubules *(7)*, and/or affect microtubule polymerization dynamics *(8,9)*. Finally, many proteins, which do not bind microtubules themselves, are tethered to them via MAPs *(10)* or motors, some of which are known to transport their cargos along microtubules *(5)*. Both MAPs and motors can be purified on microtubules. Motors association with microtubules is ATP-sensitive, whereas MAPs can be usually eluted by salt. For simplicity, in this chapter we will call all the proteins eluted by ATP ("motors"), and those eluted by NaCl ("MAPs").

Xenopus (*Xenopus laevis*) egg extracts are prepared from unfertilized eggs *(11)* and represent an abundant source of cytoskeletal proteins. Indeed, during the first 12 divisions after fertilization very little protein synthesis occurs and, thus, the egg has to supply most of the proteins needed for these rapid divisions. Freshly prepared egg extracts are in the M-phase of the cell cycle (cytostatic factor-arrested), but their status can be easily changed to interphase by addition of $Ca^{2+}$, which triggers cyclin B destruction *(12)*. This feature of egg extracts is extremely important for the studies of microtubule cytoskeleton as many accessory proteins are regulated by phosphorylation/dephosphorylation *(13,14)* and/or through inhibition by importins during the interphase/M phase transition *(15)*.

Here, we describe methods to isolate and identify a number of proteins, which bind to microtubules in Xenopus egg extracts. Sodium dodecyl sulfate (SDS)-gel resolved proteins are identified using NanoLC MS/MS sequencing and database searching. Described methods can be applied to the isolation and identification of microtubule-binding proteins from other sources and model organisms. Of note, sequence-similarity searches make it possible to identify proteins from organisms from yet unsequenced genomes.

## 2. Materials

### 2.1. Xenopus Egg Extracts

1. *X. laevis* females are from African Reptile Park, Tokai, South Africa. Pregnant mare serum gonadotropin (PMSG) and human chorionic gonadotropin (HCG) are from Sigma-Aldrich (cat. nos. G4877 and CG-10, *see* **Note 1**).
2. Cytostatic-factor (CSF)–arrested Xenopus egg extracts are prepared as described in **ref.** *16* with minor modifications. Extracts are snap-frozen in liquid nitrogen in 200-µL aliquots in thin-walled PCR tubes followed by storage at –80°C. Prior to use, tubes with extracts are thawed under hot tap water and immediately put on ice.
3. Cytochalasin B is from Sigma-Aldrich (cat. no. 30380) (*see* **Note 2**).

4. MMR buffer: 100 m$M$ NaCl, 2 m$M$ KCl, 1 m$M$ MgCl$_2$, 2 m$M$ CaCl$_2$, 0.1 m$M$ EDTA, 5 m$M$ HEPES, titrate to pH 7.8 with saturated solution of NaOH. Autoclave and store at room temperature (RT). This buffer can be also prepared as 20X stock.

5. XB buffer: 100 m$M$ KCl, 1 m$M$ MgCl$_2$, 0.1 m$M$ CaCl$_2$, 10 m$M$ HEPES, 50 m$M$ sucrose, titrate to pH 7.7 with saturated solution of KOH. Autoclave and store at RT.

6. Dejelling buffer: 2% L-cystein (Fluka, cat. no. 30089), 1 m$M$ EGTA, titrate to pH 7.8 with saturated solution of NaOH.

7. "Proteases inhibitors cocktail" (PIs) contains leupeptine, aprotinine, and pepstatine A (Euromedex, cat. nos. SP-04-2217, A162-C, and EI-9), make all together at 10 mg/mL in anhydrous DMSO and store at –20°C.

8. *Xenopus* sperm nuclei are prepared as described in Murray *(17)*, frozen in liquid nitrogen in 10-μL aliquots, and stored at –80°C.

9. Fix solution: 11% formaldehyde, 50% glycerol, and Hoechst 33342 or 33258 at 10 μg/mL in MMR buffer.

10. Rhodamin-labeled tubulin is prepared as described in Hyman et al. *(18)*.

## 2.2. MAPs and Motors Purification

1. Cow brain tubulin is prepared as described in Castoldi and Popov *(19)* and stored at –80°C.

2. Taxol (Molecular Probes, cat. no. P-3456) is dissolved in DMSO (Sigma-Aldrich, cat. no. 41648) at 20 m$M$ and stored at –20°C (*see* **Note 3**).

3. GTP (Roche, cat. no. 106356) is prepared as 200 m$M$ in water and stored at –20°C in 200-μL aliquots. ATP (Roche, cat. no. 127531) is prepared as 300 m$M$ in BRB80 (*see* step 4) and stored in 200- μL aliquots at –20°C. AMP-PNP (5'adenylylimidodiphosphate) is from Biochemika (cat. no. 01910).

4. Brinkley renaturing buffer 80 (BRB80) *(20)*, composition: 80 m$M$ Na-PIPES, 1 m$M$ EGTA, 1 m$M$ MgCl$_2$, 1 m$M$ DTT, titrate to pH 7.8 with saturated solution of NaOH. BRB80 is prepared and stored until use as 5X stock solution.

5. BRB80 washing buffer: 80 m$M$ Na-PIPES, 1 m$M$ EGTA, 1 m$M$ MgCl$_2$, 1 m$M$ DTT, 10 μ$M$ Taxol, 1 m$M$ GTP, titrated to pH 7.8 with saturated solution of NaOH.

6. All centrifugation procedures are carried out in the Optima TL100 tabletop centrifuge (Beckman).

## 2.3. SDS-Polyacrylamide Gel Electrophoresis

1. SDS-polyacrylamide gel electrophoresis (SDS-PAGE) is performed in the SE 400 apparatus (Hoefer Scientific Instruments, San Francisco, CA) according to manufacturer's instructions or in an equivalent model. For more information on SDS-electrophoresis, *see* in Ausubel et al. *(21)*.

2. Isoelectrofocusing is performed using the Pharmacia system Multiphor II according to manufacturer's instructions.

3. 2D SDS-PAGE is performed using Bio-Rad Protean II xi Cell system according to manufacturer's instructions.

### *2.4. Mass Spectrometry*

1. Cleland's reagent (dithiothreitol [DTT]) is from Merck (cat. no. 111474), iodoacetamide (cat. no. I-6125), $NH_4HCO_3$ (cat. no. A-6141) and acetonitrile are from Sigma-Aldrich.
2. Modified pig trypsin (Trypsin Gold) is from Promega (cat. no. V5280).
3. HPLC solvents (Lichrosolv®) ($H_2O$: cat. no. 1.15333, acetonitrile: cat. no. 1.00029), formic (cat. no. 1.00264) and trifluoroacetic (cat. no. 1.08262) acids are from Merck.
4. NanoLC setup consisted of a FAMOS autosampler, a SWITCHOS column-switching module, and an ULTIMATE Plus pump (Dionex).
5. C18 PepMAP100 (1 mm × 300 µm ID, 5 µm) (Dionex) is used as a trap column and C18 PepMAP100 (15 cm × 75 µm ID, 3 µm) (Dionex) as an analytical column.
6. LTQ linear trap mass spectrometer (ThermoElectron Corp.) interfaced to the nanoLC system (2.4.5) via a dynamic nanospray probe with a silicatip™ uncoated needle (20 µm ID, 10 µm tip ID; cat. no. FS360-20-10-N-20-C12, New Objective).

## 3. Methods

### *3.1. Xenopus Egg Extract Preparation*

1. CSF-arrested Xenopus egg extracts are prepared according to Desai et al. *(16)*. To induce egg maturation, 3 d before preparation eight frogs are injected subcutaneously with 100 U of PMSG each. PMSG-"primed" animals can be used for laying eggs up to 2 wk after PMSG injection. The day before extract preparation, frogs are injected with 500 U of hCG each and are kept individually in 500 mL MMR in small plastic containers in a 16°C incubator. Under these conditions, frogs lay eggs 16–18 h following hCG injection.
2. Collected eggs are washed with 800 mL of MMR to remove as much debris as possible (*see* **Note 4**). As much as 500 mL of dejelling buffer is added to eggs for a period of time between 5 and 7 min (*see* **Note 5**). Upon dejelling, eggs form a more compact mass. Dejelling buffer is then discarded and eggs are washed first with 200 mL of MMR, followed by four washes with XB buffer (prepare 500 mL). Finally eggs are washed four more times with CSF-XB buffer (prepare 250 mL). Last, CSF-XB wash solution is supplemented with PIs at 0.01 mg/mL (dilute 1:1000). After discarding the last wash solution, eggs are left in a small volume (~5 mL) of CSF-XB/PIs.
3. Dejelled and washed eggs are transferred into Ultra-clear centrifuges tubes (Beckman, cat. no. 344057) using a wide bore polyethylene pipet (Sigma-Aldrich, cat. no. Z350796). Take care to remove as much buffer as possible from the top of the tube. Tubes with eggs are transferred into polypropylene tubes (Greiner, cat. no. 187262, 18 × 95 mm) containing 0.5 mL of CSF-XB buffer and are then centrifuged at 800 rpm (120*g*) for 1 min, followed by 30 s at 1500 rpm (430*g*) in a swinging bucket rotor centrifuge (type Eppendorf 5804 or Beckman SPINCHRON® Series).

At this stage eggs should be densely packed in the tube but should not be lysed. Excess of buffer is removed from the top of the tube.

4. Eggs are crushed by centrifugation at 14,000$g$ (12,000 rpm) in a JS-13.1 rotor (Beckman) during 16 min at 4°C.
5. After centrifugation, tubes are transferred on ice. At this stage three distinct layers should be visible. The light yellow layer on top contains lipids and the dark layer on the bottom contains yolk and pigments. The cytoplasmic layer in the middle is called "CSF-arrested egg extract." To collect this fraction the tube is punctured with an 18-gauge needle and the extract is aspired using a 2-mL syringe. Extract is then supplemented with PIs at 0.01 mg/mL final concentration and stored on ice until use or is frozen for later use.
6. Upon addition of sperm nuclei, CSF-arrested egg extracts should be able to assemble half spindles and eventually bipolar spindles (*see* **Note 6**). To check the quality of extract, 20 μL is supplemented with 1 μL of sperm nuclei (1–5 × 10$^7$/mL) and 0.2–0.5 μL of Rhodamin-tubulin (the correct amount is determined empirically) *(18)*. After 30–60 min incubation, 1 μL of the reaction is mixed with 2 μL of the formaldehyde fix solution on a microscope slide, covered with an 18 × 18-mm cover slip and the presence of spindles is verified by fluorescence microscopy.

## *3.2. Preparation of Taxol-Stabilized Microtubules*

1. Microtubules are polymerized in a 500 μL tubulin solution at 50 μ$M$ (5 mg/mL) in BRB80 supplemented with 1 m$M$ GTP at 37°C during 30 min. Polymerized microtubules are supplemented with 10 μ$M$ Taxol and incubated for 10 min at 37°C (*see* **Notes 3** and **7**).
2. Polymerized microtubules are then pelleted by centrifugation at 103,000$g$ (50,000 rpm) for 14 min at 20°C in the TLA100.3 rotor. Supernatant is discarded and microtubules are resuspended in 500 μL of BRB80 with 10 μ$M$ Taxol. Microtubule suspension is stored at RT and used on the same day.

## *3.3. Purification of Motors and MAPs*

1. As much as 4 mL of freshly prepared (or thawed) CSF-extract are used for purification. Extract is diluted in 2 vol (8 mL) of BRB80 at 4°C and clarified by two successive 15 min centrifugations at 83,000$g$ (45,000 rpm) in a Beckman TLA100.3 rotor at 4°C through a 1-mL cushion of BRB80 buffer containing 40% glycerol (*see* **Note 8**).
2. To bind MAPs and motors to microtubules, the clarified extract is prewarmed in a water bath at 20°C. Taxol-stabilized microtubules in suspension (500 μL, prepared as previously described) is added to the clarified CSF-extract in the presence of 1 m$M$ GTP and 1.5 m$M$ AMP-PNP and the mixture is incubated at 20°C for 10 min (*see* **Notes 9** and **10**).
3. The microtubules/extract solution is overlaid onto 1 mL cushion of BRB80 buffer containing 40% glycerol and 10 μ$M$ Taxol and centrifuged for 10 min at 83,000$g$ (45,000 rpm) in a Beckman TLA100.3 rotor at 20°C.

4. Microtubule pellet containing MAPs and motors is resuspended in 3 mL of BRB80 washing buffer and centrifuged for 10 min at 83,000$g$ (45,000 rpm) in a Beckman TLA100.3 rotor at 20°C.

5. Repeat **step 4** two more times.

6. The final pellet is resuspended in 1 mL of washing buffer containing 10 m$M$ ATP and incubated for 10 min at 20°C. This step allows eluting motor proteins. After incubation, microtubules are pelleted for 10 min at 103,000$g$ (50,000 rpm) in a Beckman TLA100.3 rotor at 20°C and the supernatant containing eluted proteins ("motor proteins fraction") is immediately transferred on ice.

7. Repeat **step 6**. Pool together both elution fractions from **steps 6** and **7**.

8. The remaining microtubule pellet is resuspended in 1 mL of washing buffer containing 0.5 $M$ NaCl (add 1/10 v/v of 5 $M$ NaCl in H$_2$O) and incubated for 10 min at 20°C. This step allows eluting MAPs and all other proteins sensitive to higher ionic strength (*see* **Note 11**). After incubation, microtubules are pelleted by centrifugation for 10 min at 103,000$g$ (50,000 rpm) in a Beckman TLA100.3 rotor at 20°C and the supernatant containing eluted proteins ("MAPs fraction") is transferred on ice.

9. Both supernatants from **steps 6–8** are then concentrated using a 0.5-mL concentrator with a 10,000 MWCO cut-off polyethersulfone membrane (Vivaspin, cat. no. VS0101) to a volume of 50 µL. After this step, the motor protein fraction is ready for analysis by electrophoresis. The MAPs fraction at this stage contains 0.5 $M$ NaCl that could perturb proteins migration on the acrylamide gel. MAPs fraction is thus diluted in water 10 times (by addition of 450 µL H$_2$O) to reduce salt content to approx 50 m$M$ NaCl and concentrated one more time using Vivaspin 0.5 mL concentrator as previously described. The MAPs fraction is now ready for analysis by electrophoresis. All steps of purification are schematically shown in **Fig. 1**.

### 3.4. Protein Analysis on SDS-PAGE

#### 3.4.1. 1D-SDS Electrophoresis Gel Profile

1. Motors and MAPs fraction are loaded on a 6–18% gradient electrophoresis gel on a vertical slab gel at 25 mA/gel at 4°C.

2. After migration (until the front reached the bottom of the gel), the gels are stained with Coomassie Blue (*see* **Note 12**). Analysis of this gel is shown in **Fig. 2** (*see* **Note 13**).

#### 3.4.2. 2D-SDS Electrophoresis Gel Profile

1. Two-dimensional (2D) electrophoresis is performed with immobilised pH gradients for isoelectric focusing. Home made linear 3–10.5 gradients are used *(22)* and prepared according to published procedures *(23)*. IPG strips are cut with a paper cutter, and rehydrated in 7 $M$ urea, 2 $M$ thiourea, 4% CHAPS, 0.4% carrier ampholytes (3 to 10 range) and 5 m$M$ Tris cyanoethyl phosphine (Molecular Probes, cat. no. T6052) for 3- to 10.5-gradients *(24)*. The protein sample is cup-

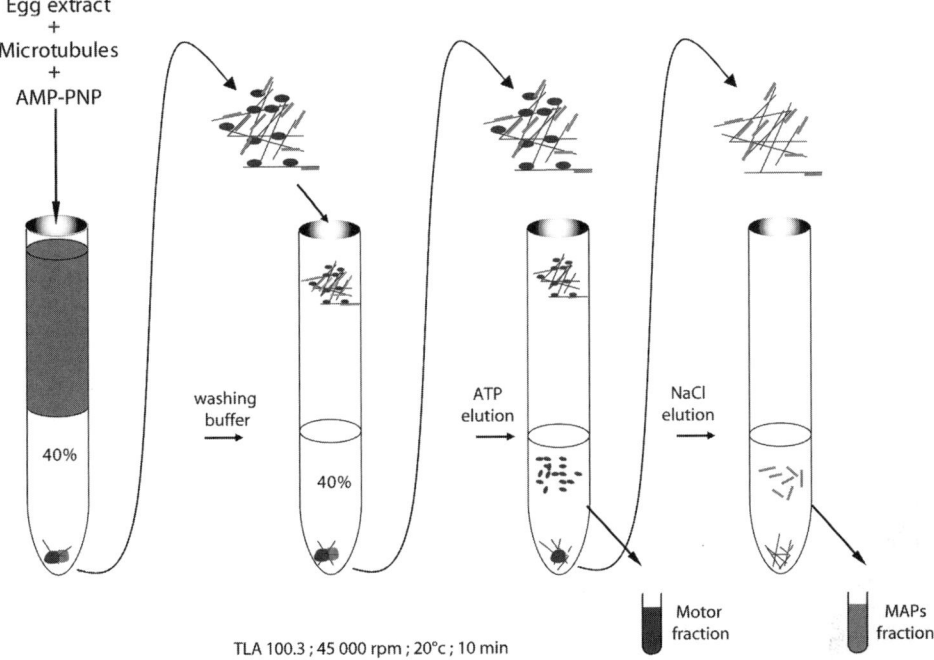

Fig. 1. Schematic view of motors and MAPs purification.

loaded at the anode. Isoelectric focusing is carried out for a total of 60,000 Vh (*see* **Note 14**).

2. After focusing, the strips are equilibrated for $2 \times 10$ min in 6 *M* urea, 2% SDS, 125 m*M* Tris-HCl pH 7.5 containing either 50 m*M* DTT (first equilibration step) or 150 m*M* iodoacetamide (second equilibration step). The equilibrated strip is loaded on the top of a 10% polyacrylamide gel, and submitted to SDS PAGE (10% gel) at 12 W/gel *(25)*.

3. After migration, the gels are stained with colloidal Coomassie Blue *(26)* (*see* **Note 15**). Analysis of this gel is shown in **Fig. 3**.

## 3.5. Mass Spectrometry Analysis of Proteins Resolved on SDS-Electrophoresis Gels

### 3.5.1. In-Gel Digestion of Protein Bands

1. Coomassie Blue-stained bands (spots) of interest are excised from 1D or 2D gels and digested in-gel as described in **refs.** *27* and *28* (*see* **Note 16**).

2. Briefly, gel pieces are cut in ca. $1 \times 1$-mm cubes and dehydrated with acetonitrile. Proteins are reduced with 10 m*M* DTT in 100 m*M* ammonium bicarbonate at 56°C and alkylated with 55 m*M* iodoacetamide. After washing with 100 m*M*

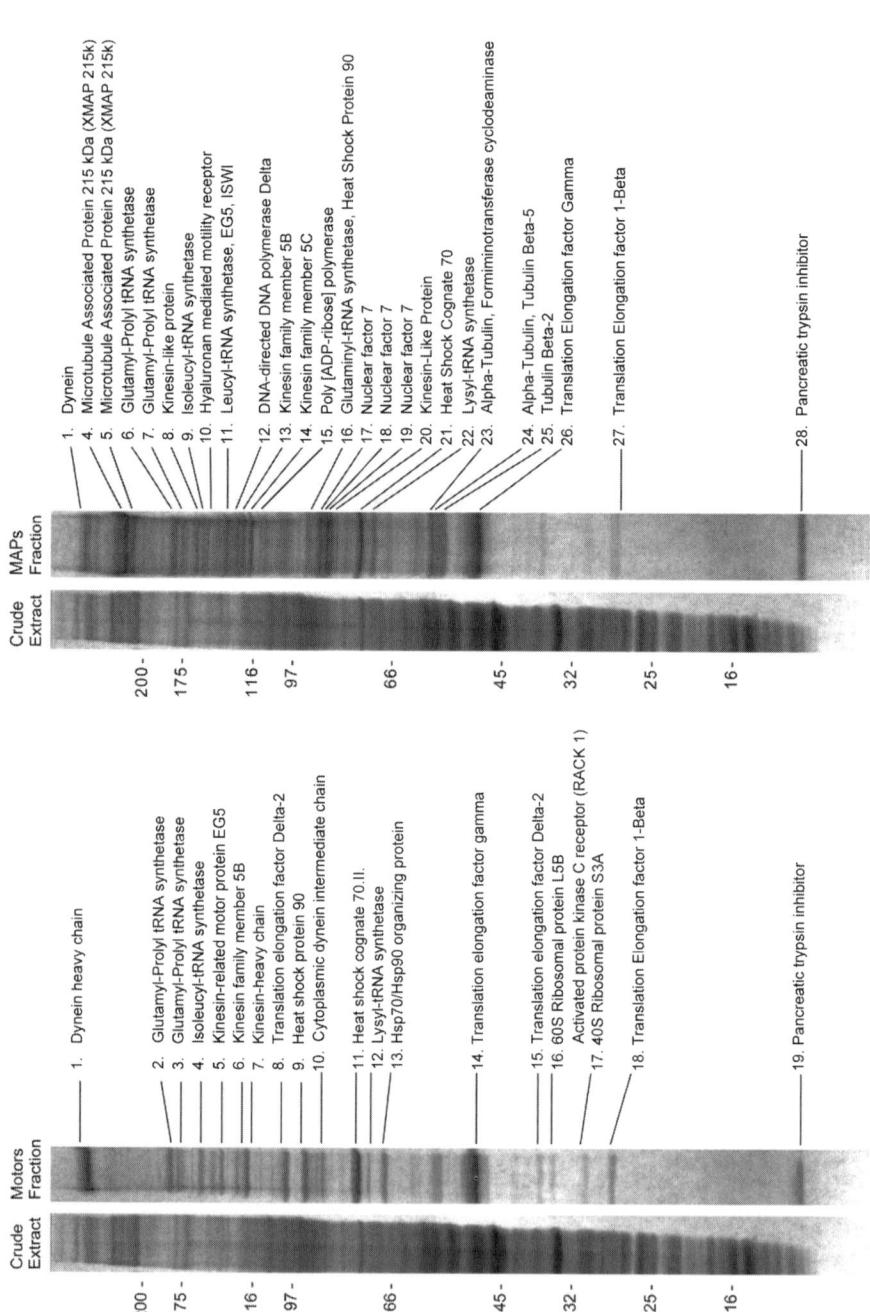

MAPs Fraction

1. Dynein
4. Microtubule Associated Protein 215 kDa (XMAP 215k)
5. Microtubule Associated Protein 215 kDa (XMAP 215k)
6. Glutamyl-Prolyl tRNA synthetase
7. Glutamyl-Prolyl tRNA synthetase
8. Kinesin-like protein
9. Isoleucyl-tRNA synthetase
10. Hyaluronan mediated motility receptor
11. Leucyl-tRNA synthetase, EG5, ISWI
12. DNA-directed DNA polymerase Delta
13. Kinesin family member 5B
14. Kinesin family member 5C
15. Poly [ADP-ribose] polymerase
16. Glutaminyl-tRNA synthetase, Heat Shock Protein 90
17. Nuclear factor 7
18. Nuclear factor 7
19. Nuclear factor 7
20. Kinesin-Like Protein
21. Heat Shock Cognate 70
22. Lysyl-tRNA synthetase
23. Alpha-Tubulin, Formiminotransferase cyclodeaminase
24. Alpha-Tubulin, Tubulin Beta-5
25. Tubulin Beta-2
26. Translation Elongation factor Gamma
27. Translation Elongation factor 1-Beta
28. Pancreatic trypsin inhibitor

Crude Extract

200 -
175 -
116 -
97 -
66 -
45 -
32 -
25 -
16 -

Motors Fraction

1. Dynein heavy chain
2. Glutamyl-Prolyl tRNA synthetase
3. Glutamyl-Prolyl tRNA synthetase
4. Isoleucyl-tRNA synthetase
5. Kinesin-related motor protein EG5
6. Kinesin family member 5B
7. Kinesin-heavy chain
8. Translation elongation factor Delta-2
9. Heat shock protein 90
10. Cytoplasmic dynein intermediate chain
11. Heat shock cognate 70.II.
12. Lysyl-tRNA synthetase
13. Hsp70/Hsp90 organizing protein
14. Translation elongation factor gamma
15. Translation elongation factor Delta-2
16. 60S Ribosomal protein L5B
17. Activated protein kinase C receptor (RACK 1)
    40S Ribosomal protein S3A
18. Translation Elongation factor 1-Beta
19. Pancreatic trypsin inhibitor

Crude Extract

200 -
175 -
116 -
97 -
66 -
45 -
32 -
25 -
16 -

Fig. 2. Analysis of proteins on a one-dimensional sodium dodecyl sulfate-electrophoresis gel (Reprinted from **ref. 32**; courtesy of *Proteomics*).

36

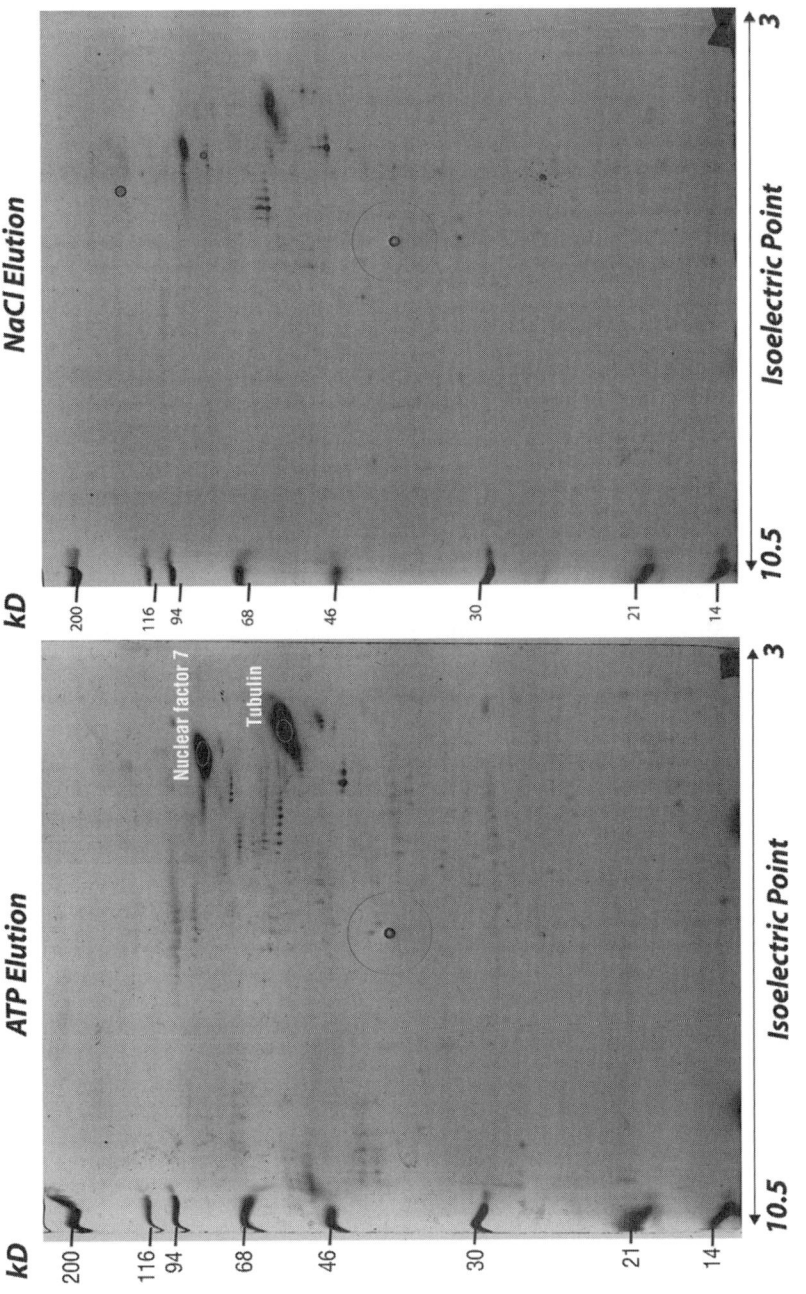

Fig. 3. Analysis of proteins on a two-dimensional sodium dodecyl sulfate-electrophoresis gel: motors fraction (ATP elution) and MAPs fraction (NaCl elution).

37

ammonium bicarbonate and dehydration with acetonitrile, a sufficient volume of digestion buffer (12.5 ng/µL of trypsin in 40 m$M$ NH$_4$HCO$_3$/10% acetonitrile) is added to cover the gel pieces. Samples are first incubated 2 h at 4°C, and the digestion is then performed overnight at 37°C, after addition of more buffer if necessary (*see* **Note 17**).

3. After digestion, peptides are extracted, successively, with 50 µL acetonitrile (equal to one to two times the volume of gel particles) and 100 µL acetonitrile: 5% formic acid (50:50). The extracts are pooled together, dried down in a vacuum centrifuge and stored at –20°C (*see* **Note 18**).

### 3.5.2. NanoLC MS/MS Sequencing

1. Dried samples are redissolved in 15–25 µL of 0.05% trifluoroacetic acid (TFA) and 4 µL are loaded onto the trap column in 0.05% TFA at the flow rate of 20 µL/min (*see* **Note 19**). After 4 min of loading and washing, peptides are eluted and separated on the analytical column at the flow rate of 200 nL/min with the following gradient: from 5 to 20% of solvent B in 20 min, 20–50% B in 16 min, 50–100% B in 5 min, 100% B during 10 min, and back to 5% B in 5 min. Solvent A: 95:5 H$_2$O:acetonitrile (v/v) with 0.1% formic acid (v/v); solvent B: 20:80 H$_2$O:acetonitrile (v/v) with 0.1% formic acid (B).

2. The eluted peptides are introduced into the mass spectrometer via a nanospray needle at the voltage of 1.8 kV, and the capillary transfer temperature is set at 200°C. The analysis is performed in data-dependent acquisition mode controlled by Xcalibur 1.4 software (ThermoElectron Corp.). The acquisition cycle consists of a survey scan covering the range of *m/z* 350 to 1500 followed by the consecutive acquisition of four MS/MS spectra from the most abundant precursor ions at the relative collision energy 35%, isolation window 4.0 amu, in three microscans with maximum ion injection time of 100 ms. The *m/z* of fragmented precursor ions are dynamically excluded for further 60 s, but otherwise no predefined exclusion lists is applied. Individual MS/MS spectra are exported into dta files by BioWorks 3.1 software from the same company

## 3.6. Bioinformatic Tools for Protein Identification

1. Identification by Mascot software. For protein identification, data files representing individual tandem mass spectra are converted into a single mgf-file and submitted to database searches using Mascot software v2.1 (Matrix Science, Ltd.) installed on a local server. Typical database searching settings: mass tolerance for precursor and fragment ions: 2.0 and 0.5 Da, respectively; instrument profile: ESI-Trap; database: MSDB; fixed modification: carbamidomethyl (cysteine); variable modification: oxidation (methionine). Protein identified with at least two peptides and a Mascot score >100 are considered as significant hits.

2. Protein identification by MS BLAST. Selected dta files are interpreted *de novo* using appropriate software, such as DeNovoX (ThermoElectron Corporation) or PepNovo *(29)*. The interpretation of each dta file results in a few peptide sequence proposals. The degenerate, redundant, and partially inaccurate and

incomplete candidate sequences obtained by the interpretation of all selected dta files are assembled into a single query for MS BLAST *(30)* search as was described in great detail in *(28,31)*. The string can contain several thousands of peptide sequences assembled in arbitrary order. The string is then submitted to MS BLAST search at the servers at EMBL, Heidelberg (http://dove.embl-heidelberg.de/Blast2/msblast.html) or at Brigham and Women's Hospital, Boston (http://genetics.bwh.harvard.edu/msblast/). The statistical confidence of hits is evaluated and hits sorted according to MS BLAST scoring scheme *(31)*. In this way it is possible to identify Xenopus proteins that are not present in a database by their similarity to available protein sequences from other species *(32)*.

## 4. Notes

1. PMSG and HCG can be acquired from any other provider but has to be checked for efficiency.
2. Cytochalasin B is an inhibitor of actin polymerisation. Use of cytochalasin B in Xenopus egg extract allows avoiding the contamination of microtubules with actin and actin-binding proteins *(33)*.
3. Taxol is a potent microtubule-stabilizing agent *(34)*. Taxol quality has to be tested. We observed that poor quality Taxol leads to partial microtubules depolymerization. This, in turn, decreases the yield of purified on microtubules proteins and results in the excessive contamination of the eluted proteins with tubulin dimers. We dilute Taxol in anhydrous DMSO, aliquot it in 10–50 µL and store at –20°C. Once thawed, aliquots of Taxol are either used up or discarded.
4. Eggs quality: egg quality is more important than egg quantity. Check and avoid lysed eggs or "activated" eggs (white eggs).
5. During and after dejelling, eggs become progressively more and more fragile and lyse easily if treated roughly. During and after this step, eggs must be manipulated carefully.
6. Extract: fresh or thawed? Before freezing extracts, we routinely test them for their competence to assemble spindles as previously described. Only extracts that can assemble spindles are considered to be in the M-phase. Extracts that contain long microtubules not associated with sperm nuclei and/or decondensed DNA (round nuclei) are considered to be in "interphase" and are discarded. Freezing extracts considerably reduces their capacity to form bipolar spindles, but MAPs and motors can be purified from both freshly prepared and frozen extracts. We did not notice significant differences in the electrophoresis spectra of proteins isolated from fresh or thawed extracts (although we cannot exclude this for some proteins). Frozen extracts offer the advantage of knowing exactly the amount of extract available for purification, which is difficult to predict when starting with freshly laid eggs. Moreover, extract preparation and testing takes time, whereas thawing extracts allows starting the purification in the morning.
7. Tubulin quality is as important as poor quality tubulin does not assemble well into microtubules. Usually about 70% of tubulin of freshly thawed tubulin should be able to assemble into microtubules.

8. Before centrifugation through glycerol cushion, mark the top of the cushion on the tube to visualize the border between cushion and extract after centrifugation. Extract is poured carefully along the tube wall on the top of the cushion to avoid mixing with 40% glycerol.

9. To scale up or down the purification procedure it is important to keep the amount of microtubules constant in respect to MAPs and motors that are to be purified on them. Generally speaking, microtubules must be in excess to avoid competition between the proteins for binding sites on microtubules.

10. The nonhydrolyzable analogue of ATP, AMP-PNP was previously shown to stabilize motors interaction with microtubules *(35)*. The use of the reagent significantly increases the yield of proteins whose association with microtubules is ATP-sensitive.

11. At 0.5 *M* NaCl, there is a slight depolymerization of microtubules. This concentration is a compromise between the goal to elute all MAPs and keep microtubules intact.

12. For scanning, we use an UMAX Powerlook 1120 scanner. We suggest scanning the gel at a resolution of at least 600 dpi.

13. Analysis of identified proteins shows that many of them are already known motors (dynein, eg5, kinesin 5B, and so on) or MAPs (XMAP215, XNF7, RHAMM, and so on), other proteins like HSP90 or poly(ADP-ribose) polymerase (PARP) were previously shown to have a microtubule localization. Last, a number of identified proteins without a known association with microtubules should be handled with care because they could be genuine contaminants or yet unknown microtubule cytoskeleton-associated proteins.

14. Large proteins (with molecular mass greater than 120 kD) do not enter the isoelectric focusing gel. This represents a serious limitation of the 2D gel analysis, especially evident for MAPs and motors, many of which are rather large proteins. Therefore, electrophoretic analysis of isolated proteins is a compromise between the high resolution of the 2D gels and the desire to have as many proteins as possible resolved on a single gel (1D gel).

15. Handling of gels intended for mass spectrometry analysis: plates for gels are washed using deionized water and stored in a clean, dust-free environment. Acrylamide solutions are filtered through 20-μm filter before pouring.

16. For cutting out protein spots and bands, we place the gels on a clean transluminator table and use a clean scalpel blade. It is not necessary to use a new blade for each band, but we wipe the blade clean after each band using an ethanol-wetted paper towel.

17. As plastic tubes accumulate static charge, they attract dust (a major source of keratin contamination). To avoid contamination during sample preparation and digestion, we work in a laminar flow hood with gloves that are frequently rinsed with deionized water, and we use tubes stored in a clean, dust-free environment.

18. Centrifugation of pooled peptide extracts is recommended to eliminate eventual remaining gel particles.

19. Cross-contamination of samples: based on staining intensity, appropriate dilution and injection order should be carried out to avoid cross-contamination by column memory effect in LC-MS/MS analyses.

## Acknowledgments

Research in the group of A.P. is funded by the "Avenir" grant of Inserm, ACI BCMS of the French Research Ministry, grant "Emergence" of the Department of Rhône-Alpes, "La Ligue contre le Cancer" (Comité de l'Isère), and Association pour la Recherche sur le Cancer. Work in the laboratory of A. S. was supported by grant PTJ-BIO/0313130 from BMBF and 1R01GM070986-01A1 from NIH NIGMS.

## References

1. Borisy, G. G. and Taylor, E. W. (1967) The mechanism of action of colchicine. Colchicine binding to sea urchin eggs and the mitotic apparatus. *J. Cell Biol.* **34,** 535–548.
2. Borisy, G. G. and Taylor, E. W. (1967) The mechanism of action of colchicine. Binding of colchincine-3H to cellular protein. *J. Cell Biol.* **34,** 525–533.
3. Sloboda, R. D., Rudolph, S. A., Rosenbaum, J. L., and Greengard, P. (1975) Cyclic AMP-dependent endogenous phosphorylation of a microtubule-associated protein. *Proc. Natl. Acad. Sci. USA* **72,** 177–181.
4. Weingarten, M. D., Lockwood, A. H., Hwo, S. Y., and Kirschner, M. W. (1975) A protein factor essential for microtubule assembly. *Proc. Natl. Acad. Sci. USA* **72,** 1858–1862.
5. Hirokawa, N., Noda, Y., and Okada, Y. (1998) Kinesin and dynein superfamily proteins in organelle transport and cell division. *Curr. Opin. Cell Biol.* **10,** 60–73.
6. Morejohn, L. C. (1994) Microtubule binding proteins are not necessarily microtubule-associated proteins. *Plant Cell* **6,** 1696–1699.
7. Dustin, P. (1980) Microtubules. *Sci. Am.* **243,** 66–76.
8. Hirokawa, N. (1994) Microtubule organization and dynamics dependent on microtubule-associated proteins. *Curr. Opin. Cell Biol.* **6,** 74–81.
9. Cassimeris, L. and Spittle, C. (2001) Regulation of microtubule-associated proteins. *Int. Rev. Cytol.* **210,** 163–226.
10. Ookata, K., Hisanaga, S., Bulinski, J. C., et al. (1995) Cyclin B interaction with microtubule-associated protein 4 (MAP4) targets p34cdc2 kinase to microtubules and is a potential regulator of M-phase microtubule dynamics. *J. Cell Biol.* **128,** 849–862.
11. Lohka, M. J. and Masui, Y. (1983) Formation in vitro of sperm pronuclei and mitotic chromosomes induced by amphibian ooplasmic components. *Science* **220,** 719–721.
12. Murray, A. W. and Kirschner, M. W. (1989) Cyclin synthesis drives the early embryonic cell cycle. *Nature* **339,** 275–280.

13. Andersen, S. S. (1998) Xenopus interphase and mitotic microtubule-associated proteins differentially suppress microtubule dynamics in vitro. *Cell Motil. Cytoskeleton.* **41,** 202–213.
14. Andersen, S. S. L. (1999) Balanced regulation of microtubule dynamics during the cell cycle: a contemporary view. *BioEssays* **21,** 53–60.
15. Nachury, M. V., Maresca, T. J., Salmon, W. C., Waterman-Storer, C. M., Heald, R., and Weis, K. (2001) Importin beta is a mitotic target of the small GTPase Ran in spindle assembly. *Cell* **104,** 95–106.
16. Desai, A., Murray, A., Mitchison, T. J., and Walczak, C. E. (1999) The use of Xenopus egg extracts to study mitotic spindle assembly and function in vitro. *Methods Cell Biol.* **61,** 385–412.
17. Murray, A. W. (1991) Cell cycle extracts. *Methods Cell Biol.* **36,** 581–605.
18. Hyman, A., Drechsel, D., Kellogg, D., et al. (1991) Preparation of modified tubulins. *Methods Enzymol.* **196,** 478–85.
19. Castoldi, M. and Popov, A. V. (2003) Purification of brain tubulin through two cycles of polymerization-depolymerization in a high-molarity buffer. *Protein Expr. Purif.* **32,** 83–88.
20. Brinkley, B. R. (1985) Microtubule organizing centers. *Annu. Rev. Cell Biol.* **1,** 145–172.
21. Ausubel, F. M., Brent, R., Kingston, R. E., et al. (2005) *Current Protocols in Molecular Biology.* John Wiley & Sons, Hoboken, NJ.
22. Gianazza, E., Celentano, F., Magenes, S., Ettori, C., and Righetti, P. G. (1989) Formulations for immobilized pH gradients including pH extremes. *Electrophoresis* **10,** 806–808.
23. Rabilloud, T., Valette, C., and Lawrence, J. J. (1994) Sample application by in-gel rehydration improves the resolution of two-dimensional electrophoresis with immobilized pH gradients in the first dimension. *Electrophoresis* **15,** 1552–1558.
24. Rabilloud, T., Adessi, C., Giraudel, A., and Lunardi, J. (1997) Improvement of the solubilization of proteins in two-dimensional electrophoresis with immobilized pH gradients. *Electrophoresis* **18,** 307–316.
25. Tastet, C., Lescuyer, P., Diemer, H., Luche, S., van Dorsselaer, A., and Rabilloud, T. (2003) A versatile electrophoresis system for the analysis of high- and low-molecular-weight proteins. *Electrophoresis* **24,** 1787–1794.
26. Neuhoff, V., Arold, N., Taube, D., and Ehrhardt, W. (1988) Improved staining of proteins in polyacrylamide gels including isoelectric focusing gels with clear background at nanogram sensitivity using Coomassie Brilliant Blue G-250 and R-250. *Electrophoresis* **9,** 255–262.
27. Shevchenko, A., Wilm, M., Vorm, O., and Mann, M. (1996) Mass spectrometric sequencing of proteins silver-stained polyacrylamide gels. *Anal. Chem.* **68,** 850–858.
28. Shevchenko, A., Sunyaev, S., Liska, A., Bork, P., and Shevchenko, A. (2003) Nanoelectrospray tandem mass spectrometry and sequence similarity searching

for identification of proteins from organisms with unknown genomes. *Methods Mol. Biol.* **211,** 221–234.

29. Frank, A. and Pevzner, P. (2005) PepNovo: de novo peptide sequencing via probabilistic network modeling. *Anal. Chem.* **77,** 964–973.
30. Shevchenko, A., Sunyaev, S., Loboda, A., et al. (2001) Charting the proteomes of organisms with unsequenced genomes by MALDI-quadrupole time-of-flight mass spectrometry and BLAST homology searching. *Anal. Chem.* **73,** 1917–1926.
31. Habermann, B., Oegema, J., Sunyaev, S., and Shevchenko, A. (2004) The power and the limitations of cross-species protein identification by mass spectrometry-driven sequence similarity searches. *Mol. Cell. Proteomics.* **3,** 238–249.
32. Liska, A. J., Popov, A. V., Sunyaev, S., et al. (2004) Homology-based functional proteomics by mass spectrometry: application to the Xenopus microtubule-associated proteome. *Proteomics* **4,** 2707–2721.
33. Spudich, J. A. and Lin, S. (1972) Cytochalasin B, its interaction with actin and actomyosin from muscle (cell movement-microfilaments-rabbit striated muscle). *Proc. Natl. Acad. Sci. USA* **69,** 442–446.
34. Schiff, P. B., Fant, J., and Horwitz, S. B. (1979) Promotion of microtubule assembly in vitro by taxol. *Nature* **277,** 665–667.
35. Brady, S. T. and Lasek, R. J. (1984) Adenylyl imidodiphosphate (AMPPNP), a nonhydrolyzable analogue of ATP, produces a stable intermediate in the motility cycle of fast axonal transport. *Biol. Bull.* **167,** 503

# 4

# Preparation and Characterization of Posttranslationally Modified Tubulins From *Artemia franciscana*

## Paul A. O'Connell and Thomas H. MacRae

## Summary

Tubulin heterogeneity within eukaryotic cells is generated by differential gene expression and posttranslational modification of α- and β-tubulin gene products, either as heterodimers or when polymerized into microtubules. The characterization of posttranslationally modified tubulins from the crustacean *Artemia franciscana* is presented, although tubulins from other sources can be studied with these procedures. Tubulin is prepared from cell free extracts by taxol-induced assembly and centrifugation of microtubules through sucrose cushions, which also yields microtubule-associated proteins, or it is purified to apparent homogeneity by relatively simple chromatographic procedures and assembly/disassembly steps. To detect posttranslationally modified tubulins protein samples are electrophoresed in sodium dodecyl sulfate (SDS) polyacrylamide gels, blotted to nitrocellulose membranes and probed with isoform-specific antibodies. Isotubulins, for which gene-encoded amino acid differences and posttranslational modifications generate charge variations, are resolved in two-dimensional gels using isoelectric focusing followed by SDS polyacrylamide gel electrophoresis, a procedure useful for resolution of microtubule-associated proteins. Isoforms patterns are visualized by Coomassie blue and/or silver staining and individual isoforms are identified by antibody reactivity on Western blots. Tubulin isoforms are localized in *Artemia* by immunofluorescent staining of larvae. The focus of this chapter is the purification of tubulin from a nonneural source and characterization of tyrosinated, detyrosinated, and nontyrosinatable α-tubulins using polyclonal antibodies made to carboxy-terminal peptides of each isoform.

**Key Words:** Tubulin purification; tubulin posttranslational modification; tubulin tyrosination/detyrosination; nontyrosinatable tubulin; taxol; two-dimensional gel electrophoresis; ELISA; immunoprobing of Western blots; immunofluorescent staining of *Artemia* larvae.

From: *Methods in Molecular Medicine, Vol. 137: Microtubule Protocols*
Edited by: Jun Zhou © Humana Press Inc., Totowa, NJ

## 1. Introduction

The main structural protein of microtubules is tubulin, a heterodimer formed of α- and β-tubulin, each usually encoded by small gene families *(1–4)*. Tubulins other than α- and β-isotypes exist, including γ, δ, ε, ζ, and η, but their distribution within cells is restricted mainly to centrioles and basal bodies *(5,6)*. Tubulin heterogeneity arises as a result of differential gene expression and by posttranslational modification, producing among others the tyrosinated, detyrosinated, and nontyrosinatable tubulin isoforms *(4,6–11)*. Most α-tubulins are synthesized with a carboxy-terminal tyrosine and upon polymerization this amino acid residue is removed by tubulin tyrosine carboxypeptidase, frequently exposing glutamic acid and producing detyrosinated tubulin. Tyrosine is reattached by tubulin tyrosine ligase when microtubules disassemble, giving tyrosinated tubulin which is now primed to reenter the cycle of polymerization, detyrosination, depolymerization, and tyrosination. Detyrosinated tubulin may irreversibly lose the carboxy-terminal glutamic acid, presumably by enzyme action, producing nontyrosinatable tubulin, an isoform removed from the assembly/disassembly cycle.

Other isoforms include acetylated α-tubulin, the lone exception to the rule that the hyper-variable carboxy-terminus is the site of tubulin posttranslational modification. An acetyl group is attached to lysine 40 by α-tubulin acetyltransferase and removed by one or more tubulin deacetylases, which have activity toward histones *(7,12)*. Other posttranslational changes affecting α- and β-tubulin are polyglutamylation and polyglycylation, the respective addition of glutamate and glycine peptides to the γ-carboxyl group of tubulin glutamates *(13–15)*. These isoforms, along with detyrosinated α-tubulin, are thought to mediate molecular motor binding to microtubules and modulate axoneme structure *(9,13)*. The detyrosinated isoforms are often associated with stable microtubules, but do not confer stability, the latter possibly due to a kinesin-like, plus-end cap *(16)*. Detyrosinated tubulin may signal binding of vimentin intermediate filaments by a kinesin-dependent mechanism *(17)* and influence intracellular vesicle transport *(18)*. Posttranslational modifications potentially alter microtubule-associated protein binding by affecting protein–protein interactions, as occurs for microtubules containing polyglutamylated tubulin *(19)*. The consequences of posttranslational modifications may be interdependent, resulting in sequential, hierarchical tubulin modifications on a single microtubule *(10)*, and posttranslational modifications have the potential to reduce microtubule heterogeneity by obscuring isotype-specific differences within tubulins *(8)*.

In this chapter, procedures for tubulin preparation from the crustacean *Artemia franciscana*, including taxol-induced tubulin assembly, are described. The methods are representative of those used to purify tubulin from nonneural sources where levels of the protein are much lower than in brain. Preparation of antibodies for detection of tyrosinated, detyrosinated, and nontyrosinatable tubulins on Western blots and in larvae are presented, as are procedures for protein resolution and staining in one and two-dimensional gels.

## 2. Materials

### 2.1. Purification of Artemia Tubulin

1. *A. franciscana* cysts (Sanders Brine Shrimp Co., Ogden, UT).
2. Miracloth (Calbiochem, La Jolla, CA).
3. Hatch medium: 422 m$M$ NaCl, 9.4 m$M$ KCl, 25.4 m$M$ MgSO$_4$·22.7 m$M$ MgCl$_2$·1.4 m$M$ CaCl$_2$·0.5 m$M$ NaHCO$_3$. Disodium tetraborate may be added to 0.1% (w/v) but is not essential. Store at 4°C.
4. PIPES buffer: 100 m$M$ 1,4-piperazine-N,N'-bis(2-ethanesulfonic acid) as free acid (Sigma-Aldrich, Oakville, Ontario) 1 m$M$ ethylene glycol-bis(amino-ethylether)-tetraacetic acid (EGTA), 1 m$M$ MgCl$_2$, pH 6.5 with 10 $N$ NaOH (*see* **Note 1**). Store PIPES buffers at 4°C.
5. PIPES buffer with 4 $M$ glycerol: after adjusting buffer pH add 146 mL of glycerol and bring the final volume to 500 mL.
6. PIPES buffer with 8 $M$ glycerol: after adjusting buffer pH add 292 mL of glycerol and bring the final volume to 500 mL.
7. 0.5X PIPES buffer with 1 $M$ NaCl.
8. PIPES buffer with 0.2 $M$ NaCl.
9. PIPES buffer with 0.3 $M$ NaCl.
10. 2X PIPES buffer.
11. PIPES buffer with 10 m$M$ guanosine triphosphate (GTP) (Sigma-Aldrich).
12. Cellulose phosphate ion exchanger, Phosphocellulose P11, fibrous form (Whatman, Mandel Scientific, Guelph, Ontario). Suspend 30 g dry weight of phosphocellulose in 1000 mL of 0.5 $N$ NaOH for 5 min stirring gently with a glass rod, then wash on a Buchner funnel under suction with water until the pH of the wash water is 10 or lower. Adding a few drops of concentrated HCl to the final washes speeds the process. Suspend the phosphocellulose in 1000 mL of 0.5 $N$ HCl for 5 min followed by washing with H$_2$O until the pH is approx 3.0. Add 250 mL of PIPES buffer, adjust the pH to 6.5 with 10 $N$ NaOH and leave the slurry overnight at 4°C. Adjust the pH to 6.5, filter on a Buchner funnel and suspend the phosphocellulose in 250 mL of fresh PIPES buffer. Store at 4°C, adding 0.02% (w/v) NaN$_3$ if long term storage is anticipated. NaN$_3$ is toxic and must be used with caution.
13. Advanced ion exchange cellulose, diethylaminoethyl cellulose, preswollen ion-exchange cellulose, DE 52, Whatman (DE 52 cellulose) (Mandel Scientific). Suspend 100 g of DE 52 cellulose in 500 mL of 2X PIPES buffer, stir gently with a glass rod for 5 min, leave 20 min at room temperature without stirring and decant fines. Add 300 mL of PIPES buffer, mix gently, adjust pH to 6.5 and leave for 10–15 min without stirring before decanting fines. Add 200 mL of PIPES buffer and repeat the procedure just described, leaving enough buffer after de-

canting to cover the surface. Store at 4°C, adding 0.02% (w/v) NaN$_3$ if long term storage is anticipated.

14. Guanosine triphosphate (GTP) (Sigma-Aldrich) 18 m$M$ stock solution in either PIPES buffer or H$_2$O. Store at –20°C.
15. MgCl$_2$ solution: 100 m$M$ MgCl$_2$ in H$_2$O.

## 2.2. Taxol-Induced Tubulin Assembly

1. Cell-free extract or purified tubulin from *Artemia* cysts or larvae (*see* **Subheading 3.1.**).
2. PIPES buffer (*see* **Subheading 2.1.**).
3. Taxol stock solution: 10 m$M$ paclitaxel (Drug Synthesis and Chemistry Branch, Developmental Therapeutics Program, Division of Cancer Treatment and Diagnosis, National Cancer Institute, Bethesda, MD) dissolved in dimethyl sulfoxide (DMSO). Store at –70°C.
4. Taxol working solution: 0.25 m$M$ paclitaxel, immediately before use dilute 5 µL of 10 m$M$ taxol in 95 µL DMSO followed by addition of 100 µL PIPES buffer at room temperature. The taxol may precipitate if cold PIPES buffer is added or the diluted solution is stored.
5. PIPES buffer with 15% (w/v) sucrose. Store at –20°C.

## 2.3. SDS-Polyacrylamide Gel Electrophoresis

1. Running buffer: 0.025 $M$ Tris, 0.2 $M$ glycine, 0.04% SDS, pH 8.1–8.3 before addition of SDS (*see* **Note 2**). Unless otherwise noted, store all electrophoresis solutions at room temperature.
2. Acrylamide solution: 40% acrylamide/bis solution, 37.5:1 (Bio-Rad, Hercules, CA). Acrylamide is a neurotoxin and must be handled with care. Store at 4°C.
3. Running gel buffer: 1.5 $M$ Tris-HCl, 0.005% (w/v) SDS, pH 8.8 with HCl before adding SDS.
4. Stacking gel buffer: 0.5 $M$ Tris-HCl, 0.005% (w/v) SDS, pH 6.8 with HCl before adding SDS.
5. N,N,N,N'-Tetramethyl-ethylenediamine (TEMED), Bio-Rad: 0.2% (v/v) in H$_2$O.
6. TEMED: 2.0% (v/v) in H$_2$O.
7. Ammonium persulfate: 0.5% (w/v) in H$_2$O, made immediately before use.
8. Butanol saturated with H$_2$O.
9. Treatment buffer, 2X: 0.124 $M$ Tris, 0.139 $M$ SDS, 0.2% (v/v) glycerol, 0.1% (v/v) mercaptoethanol, 0.2% (w/v) bromophenol blue, pH the Tris to 6.8 with HCl before adding the remaining components.
10. Coomassie blue staining solution: 0.5% (w/v) Coomassie Brilliant Blue R, 40% (v/v) methanol, 7% (v/v) acetic acid. Stir for 1 h and filter through two layers of Miracloth.
11. Destaining solution: 20% (v/v) methanol, 7% (v/v) acetic acid, 10% (v/v) glycerol.
12. Acetic acid: 7% (v/v) in H$_2$O.

## 2.4. Two-Dimensional Gel Electrophoresis

1. Immobiline DryStrip gels, 13 cm, pH 3.0–10.0 (Amersham Biosciences, Piscataway, NJ). Other lengths and pH ranges are available. Store at –20°C.
2. IPG buffer, pH 3.0–10.0 (Amersham Biosciences). IPG buffer must be the same pH range as the Immobiline DryStrip gels used for isoelectric focusing. Store at 4°C.
3. Dithiothreitol (DTT) (Sigma-Aldrich).
4. Mineral oil (Sigma-Aldrich).
5. Iodoacetamide (Sigma-Aldrich).
6. Acrylamide: 30% acrylamide/bis solution, (Bio-Rad) 37.5:1. Store at 4°C.
7. Rehydration solution: 8 $M$ urea, 2% CHAPS (w/v), 0.002% (w/v) bromophenol blue. Store in 2.5-mL samples at –20°C. Immediately before use, add DTT to 0.2% (w/v) and IPG buffer to 0.5% (v/v).
8. SDS equilibration buffer: 50 m$M$ Tris-HCl, pH 8.8, 6 $M$ urea, 30% (v/v) glycerol, 2% (w/v) SDS, 0.002% (w/v) Bromophenol blue. Add DTT to 0.06 $M$ and iodoacetamide to 0.135 $M$ just before use. Store at –20°C.
9. Resolving gel buffer, 4X: 1.5 $M$ Tris-HCl, pH 8.8. Store at 4°C.
10. SDS: 10% (w/v) SDS in $H_2O$. Store at room temperature.
11. Ammonium persulfate: 10% (w/v) ammonium persulfate in $H_2O$. Prepare immediately before use.
12. SDS electrophoresis buffer: 0.025 $M$ Tris, 0.2 $M$ glycine, 0.1% (w/v) SDS, pH 8.1–8.3 before addition of SDS (*see* **Note 2**). Store at room temperature.
13. Gel storage solution: 0.375 $M$ Tris-HCl, pH 8.8, 0.1% (w/v) SDS. Store at 4°C.
14. Agarose sealing solution: combine 0.5% (w/v) agarose and 0.002% (w/v) bromophenol blue in $H_2O$, mix, and microwave until agarose dissolves. Store at room temperature in 2-mL samples.
15. Two-dimensional (2D) Clean-Up Kit: Ettan™ Sample Preparation Kits and Reagents (Amersham Biosciences, cat. no. 80-6484-51).

## 2.5. Silver Staining of Two-Dimensional Gels

1. Fixation solution 1: 50% (v/v) methanol, 10% (v/v) acetic acid in $H_2O$. Unless noted otherwise, all solutions are stored at room temperature.
2. Fixation solution 2: 50% (v/v) methanol in $H_2O$.
3. Sensitization solution: 1.2 m$M$ $Na_2S_2O_3$ in $H_2O$.
4. Silver staining solution: 11 m$M$ $AgNO_3$ in $H_2O$. Chill to 4°C prior to use.
5. Developing solution: 0.28 $M$ $Na_2CO_3$, 0.025% (v/v) formalin. Add formalin immediately before use.
6. Stop solution: 0.038 $M$ (ethylenedinitrilo)tetraacetic acid (EDTA).

## 2.6. Preparation of Antibodies to Posttranslationally Modified Tubulins

1. Peptides corresponding to the carboxy terminus of tyrosinated ($^+H_3N$-Gly-Glu-Glu-Glu-Gly-Glu-Glu-Tyr-COO$^-$), detyrosinated ($^+H_3N$-Gly-Glu-Glu-Glu-Gly-Glu-Glu-COO$^-$) and nontyrosinatable ($^+H_3N$-Gly-Glu-Glu-Glu-Gly-Glu-COO$^-$) tubulins.

2. Peptide stock solutions: dissolve 3 mg of peptide in 1 mL of 0.2 $M$ Na acetate, pH 5.5. Store at –20°C.
3. Keyhole limpet hemocyanin, stock solution: dissolve 10 mg of Keyhole limpet hemocyanin (Sigma-Aldrich) in 1 mL of $H_2O$. Store at –20°C.
4. Na acetate stock solution: 2 $M$ Na acetate, pH 5.5. Store at room temperature.
5. Glutaraldehyde: 25% aqueous solution of EM grade glutaraldehyde stored in ampoules under nitrogen (Marivac, Halifax, Nova Scotia).
6. Na borohydride.
7. Dialysis buffer: 10 m$M$ Tris-HCl, 140 m$M$ NaCl, pH 7.4. Store at room temperature.
8. Rabbits obtained from Charles River Canada, St. Constant, Quebec, and cared for in accordance with guidelines in Guide to the Care and Use of Experimental Animals available from the Canadian Council on Animal Care.
9. Titermax Gold Adjuvant (Sigma-Aldrich).
10. Nitrocellulose membranes, Protran™ (Schleicher & Schuell, Mandel Scientific).
11. Ponceau S: 0.2% (w/v) Ponceau S (Sigma-Aldrich) in 3% (w/v) trichloroacetic acid. Store at room temperature.
12. Skim milk solution: 5% (w/v) Carnation skim milk powder in TBS buffer made immediately before use.
13. TBS buffer: 10 m$M$ Tris-HCl, 140 m$M$ NaCl, pH 7.4. Store at room temperature.
14. TBS-Tween buffer: TBS buffer with 0.1% (v/v) Tween-20. Store at room temperature.
15. Glycine buffer: 5 m$M$ glycine, 100 m$M$ NaCl, 0.5% (v/v) Tween-20, 10% (w/v) bovine serum albumin, pH 2.3. Store at –20°C.
16. Tris buffer: 1 $M$ Tris-HCl, pH 8.5.
17. PMSF-treated carboxypeptidase A (type II) (carboxypeptidase A): 10 µg/mL (0.6 U/mL) carboxypeptidase A (Sigma-Aldrich) in TBS.

## 2.7. Determining Antibody Titer and Specificity by ELISA

1. Costar 96-well, polystyrene, nontreated, flat-bottom assay plates with low evaporation lids (Corning Inc., NY).
2. Coating buffer: 15 m$M$ $Na_2CO_3$, 35 m$M$ $NaHCO_3$, 0.02% (w/v) $NaN_3$, pH 9.6.
3. Purified *Artemia* tubulin (*see* **Subheadings 2.1.** and **3.1.**).
4. Gelatin (Bio-Rad) 3% (w/v) in TBS-Tween.
5. PMSF-treated carboxypeptidase A (type II) (carboxypeptidase A): 10 µg/mL (0.6 U/mL) carboxypeptidase A (Sigma-Aldrich) in TBS.
6. TBS-Tween buffer (*see* **Subheading 2.6.**).
7. Alkaline phosphatase-conjugated anti-rabbit IgG: dilute the alkaline phosphatase-conjugated anti-rabbit IgG (Jackson ImmunoResearch Lab., Inc., Bio/Can Scientific, Etobicoke, Ontario) in TBS-Tween buffer.
8. AP buffer: 0.1 $M$ glycine, 1 m$M$ $MgCl_2$, pH 10.4.
9. ρ-Nitrophenol phosphate solution: 0.08% (w/v) ρ-nitrophenol phosphate (Sigma-Aldrich) in AP buffer.
10. Peptides corresponding to the carboxy terminus of tyrosinated, detyrosinated, and nontyrosinatable tubulins (*see* **Subheading 2.6.**).

## 2.8. Determining Antibody Specificity by Immunoprobing of Western Blots

1. Blotting buffer: 0.024 $M$ Tris, 0.192 $M$ glycine, 20% (v/v) methanol, pH as mixed. Store at 4°C.
2. Nitrocellulose membrane (*see* **Subheading 2.6.**).
3. Ponceau S (*see* **Subheading 2.6.**).
4. Skim milk solution (*see* **Subheading 2.6.**).
5. TBS buffer (*see* **Subheading 2.6.**).
6. TBS-Tween buffer (*see* **Subheading 2.6.**).
7. HST buffer: 10 m$M$ Tris, 1 $M$ NaCl, 0.5% (v/v) Tween-20, pH 7.4 with HCl before addition of Tween-20. Store at room temperature.
8. Horseradish peroxidase (HRP)-conjugated goat anti-rabbit IgG antibody: dilute the HRP conjugated antibody (Jackson ImmunoResearch Lab., Inc.) in either TBS-Tween or HST.
9. Western Lightning™ Chemiluminescence (ECL) Reagent Plus kit (PerkinElmer, Boston, MA).
10. Fuji medical X-ray film (Fisher Scientific, Ltd.).

## 2.9. Immunofluorescent Staining of Artemia

1. *A. franciscana* cysts (*see* **Subheading 2.1.**).
2. Phosphate-buffered saline (PBS): 0.137 $M$ NaCl, 0.003 $M$ KCl, 0.01 $M$ Na$_2$HPO$_4$, 0.001 $M$ KH$_2$PO$_4$, pH 7.4 as mixed. Store either at room temperature or 4°C.
3. Paraformaldehyde fixative: freshly made 4% (w/v) paraformaldehyde (Sigma-Aldrich) in phosphate-buffered saline, pH 7.4.
4. PBSAT: PBS containing 0.5% (w/v) bovine serum albumin (Sigma-Aldrich) and 0.5% (v/v) Triton X-100 (Sigma-Aldrich).
5. Vectashield™ Mounting Medium (Vector Laboratories, Burlingame, CA).
6. Phosphate-buffered saline II (PBS II): 160 m$M$ NaCl, 6 m$M$ KCl, 2 m$M$ CaCl$_2$, 0.4 m$M$ NaH$_2$PO$_4$, 29 m$M$ NaHCO$_3$, pH 8.2.
7. PBS II containing 0.075% (w/v) saponine (Sigma-Aldrich).
8. Paraformaldehyde fixative: freshly made 2% (w/v) paraformaldehyde (Sigma-Aldrich) in PBS II containing 0.075% saponine.
9. PBS II containing 11% (w/v) sucrose.
10. Methylcellulose solution: 3% (w/v) carboxymethylcellulose sodium salt (Fluka AG Buchs SG) in H$_2$O. It may be necessary to find another supplier.
11. Poly(L)-lysine solution: 1 mg/mL of poly(L)-lysine (Sigma-Aldrich) in H$_2$O. Place clean glass slides in the poly(L)-lysine solution for 5 min, dip in H$_2$O twice and air-dry.

## 3. Methods

*Artemia* embryos undergoing oviparous development are released from females as encysted gastrulae and enter diapause, an extreme state of hibernation characterized by virtual cessation of metabolic activity *(20,21)*. Diapause is broken by desiccation and embryos resume development, emerging from the cyst enclosed in a membrane that ruptures (hatching) to yield swimming nauplii or first instar larvae. Encysted embryos immediately postdiapause possess low levels of active proteases, but these increase during development. The virtual absence of active proteases facilitates tubulin purification from encysted *Artemia* embryos, but this advantage is lost as development progresses, and may never exist for other organisms, which lack a resting stage such as diapause. This has significance because spurious removal of one or more carboxy-terminal amino acid residues by nonspecific exoproteases may lead to the incorrect conclusion that detyrosinated and/or nontyrosinatable tubulins are present in cells.

### 3.1. Purification of Artemia Tubulin

1. Hydrate 50 g (dry weight) of *Artemia* cysts in 300 mL of cold $H_2O$ for at least 3 h. Collect the cysts by suction on a Buchner funnel, wash twice with cold $H_2O$ and twice with cold hatch medium. Suspend 125 g (wet weight) of cysts in 1800 mL of cold hatch medium and incubate at 27–28°C for 3–6 h with shaking at 200 rpm.

2. Harvest cysts by suction on a Buchner funnel, wash twice with cold $H_2O$ and twice with cold PIPES buffer containing 4 $M$ glycerol. Homogenize the cysts in 30-g batches for 8 min in a Retsch motorized mortar and pestle (Brinkman Instruments Canada, Rexdale, Ontario) at 4°C and then transfer to a beaker chilled on ice. Add 1 mL of PIPES buffer containing 4 $M$ glycerol to 30 g of cysts prior to homogenization and rinse the pestle subsequent to homogenate recovery with 1 mL of the same buffer.

3. Add 10 mL of PIPES buffer containing 4 $M$ glycerol to the total cyst homogenate and stir for 15 min at 4°C. The suspension is very thick.

4. Centrifuge the homogenate at 16,000$g$ for 10 min and filter the supernatant through a double layer of Miracloth into a chilled beaker. Squeeze the Miracloth while wearing gloves to extract as much liquid as possible.

5. Centrifuge the filtered supernatant at 40,000$g$ for 30 min at 4°C, transfer the upper 80% of the supernatant to a fresh tube and centrifuge under the same conditions for 20 min. Collect the upper 80% of the supernatant and either use immediately or freeze at –80°C.

6. Thaw frozen supernatants quickly in warm water with gentle mixing and centrifuge at 40,000$g$ for 30 min at 4°C. The supernatant protein concentration should be 30 mg/mL or higher.

7. Apply 400–500 mg of protein to a Phosphocellulose P11 column (length 18.5 cm and diameter 2.5 cm) washed with 0.5X PIPES buffer containing 1 $M$ NaCl and equilibrated with PIPES buffer. Elute protein with PIPES buffer at a flow rate of 35 mL/h, collecting 5 mL fractions. Read the $A_{280}$ of each fraction, pool protein containing samples, usually tubes nos. 6–25, and freeze at –70°C.

8. Thaw the phosphocellulose fraction quickly in warm water and centrifuge at 40,000$g$ for 20 min at 4°C. The supernatant should contain approx 250 mg of protein.

9. Apply the entire sample to a DE 52 column (18.5 cm length, 2.5 cm diameter) washed with 0.5X PIPES buffer containing 1 $M$ NaCl and equilibrated with PIPES buffer. Wash the column successively at 60 mL/h with PIPES buffer and PIPES buffer containing 0.2 $M$ NaCl until the $A_{280}$ decreases to less than 0.1. Elute tubulin with PIPES buffer containing 0.3 $M$ NaCl, collecting 5-mL fractions and reading the $A_{280}$ of each fraction. Pool the protein containing fractions, which should yield approx 50 mL.

10. Slowly add powdered $(NH_4)_2SO_4$ to 50% saturation (29.1 g per 100 mL) with gentle stirring at 4°C, followed by stirring for 20 min after all $(NH_4)_2SO_4$ has dissolved. Collect the precipitate by centrifugation at 12,000$g$ for 15 min at 4°C, dissolve in 1 mL of PIPES buffer containing 1.0 m$M$ GTP, and dialyze overnight at 4°C in 100 mL of PIPES buffer containing 8 $M$ glycerol.

11. Dilute the dialyzed sample 1:1 with cold PIPES buffer and centrifuge at 40,000$g$ for 30 min at 4°C. Collect the supernatant, add 100 m$M$ $MgCl_2$ to a final concentration of 10 m$M$ and 18 m$M$ GTP to a final concentration of 1.8 m$M$, mix gently, incubate at 37°C for 30 min and centrifuge at 40,000$g$ for 30 min at 28°C. Discard the supernatant and rinse the tube and pellet with a small amount of PIPES buffer at 37°C. Suspend the pellet in 0.3 mL of cold PIPES buffer and incubate on ice for 30 min with occasional gentle vortexing. Centrifuge at 40,000$g$ for 30 min at 4°C, place the supernatant in a fresh tube and centrifuge for 20 min. The supernatant of approx 0.3 mL contains the purified tubulin at 3–5 mg/mL and it can either be used immediately or stored at –70°C.

## 3.2. Taxol-Induced Tubulin Assembly

1. Add 20 µL of taxol working solution to 480 µL of cyst cell free protein extract and incubate for 30 min at 37°C. GTP is not required for efficient tubulin assembly.

2. Pipet reaction mixtures onto 2.5 mL sucrose cushions and centrifuge at 40,000$g$ for 30 min at 22°C

3. Discard supernatants and sucrose cushions, rinse tubes and pellets twice with 1 mL of PIPES buffer and suspend in 200 µL of PIPES buffer (*see* **Note 3**).

4. Use samples immediately for SDS polyacrylamide gel electrophoresis **(Fig. 1)** or store at –20°C.

## 3.3. SDS-Polyacrylamide Gel Electrophoresis

1. The Mighty Small™ SE 245 dual gel caster (Hoefer, Inc., San Francisco, CA) and SE 250 minigel electrophoresis apparatus (Hoefer) are used.

2. Prepare a 10% running gel solution by mixing 4 mL of 40% acrylamide/bis solution, 4 mL of running gel buffer, 2 mL of 0.2% TEMED, 4 mL of $H_2O$, and 2 mL of 0.5% ammonium persulfate. Pipet into the gel casting assembly leaving sufficient space for the stacking gel. Overlay the gels with water-saturated *n*-butanol

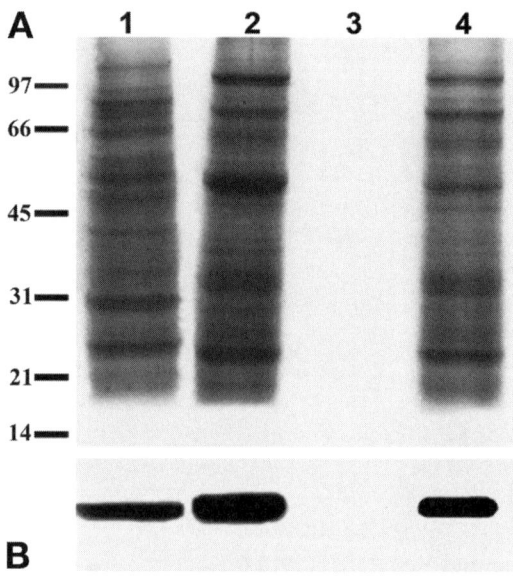

Fig. 1. Taxol-induced tubulin assembly and MAP extraction. Tubulin assembly was induced in *Artemia* cell free extracts, microtubules were collected by centrifugation through sucrose gradients and MAPs were extracted with PIPES buffer containing 0.5 *M* NaCl. Protein samples were electrophoresed in 12.5% sodium dodecyl sulfate polyacrylamide gels and either stained with Coomassie blue (**A**) or blotted to nitrocellulose and immunostained with anti-tubulin antibody (**B**). Lane 1, 5 μg of *Artemia* cell free extract protein; lane 2, microtubule proteins obtained upon incubation of *Artemia* cell-free extracts in the presence of taxol; lane 3, microtubule proteins obtained upon incubation of *Artemia* cell free extract in the absence of taxol; lane 4, 5 μg of MAPs from taxol induced microtubules. Molecular weight markers $10^{-3}$X are on the left side of the figure.

and polymerize for 0.5–1 h. Gels can be stored for at least 2 wk at 4°C by sealing with plastic wrap.

3. Prepare the stacking gel solution by mixing 1 mL of 40% acrylamide/bis solution, 2.5 mL of stacking gel solution, 1.25 mL of 2.0% TEMED, 4 mL of $H_2O$, and 1.25 mL of 0.5% ammonium persulfate. Immediately remove the overlay solution from the running gel, apply the stacking gel solution, insert a comb, and polymerize for 15 min.

4. Remove the comb, install gels in the SE 250 gel apparatus, and fill both chambers with running buffer.

5. Combine equal volumes of protein sample and 2X treatment buffer, place in a boiling water bath for 3 min and either use immediately or store at –20°C. Silver stained one-dimensional SDS polyacrylamide gels require 10- to 100-fold less protein per lane than those stained with Coomassie blue.

6. Gently pipet protein samples beneath the buffer in each well and load unused wells with sample buffer lacking protein.
7. Cool the gel apparatus with running water and apply a constant current of 30 mA per gel for approx 1 h or until the tracking dye reaches the gel bottom.
8. Place gels in a tray containing Coomassie blue staining solution and incubate with gentle agitation for 20–30 min. For silver staining (*see* **Subheading 3.4.**).
9. Replace Coomassie blue with destaining solution, cover and shake slowly until desired band intensity is attained. Kimwipe tissues packed in a corner of the tray shorten destaining time.
10. Store the gels in 7% acetic acid at room temperature.

## *3.4. Two-Dimensional Gel Electrophoresis*

1. Prepare protein samples with the 2D Clean-Up Kit: Ettan™ Sample Preparation Kit.
2. A Hoefer SE 600 Dual Cooled Vertical Slab Unit with 18 × 16-cm glass plates and 1.5-mm thick spacers is used for SDS polyacrylamide gel electrophoresis when running two-dimensional gels. Clean the glass plates with 70% ethanol and dry with Kimwipe tissues. Wear gloves when handling plates.
3. Prepare a 10% gel solution by mixing 33.3 mL of acrylamide/bis solution, 25 mL of 4X resolving gel buffer, 1 mL of 10% SDS, 40.2 mL of $H_2O$, 500 µL of 10% ammonium persulfate and 33 µL of TEMED.
4. Pipet the gel solution into the casting assembly, overlay immediately with water-saturated *n*-butanol and polymerize for 1 h.
5. The gel can be used immediately or stored at 4°C for 2 wk.
6. The Ettan™ IPGphor™ Isoelectric Focusing System and Immobiline DryStrip gels are used for isoelectric focusing.
7. Pipet 250 µL of rehydration solution containing up to 400 µg of protein along the full length of a strip holder.
8. Remove the protective covering from a 13-cm DryStrip gel and place it gel side down into the holder, ensuring the rehydration solution is distributed evenly across the entire surface and that the gel contacts electrodes embedded in the strip holders.
9. Layer approx 400 µL of mineral oil over each DryStrip gel and place covers on the strip holders.
10. Program the Ettan™ IPGphor™ Isoelectric Focusing System using conditions empirically determined for each sample and which depend on DryStrip gel length, pH gradient, protein load and rehydration solution composition. For a 13-cm, pH 3.0–10.0 DryStrip gel, rehydration is at 0 V for 12 h, 500 V for 1 h, 1000 V for 1 h, and 8000 V for 2 h.
11. Turn off the power upon completion of IEF and remove the DryStrip gels from strip holders with forceps. Gently rinse the gels, place in a screw cap tube and either equilibrate or store at –80°C.
12. Incubate each DryStrip gel with gentle agitation for 15 min in 10 mL of SDS equilibration buffer containing 100 mg of DTT immediately before SDS polyacrylamide gel electrophoresis.

13. Decant the SDS equilibration solution, add 10 mL of SDS equilibration buffer containing 250 mg of iodoacetamide to each DryStrip gel and agitate gently for 15 min.
14. Fill the tank with SDS electrophoresis buffer cooled to 10–15°C.
15. Position each equilibrated DryStrip gel on the surface of a SDS polyacrylamide gel with forceps, ensuring the entire bottom edge of the strip contacts the slab gel surface. Overlay each gel with 1 mL of melted agarose cooled to 40–50°C.
16. After the agarose solidifies fill the upper buffer chamber with SDS electrophoresis buffer and apply a constant current of 30 mA per gel for approx 3 h, or until the bromophenol blue tracking dye is approx 0.5 cm from the gel bottom.
17. Recover gels while wearing gloves and either stain with Coomassie blue (*see* **Subheadings 2.3.** and **3.3.**) or silver stain.

## 3.5. Silver Staining of Two-Dimensional Gels

1. For silver staining place each gel in 250 mL of fixation solution 1 and either agitate for 30 min or leave overnight at room temperature if the staining protocol cannot be completed (*see* **Note 4**).
2. Remove fixation solution 1, add 250 mL of fixation solution 2 and agitate for 15 min.
3. Discard fixation solution 2, wash gels five times for 5 min with 250 mL of $H_2O$, add 250 mL of sensitization solution and agitate gently for 1 min.
4. Rinse gels twice for 1 min with 250 mL of $H_2O$, add 250 mL of silver staining solution cooled to 4°C and agitate for 25 min.
5. Rinse gels twice for 1 min with 250 mL of $H_2O$, add 250 mL of developing solution and agitate gently for 5–10 min.
6. Discard the developer, quickly add 250 mL of stop solution and agitate for 10 min.
7. Drain the stop solution and rinse each gel twice for 1 min with 250 mL of $H_2O$. The gel is ready to scan and it can also be used to recover proteins for mass spectrometry (**Fig. 2**).

## 3.6. Preparation of Antibodies to Posttranslationally Modified Tubulins

1. To conjugate tubulin peptides to Keyhole limpet hemocyanin with an estimated coupling efficiency of 25% mix 100 µL of stock peptide solution, 170 µL of Keyhole limpet hemocyanin stock solution, 21 µL of 2 *M* Na acetate, pH 5.5 and 25 µL of 8% glutaraldehyde in $H_2O$. Incubate at room temperature for 24 h. Add 4 mg of Na borohydride and incubate for 1.5 h at room temperature. Dialyze against three changes of 150 mL of dialysis buffer for approx 40 h at 4°C (*see* **Note 5**).
2. Emulsify each conjugated peptide in an equal volume of TitreMax Gold adjuvant by vortexing vigorously and warming in tap water. Inject rabbits subcutaneously four times at 2 wk intervals, preparing fresh conjugated peptides for each injection.
3. Exsanguinate rabbits by cardiac puncture, collecting blood in glass tubes. Incubate the blood for 1 h at room temperature, separate clots from tube walls with a thin glass rod and incubate overnight at 4°C. Remove the straw colored serum

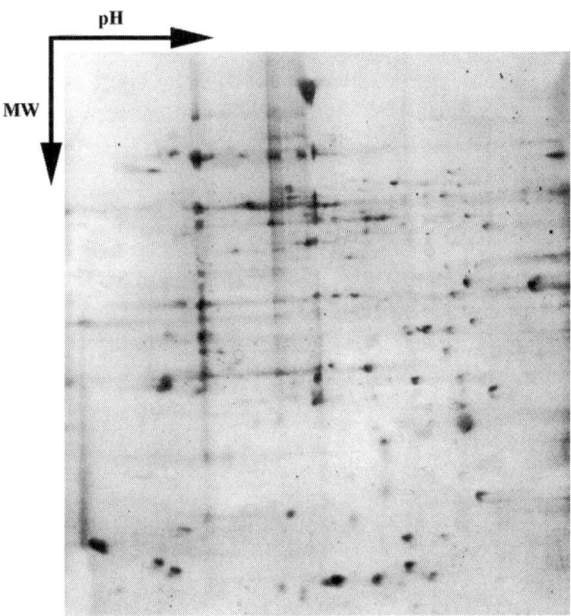

Fig. 2. Silver stained two-dimensional gel of *Artemia* MAPs. Four hundred micrograms of *Artemia* MAPs were resolved by two-dimensional gel electrophoresis and silver stained. The directions of isoelectric focusing (pH) and sodium dodecyl sulfate polyacrylamide gel electrophoresis (MW) are indicated by labeled arrows.

with a Pasteur pipet, avoiding red blood cells, and centrifuge the clot at 3000*g* for 5 min at 4°C. Remove residual serum, pool with the previous sample and centrifuge at 3000*g* for 5 min at 4°C. The serum can be used immediately or stored at –20°C.

4. For affinity purification of antitubulin antibody electrophorese 100 µg of *Artemia* tubulin in an SDS polyacrylamide gel, transfer to nitrocellulose, stain with Ponceau S and cut the tubulin band from the blot (*see* **Subheadings 3.1.**, **3.3.**, and **3.7.**). Tubulin from *Artemia* cysts developed as long as 6 h is completely tyrosinated and is used for purification of antibody to this posttranslationally modified isoform. Tubulin from *Artemia* grown 24 h (early larvae) is almost completely detyrosinated. To ensure full detyrosination incubate blots in carboxypeptidase A at 10 µg/mL (0.6 U/mL) in TBS for 30 min at 37°C, wash several times with TBS and with TBS-Tween at room temperature.

5. Incubate nitrocellulose strips in 5% skim milk powder for 30 min, wash three times for 10 min each in TBS-Tween and once in TBS. Cut strips into pieces and incubate for 1 h at room temperature with gentle shaking in 100 µL of serum diluted with 400 µL of TBS-Tween.

6. Wash the membrane as described, elute the bound antibody with 350 μL of glycine buffer and neutralize immediately with Tris buffer once it is removed from the membrane. The antibody is either used immediately or stored at –70°C. The membrane is washed with TBS-Tween and stored in the same buffer for repeated use.

## 3.7. Determining Antibody Titer and Specificity by ELISA and Immunoprobing of Western Blots

1. Coat assay plates by adding 100 μL of either tyrosinated or detyrosinated *Artemia* tubulin (*see* **Subheading 3.6.**) at 0.01 mg/mL in coating buffer to each well followed by incubation at 37°C for 1 h and overnight at 4°C.
2. Remove the tubulin-containing buffer, add 100 μL of gelatin to each well, incubate at 37°C for 15 min, remove the gelatin and rinse five times with TBS-Tween.
3. Prepare a twofold serial dilution of antibody, add to assay wells, incubate at 37°C for 1 h and wash five times for 2 min each with TBS-Tween.
4. Add 100 μL of alkaline phosphatase-conjugated anti-rabbit IgG antibody diluted 1:500 in TBS-Tween, incubate 30 min at 37°C, wash five times in TBS-Tween and rinse with AP buffer.
5. Add 100 μL of ρ-nitrophenol phosphate solution and measure the absorbance at 410 nm in a microplate reader.
6. Competitive ELISAs are as just described except before addition to tubulin-containing wells antibody at a predetermined dilution is incubated for 1 h at room temperature with twofold serial dilutions of peptides corresponding to tyrosinated, detyrosinated, and nontyrosinatable tubulins (*see* **Note 6**).
7. Antibody specificity is also determined by immunoprobing of Western blots (*see* **Subheading 3.8.**) containing tyrosinated and detyrosinated tubulin (*see* **Subheading 3.6.**).

## 3.8. Immunoprobing of Western Blots

1. Transfer proteins overnight at room temperature from SDS polyacrylamide gels to nitrocellulose membranes using a Bio-Rad Trans-Blot Cell and a Bio-Rad Model 200/2.0 Power Supply at 100 mA.
2. Remove membranes from the blotting apparatus, rinse with $H_2O$, stain with Ponceau S for 2 min and wash briefly with $H_2O$. Label molecular weight markers and other bands of interest with pencil.
3. Incubate membranes in skim milk solution at room temperature for 30 min with gentle shaking.
4. Remove the skim milk solution and incubate membranes for 15 min with gentle shaking at room temperature in primary antibody diluted in TBS-Tween (*see* **Note 7**).
5. Discard the primary antibody and wash membranes twice with gentle agitation at room temperature for 3 min in TBS-Tween, once in HST and twice in TBS-Tween.

6. Incubate the membranes for 15 min with gentle shaking at room temperature in horseradish peroxidase-conjugated goat anti-rabbit IgG secondary antibody diluted in HST.
7. Remove the secondary antibody and wash membranes as previously described with the addition of a final 5-min wash in TBS.
8. Detect antibody-reactive proteins with the Western Lightning™ Chemiluminescence (ECL) Reagent Plus kit, or an equivalent product, following manufacturer's instructions and exposing the membrane to X-ray film (*see* **Note 8**) (**Fig. 1B**).

### 3.9. Immunofluorescent Staining of Artemia

Two procedures are provided for immunofluorescent staining of *Artemia*.

1. In the first procedure, incubate *A. franciscana* larvae at the desired stage of development overnight at room temperature in freshly prepared paraformaldehyde fixative.
2. Wash the samples twice by suspending in PBS, dissect on a clean glass slide by chopping with a fresh razor blade and place in an Eppendorf tube containing PBSAT.
3. Add primary antibody, incubate at room temperature for 1 h and overnight at 4°C.
4. Rinse samples twice with PBSAT, expose to secondary antibody for 2–4 h at room temperature, rinse two times with PBS and mount in Vectashield™ Mounting Medium.
5. In the second procedure, thorax and abdomen from *Artemia* adults are fixed overnight in 2% paraformaldehyde dissolved in PBS II containing 0.075% saponine.
6. Rinse the samples with PBS II and incubate in 11% sucrose in PBS II for at least 4 h. Samples may be stored in 11% sucrose at –18°C and they are brought to room temperature before subsequent processing.
7. Embed the samples in 3% methylcellulose, freeze, cut 14-μm cryosections, collect on poly(L)-lysine-coated glass slides and incubate with antibody as previously described.
8. To localize actin either stain with an actin-specific antibody raised in mouse, using the procedure just described, or stain for 1 h with fluorochrome-labeled phalloidin (Molecular Probes, Eugene, OR or Sigma-Aldrich) at 2.0 μg/mL (**Fig. 3**).

## 4. Notes

1. 1,4-Piperazine-N,N'-bis(2-ethanesulfonic acid) does not dissolve until the pH is raised by addition of NaOH.
2. It should not be necessary to adjust the pH of the running buffer and doing so will cause gels to run slowly.
3. Centrifugation of cell free extracts after incubation with taxol is a convenient way to obtain both tubulin and microtubule-associated proteins.
4. The silver staining method yields protein samples suitable for mass spectrometry. Silver staining is very susceptible to interference so plastic wrap or a

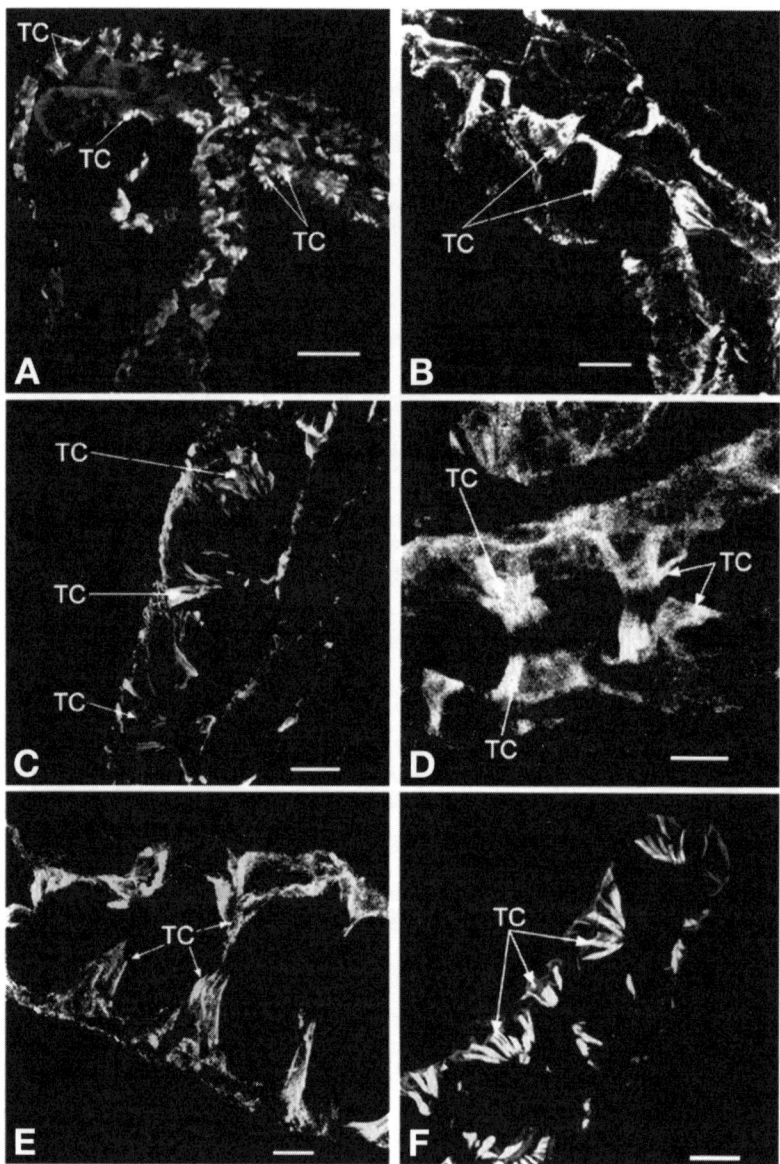

Fig. 3. Immunofluorescent staining of posttranslationally modified *Artemia* tubulin. Hindguts from *Artemia* adults were fixed in 2% glutaraldehyde in PBS II and stained with antibodies to tubulin including TAT, a general anti-α-tubulin antibody obtained from Dr. Keith Gull, Oxford University, UK *(24)*, **A** and **B**; Anti-Y, tyrosinated α-tubulin *(25)*, **C**; Anti-E, detyrosinated α-tubulin *(25)*, **D**; 6-11B-1, acetylated α-tubulin from Sigma-Aldrich, **E**. The secondary antibodies were FITC-conjugated antimouse IgG, (**A,B,E**) and FITC-conjugated antirabbit IgG (**C,D**). The tendon

polycarbonate sheet should be placed between gloved hands and the gel when manipulating. Do not place pressure on gels. Wear powder-free gloves, ensure all containers are clean and prepare solutions immediately prior to use. Stain only one gel per dish.

5. The procedure is based on **refs. *22*** and ***23***.
6. Peptide-based ELISAs and competitive immunoblotting are used to determine specificity and titer of antibodies to nontyrosinatable tubulin because this isoform cannot be obtained separate from tyrosinated and detyrosinated tubulin.
7. Minimizing primary and secondary antibody concentrations, as well as the time membranes are in contact with antibodies and the amount of protein on membranes, will reduce/eliminate non-specific staining. These conditions are determined empirically using samples in which proteins, in addition to the protein of interest, are present.
8. If membranes containing complex mixtures of proteins are exposed to film too long many bands in addition to the protein of interest will appear. Exposure times are determined empirically.

## Acknowledgments

The work was supported by a Natural Sciences and Engineering Research Council of Canada Discovery Grant to THM and a Nova Scotia Health Research Foundation Student Award to PAO.

## References

1. Krebs, A., Goldie, K. N., and Hoenger, A. (2005) Structural rearrangements in tubulin following microtubule formation. *EMBO Rep.* **6,** 227–232.
2. Amos, L. A. (2004) Microtubule structure and its stabilization. *Org. Biomol. Chem.* **2,** 2153–2160.
3. Löwe, J., Li, H., Downing, K. H., and Nogales, E. (2001) Refined structure of αβ-tubulin at 3.5 Å resolution. *J. Mol. Biol.* **313,** 1045–1057.
4. Ludueña, R. F. (1998) Multiple forms of tubulin: different gene products and covalent modifications. *Int. Rev. Cytol.* **178,** 207–275.
5. Dutcher, S. K. (2003) Long-lost relatives reappear: identification of new members of the tubulin superfamily. *Curr. Opin. Microbiol.* **6,** 634–640.
6. McKean, P. G., Vaughan, S., and Gull, K. (2001) The extended tubulin superfamily. *J. Cell Sci.* **114,** 2723–2733.

---

(Fig. 3. *continued*) cells were stained with FITC-conjugated phalloidin, **F**. The pyramid-shaped tendon cell microtubule bundles contain tyrosinated, detyrosinated and acetylated tubulins, with the most extensive staining evident for the detyrosinated isoform, followed by the acetylated and then the tyrosinated isoforms, although it is difficult to accurately quantitate staining by this method. TC, tendon cell. Bar = 50 μm in **A**, bar = 20 μm in **D,F**, bar = 10 μm in **B,C,E**. (From **ref. *26*** with permission.)

7. Westermann, S. and Weber, K. (2003) Post-translational modifications regulate microtubule function. *Nat. Rev. Mol. Cell Biol.* **4,** 938–947.

8. MacRae, T. H. (1997) Tubulin post-translational modifications. Enzymes and their mechanisms of action. *Eur. J. Biochem.* **244,** 265–278.

9. Rosenbaum, J. (2000) Functions for tubulin modifications at last. *Curr. Biol.* **10,** R801–R803.

10. Idriss, H. T. (2000) Man to trypanosome: the tubulin tyrosination/detyrosination cycle revisited. *Cell Motil. Cytoskel.* **45,** 173–184.

11. Erck, C., Peris, L., Andrieux, A., et al. (2005) A vital role of tubulin-tyrosine-ligase for neuronal organization. *Proc. Natl. Acad. Sci. USA* **102,** 7853–7858

12. Hubbert, C., Guardlola, A., Shao, R., et al. (2002) HDAC6 is a microtubule-associated deacetylase. *Nature* **417,** 455–458.

13. Redeker, V., Levilliers, N., Vinolo, E., et al. (2005) Mutations of tubulin glycylation sites reveal cross-talk between the C termini of $\alpha$- and $\beta$-tubulin and affect the ciliary matrix in *Tetrahymena. J. Biol. Chem.* **280,** 596–606.

14. Thazhath, R., Jerka-Dziadosz, M., Duan, J., et al. (2004) Cell context-specific effects of the $\beta$-tubulin glycylation domain on assembly and size of microtubular organelles. *Mol. Biol. Cell* **15,** 4136–4147.

15. Janke, C., Rogowski, K., Wloga, D., et al. (2005) Tubulin polyglutamylase enzymes are members of the TTL domain protein family. *Science* **308,** 1758–1762.

16. Infante, A. S., Stein, M. S., Zhai, Y., Borisy, G. G., and Gundersen, G. G. (2000) Detyrosinated (Glu) microtubules are stabilized by an ATP-sensitive plus-end cap. *J. Cell Sci.* **113,** 3907–3919.

17. Kreitzer, G., Liao, G., and Gundersen, G. G. (1999) Detyrosination of tubulin regulates the interaction of intermediate filaments with microtubules in vivo via a kinesin-dependent mechanism. *Mol. Biol. Cell* **10,** 1105–1118.

18. Lin, S. X., Gundersen, G. G., and Maxfield, F. R. (2002) Export from pericentriolar endocytic recycling compartment to cell surface depends on stable, detyrosinated (Glu) microtubules and kinesin. *Mol. Biol. Cell* **13,** 96–109.

19. Bonnet, C., Boucher, D., Lazereg, S., et al. (2001) Differential binding regulation of microtubule-associated proteins MAP1A, MAP1B, and MAP2 by tubulin polyglutamylation. *J. Biol. Chem.* **276,** 12,839–12,848.

20. MacRae, T. H. (2003) Molecular chaperones, stress resistance and development in *Artemia franciscana. Semin. Cell Develop. Biol.* **14,** 251–258.

21. Liang, P. and MacRae, T. H. (1999) The synthesis of a small heat shock/$\alpha$-crystallin protein in *Artemia* and its relationship to stress tolerance during development. *Dev. Biol.* **207,** 445–456.

22. Bulinski, J. C., Kumar, S., Titani, K., and Hauschka, S. D. (1983) Peptide antibody specific for the amino terminus of skeletal muscle $\alpha$-actin. *Proc. Natl. Acad. Sci. USA* **80,** 1506–1510.

23. Gundersen, G. G., Kalnoski, M. H., and Bulinski, J. C. (1984) Distinct populations of microtubules: tyrosinated and nontyrosinated alpha tubulin are distributed differently in vivo. *Cell* **38,** 779–789.

24. Woods, A., Sherwin, T., Sasse, R., MacRae, T. H., Baines, A. J., and Gull, K. (1989) Definition of individual components within the cytoskeleton of *Trypanosoma brucei* by a library of monoclonal antibodies. *J. Cell Sci.* **93,** 491–500.

25. Xiang, H. and MacRae, T. H. (1995) Production and utilization of detyrosinated tubulin in developing *Artemia* larvae: evidence for a tubulin-reactive carboxypeptidase. *Biochem. Cell Biol.* **73,** 673–685.

26. Criel, G. R. J., Van Oostveldt, P., and MacRae, T. H. (2005) Spatial organization and isotubulin composition of microtubules in epidermal tendon cells of *Artemia franciscana. J. Morph.* **262,** 203–215.

# 5

## Studying the Structure of Microtubules by Electron Microscopy

### Linda A. Amos and Keiko Hirose

#### Summary

Although the structures of individual proteins and moderately sized complexes of proteins may be investigated by X-ray crystallography, the interaction between a long polymer, such as a microtubule, and other protein molecules, such as the motor domain of kinesin, need to be studied by electron microscopy. We have used electron cryo-microscopy and image analysis to study the structures of microtubules with and without bound kinesin motor domains and the changes that take place when the motor domains are in different nucleotide states. Among the microtubules that assemble from pure tubulin, we select a minor subpopulation that has perfect helical symmetry, which are the best for three-dimensional reconstruction. Gold labeling can be used to mark the positions of certain regions of protein sequence.

**Key Words:** Electron microscopy; 3D structure; helical image analysis; macromolecular complexes; microtubules; kinesin; MAPs.

## 1. Introduction

Microtubules (MTs) (**Fig. 1**) are essential structural components of eukaryotic cells and bind many other proteins (*1*). These interactions can be studied in vitro at a molecular level by electron microscopy (EM). The results help to fill the gap between images showing the distribution of fluorescently labeled proteins inside cells, obtained by light microscopy, and the near-atomic structures of individual protein molecules and multi-protein complexes, solved by X-ray crystallography or nuclear magnetic resonance (NMR). Because MTs have helical symmetry (**Fig. 2**), it is usually possible to obtain three-dimensional (3D) data from 2D EM images without resorting to collecting tilted views (unlike the case of two-dimensional [2D] crystals or asymmetric, single particles). Invaluable information has been gathered about the interaction of MTs with motor molecules, such as kinesin family proteins (*2–18*), and structural microtubule-

From: *Methods in Molecular Medicine, Vol. 137: Microtubule Protocols*
Edited by: Jun Zhou © Humana Press Inc., Totowa, NJ

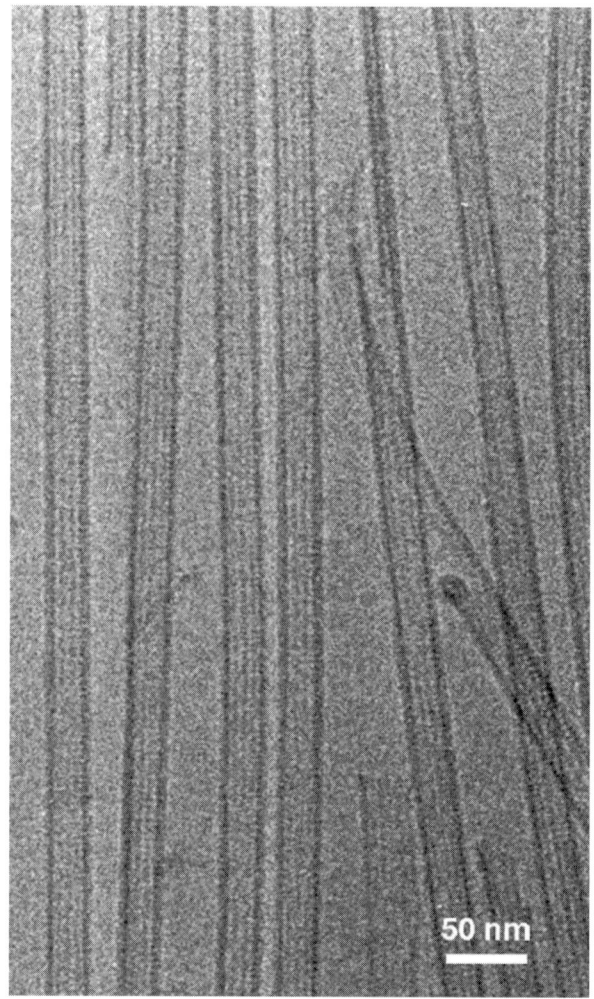

Fig. 1. Electron micrograph of a frozen hydrated array of microtubules reassembled from purified tubulin.

associated proteins (MAPs), such as tau and double-cortin *(19–20)*. Many important complexes remain to be investigated structurally in this way. The buffers and incubation conditions needed will vary depending on the properties of the accessory proteins. Here, we outline some of the conditions that have been used for kinesins and tau.

Electron micrographs of protein specimens embedded in negative stain (a heavy-metal salt solution) have quite high contrast but MTs tend to be dis-

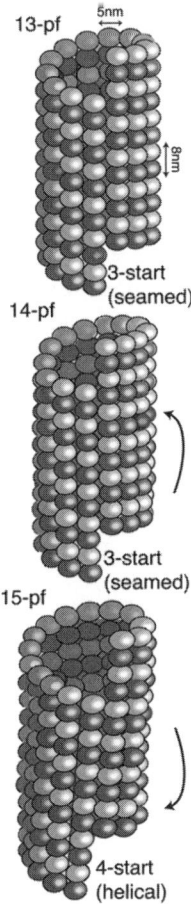

Fig. 2. Microtubules (MTs) with 13, 14, and 15 protofilaments (PFs). Each αβ-tubulin heterodimer is represented as a dark and a light sphere and is 8 nm long. Each column of heterodimers (a PF) is 5 nm wide. PFs in the 13-PF MT are straight, those in the 14-PF and 15-PF MTs follow left-handed and right-handed helices, respectively, owing to small rotations of the surface lattice (arrows *[46,47]*). The lattice of monomer subunits also forms families of three or four shallow left-handed helices; in the 15-PF MT, where there is a four-start monomer family, the heterodimers form a two-start family. The 13-PF and 14-PF dimer lattices each have a step at the "seam" *(48–50)*.

torted (flattened) when the stain dries. Also, irregularities in the dried stain and the thin layer of carbon on which the specimens are supported both contribute to a speckled background that may make it difficult to see the specimen clearly. The contributions from stain and the carbon film, as well as flattening, are avoided when frozen-hydrated specimens are suspended in a thin layer of ice supported on a holey carbon film but then the relative noise level may be similar because of the low contrast difference between protein and ice (protein is slightly denser than ice, as opposed to being significantly less dense than negative stain; hence the sign of the contrast is reversed: *see* **Subheading 3.2.1.**). Even under perfect conditions, however, it would be difficult to appreciate the structure of the specimen directly from its projected image, without some image processing, because everything throughout its depth appears superimposed. The main advantages of frozen hydrated specimens are that they are quickly frozen in near-physiological conditions and are not distorted by being dried. This is especially important for tubular structures such as MTs, which tend to collapse when dried down on a surface.

Images of MTs have been analysed in several different ways, mostly by treating them as objects with perfect helical symmetry. In this case it is not difficult to reconstruct a three-dimensional (3D) image and study single protein molecules within the complexes. Reassembled MTs vary in structure and usually include specimens with 12–16 longitudinal PFs of tubulin subunits. In vivo, the vast majority of MTs have 13 PFs, which run parallel to the axis of the tube (**Fig. 2**) but the PF numbers for MTs assembled from pure tubulin in vitro vary, with a peak at 14. 13-PF or 14-PF MTs do not have perfect helical symmetry because, after they have assembled first as 2D sheets, when they close to form a tube there is a mismatch between the heterodimers at the seam, despite the fact that the tubulin monomers form a perfect helical lattice (*see* **Fig. 2**). The addition of a 15th PF and rotation of the lattice by a few degrees allows the heterodimers to match. A smaller angle of rotation in the other direction matches up the monomer lattice in 14-PF MTs, though the dimers still mismatch as in the straight 13-PF MTs. These rotations mean that the PFs do not run straight, when the number of PFs is more (or less) than 13, but instead rotate gradually around the axis.

In the case of helically symmetrical specimens, it makes sense to work in reciprocal space, because the diffraction pattern from a helix consists of a series of lines, known as layer lines (*see* **Figs . 3** and **4**) and the spaces between these lines can be ignored. This is because the regularly repeating features throughout the whole image contribute to each layerline but random differences (noise) produce diffracted intensity that is spread everywhere. The amount of noise superimposed on the layerlines will be only a small proportion of the total. In general, a longer MT, with more copies of the features that we want to see, will produce sharper layerlines with a lower contribution of noise. We therefore aim to take images of long, straight regular specimens if possible. It is possible to straighten the images of tubes that bend smoothly (**Fig. 5**), but one does not know whether this process makes appropriate changes at a molecular level so it is desirable to minimize such corrections.

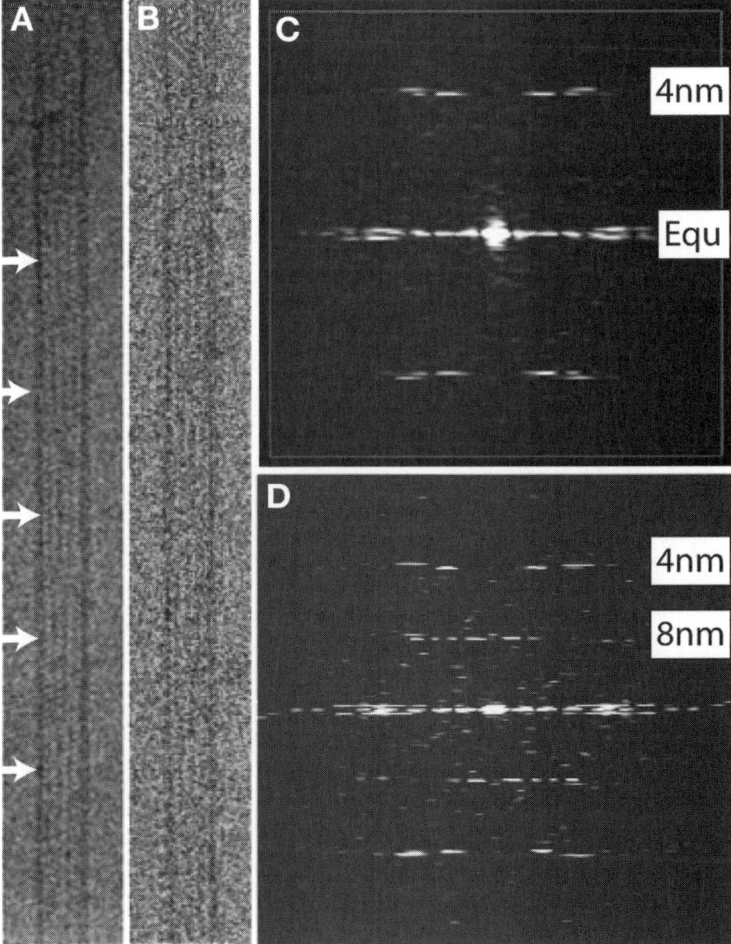

Fig. 3. Cryo- electron microscopy images of Microtubules (MTs) and their diffraction patterns (amplitudes of the computed Fourier transforms). The plain MT in **A** produces strong diffraction spots in its diffraction pattern (**C**) at a reciprocal spacing of 1/(4 nm) because of the regular lattice of monomers but shows no clear evidence for the 8 nm periodicity of the tubulin heterodimers. In contrast, a MT decorated all over with kinesin motor domains (**B**) has a strong 8-nm layerline (**D**). Both diffraction patterns have the characteristics of 15-PF MTs (*see* **Fig. 4**), with $n = 11$ reflections on the 4-nm group of layerlines at a slightly higher level than the $n = -4$ reflections (the $n = -19$ reflections that lie just below the $n = -4$ peaks are too weak to show up in these figures). Both images show Moiré patterns arising from interference between PFs on the near and far sides of the tubes; the longitudinal spacing of the alternating striped and fuzzy sections (arrows in **A**) depends on the pitch of the PF helices.

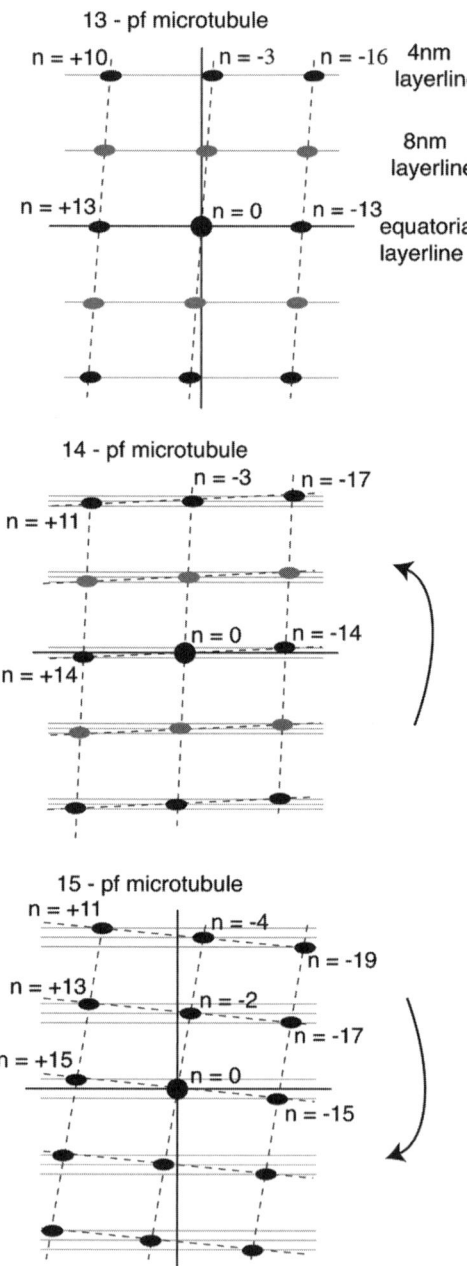

Fig. 4. The reciprocal lattices corresponding to the three helical lattices in **Fig. 2**. Reflections on the 8-nm layerlines are present only when there is a detectable difference between the two monomers of a heterodimer. Binding of kinesin produces a large difference. Structural differences between α- and β-tubulin in undecorated microtubules (MTs) produce diffracted intensity on higher orders of the 8-nm layerlines. The

Strategies for the analysis of specimens that lack perfect helical symmetry have been described *(21–23)* but the extra information required to reconstruct a seam, for example, may reduce the overall reliability of the results compared with the straight-forward analysis of 15-PF MTs. So it is usually preferable to opt for the latter, even though one has to search amongst the EM images for the 5–15% of MTs that have 15 PFs.

## 2. Materials
### *2.1. Buffers and Chemicals*

1. Buffer: BRB80 (80 m$M$ PIPES [pH 6.8], 1 m$M$ EGTA, 5 m$M$ MgCl$_2$, 1 m$M$ DTT, 0.5 m$M$ GTP).
2. MES buffer: (60 m$M$ MES, 5 m$M$ MgSO$_4$, 1 m$M$ EGTA, 1 m$M$ DTT, pH 6.5); PC (phosphocellulose) resin (Whatman P10): tubulin does not bind, even at low ionic strength but MAPs do.
3. MT stabilizers: DMSO (dimethyl sulfoxide); TMAO (trimethylamine $N$-oxide).
4. GMPCPP (guanylyl-[α,β]-methylene-diphosphonate) is a slowly hydrolysable GTP analogue *( 24)*.
5. Kinesin reagents: AMPPNP (adenylylimidodiphosphate) is a slowly hydrolysable ATP analogue that makes kinesin motor domains bind strongly to MTs; Apyrase: buy a preparation with high activity for hydrolysis of ADP to AMP; Hexakinase: used with glucose to convert traces of ATP to ADP.
6. Gold labels: nanogold and undecagold (http://nanoprobes.com).

### *2.2. EM Accessories*

1. Negative stain: 1–2% w/v uranyl acetate (UAc), pH <5.0, unbuffered, because UAc precipitates if the solution is brought to neutral pH.
2. Carbon film: as well as plain metal EM grids (e.g., www.agarscientific.com) it is possible to buy grids already coated with carbon films. Some contain a regular array of holes **(Fig. 7B)** in a choice of sizes (www.quantifoil.com).

---

Fig. 4. *(continued)* value of $n$ is the number of separate helices in the family that gives rise to the diffraction spot. Above the equatorial line, positive values of $n$ are for right-handed helices, negative values for left-handed helices. Because of the seams, values of $n$ on the 8-nm layerlines are nonintegral (therefore, shown in gray) for 13-PFs and 14-PF MTs, though the reflections may appear to be well behaved for the side of the lattice that has no seam. In an electron microscopy image, the lattice on the far side of a MT is a mirror image of that on the near side. Thus, the diffraction pattern from a MT (e.g., **Fig 3C,D**) is (approximately) a combination of one of these reciprocal lattices and its mirror image. For the 13-PF lattice, this leads to some overlap between $n = +10$ and $n = -16$ reflections, for example. Besides being fully helical (integral values of $n$ on the 8-nm layerline), the rotated lattice of 15-PF MTs has the great advantage of separating the contributions to the diffraction pattern from the protofilaments on the near and far sides of the tube (equatorial group of layerlines) and also the contributions from different families of helices with both 4- or 8-nm pitches.

Fig. 5. Straightening of a curved MT. The boxed part of **A** (1024 pixels long) is shown after straightening in **B**.

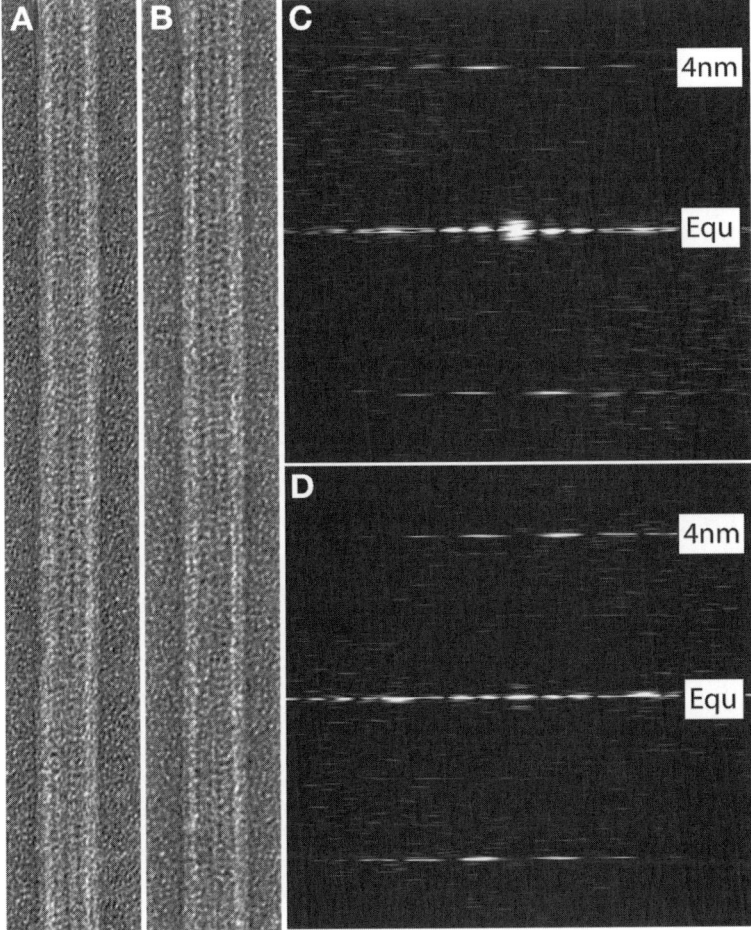

Fig. 6. Negatively stained microtubules (**A,B**) and their diffraction patterns (**C,D**). UAc stain appears black and protein (absence of stain) appears white. This is the reverse of the previous figures, where protein appears darker than the background of ice. As often seen with negative stain, one side of each tube shows better contrast than the other; hence the diffraction patterns are not symmetrical in intensity.

## 3. Methods

### 3.1. Protein Specimens

#### 3.1.1. Microtubule Assembly

1. Tubulin is purified as described in detail in other chapters. We start from fresh pig brain tissue and carry out two cycles of polymerization and depolymerization in BRB80 assembly buffer with 25% glycerol and then remove endogenous MAPs with a PC column. Alternatively, the higher-salt method of Castoldi and

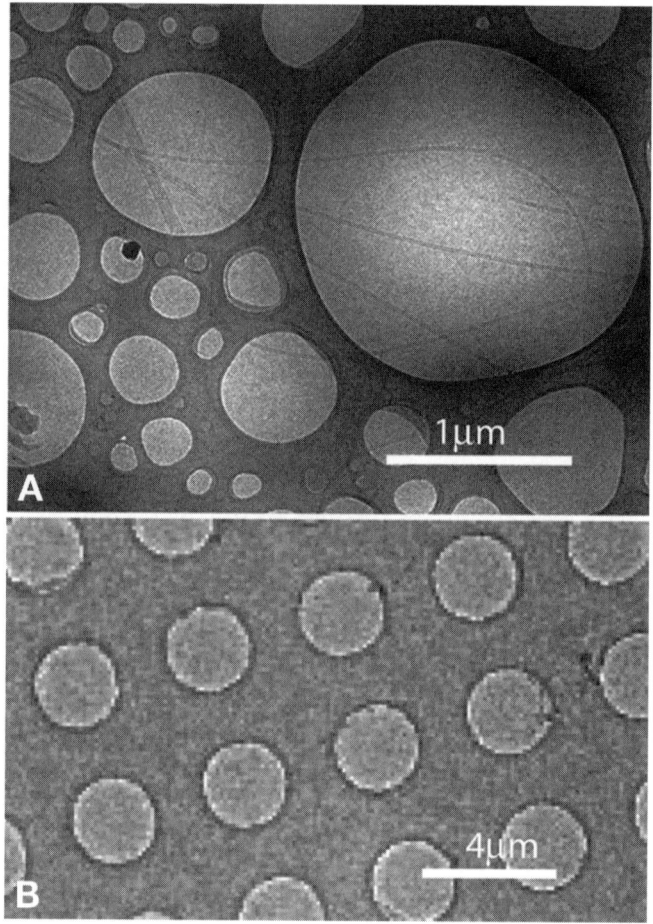

Fig. 7. Holey carbon films viewed by electron microscopy at low magnification. Film **A** was formed as described in the text and a microtubule sample applied to it. The holes in the carbon are filled with amorphous ice. **B** is part of a Quantifoil™ grid.

Popov *(25)* is very efficient. Purified tubulin can be stored in liquid nitrogen, either before PC column purification (in glycerol) or after (in small aliquots), or even as assembled MTs (usually with Taxol).

2. Before polymerization into MTs for EM, frozen soluble tubulin is left to thaw on ice for approx 0.5 h, and centrifuged to remove denatured protein. The solution is incubated at 37°C for 20 min in BRB80 with 5–10% DMSO, which is added to increase the proportion of 15-PF MTs *(26)*. The MTs may then be stabilized with 20–100 μ*M* Taxol. When removal of GTP is required, the MTs are centrifuged, and resuspended in a Taxol solution without GTP.

### 3.1.2. Decoration With Kinesin

Kinesin family motor domains bind to individual tubulin heterodimers in a microtubule *(27)* and thereby emphasize the 8-nm longitudinal periodicity (**Fig. 3B,D**). Motor domain constructs can usually be expressed in *Escherichia coli* and purified either with the help of a tag (e.g., 3xHis) or without one, as described for dimeric kinesin KΔ401 *(28)*. It is worth trying different buffers to find the best conditions for forming the most regular complexes (these can be assayed fairly quickly by negative stain EM—*see* **Subheading 3.2.1.**). For conventional kinesin, taxol-stabilized MTs were diluted in MES buffer (with 20 μ*M* Taxol, pH 6.5), applied to an EM grid coated with a holey carbon film, and KΔ430 was added to give a final concentration of 10 μ*M*. One micromolar AMPPNP was added to put the motors into a state resembling ATP-bound. For freezing in the no-nucleotide state, the MT-KΔ430 mixture was incubated on the grids with 2 U/mL apyrase and then rapidly frozen by plunging the grids into an ethane slush. For the ADP state, the MT-kinesin mixture was put on to the grid first and 1 m*M* ADP, 1 U/mL hexokinase, and 0.01% glucose were added just before freezing. For *Drosophila*, a phosphate or HEPES buffer made the motor proteins more soluble than BRB80 or MES and gave better decoration.

### 3.1.3. Tau Purification

Tau protein, for stabilizing MTs, can also be expressed in *E. coli*. The bacterial pellet is lysed, spun at 15,300*g* and the supernatant loaded on to a PC column *(29)*. The column is washed with 0.1 *M* NaCl in 50 m*M* PIPES, and the protein eluted out with 0.3 *M* NaCl in 50 m*M* PIPES, passed through a 0.20-μm filter and dialysed overnight against saturated ammonium sulfate. The protein is precipitated twice in ammonium sulfate, resuspended in BRB80 and spun at 235,000*g*, then boiled with 0.5 *M* NaCl/2% β-mercaptoethanol in BRB80 for 5 min, followed by chilling on ice, and then gel-filtration through a Sephacryl S200 column prewashed with BRB80.

### 3.1.4. Assembly of Microtubules With Tau/Without Taxol

For the experiments reported by Kar et al. *(19)*, different concentrations of tau (from 1 to 12 μ*M*) were incubated with 10 or 20 μ*M* tubulin in BRB80 buffer with 0.1 *M* TMAO, 1 m*M* GMPCPP and 5% DMSO and incubated for 20 min at 37°C. The presence of TMAO or other osmolytes *(30)* appears to make tau fold up into a better conformation next to tubulin and thus promotes assembly. Assembly in the presence of both TMAO and GMPCPP was designed to help make well-ordered specimens for computer analysis, without adding Taxol, which may interfere with the binding of MAPs.

### 3.1.5. Labeling Tau With Nanogold

1. Electron microscopy can be used to investigate cryo specimens labeled with nanogold or the smaller undecagold particles, because gold has a sufficiently higher density than protein or ice for even quite small amounts to show up in processed images. Nanogold can often be seen directly in unprocessed images but was not immediately evident when incorporated into MTs *(19)*. For this reason and because every tau molecule binds across several tubulin dimers, 3D analysis was carried out for tau labeled with nanogold rather than undecagold. The smaller label is preferable for 1:1 complexes, such as kinesin decorated MTs *(13)*, because it is less likely to interfere with protein–protein binding.

2. Tau protein, mutated to have just a single cysteine residue, was labeled with 1.4 nm maleimido-gold reagent obtained from Nanoprobes, following the supplier's protocol. Labeled tau is separated from an excess of free gold on a Superdex-200 column and labeling is detected by silver-stained and protein-assayed dot blots of each column fraction. Spectroscopic measurements at 280 and 420 nm indicate the 80–95% of molecules that are labeled, as described in the protocol. After separation from free gold, the labeled protein is incorporated immediately into MTs by incubation with tubulin under the previously listed conditions and EM grids are prepared at once, if possible. The presence of gold in the MTs is checked spectroscopically or by dot-blotting. The gold label has been shown to be unstable after 3 d for some proteins but seemed to remain attached to MTs flash-frozen and stored in liquid nitrogen for a few weeks.

## 3.2. EM Specimen Preparation

### 3.2.1. Negative Staining

A solution containing approx 0.1 mg/mL protein is applied to an EM grid coated with a plain (not holey) carbon film. MTs stick well to the carbon unless it has become hydrophobic with time. They are usually negatively stained with 1–2% w/v uranyl acetate (**Fig. 6**), whose low pH fixes MTs; this contrasts with higher pH stains such as sodium phosphotungstate, which cause depolymerization of unfixed MTs. Holey carbon film can also be used to avoid flattening of MTs. However, a stain film stretched over the holes easily breaks during drying of grids and by irradiation with the electron beam.

### 3.2.2. Holey Carbon Film

1. There are at least two methods for preparation of homemade holey-grids (**Fig. 7A**). In one method, a mixture of 50 mL Formvar in chloroform and 20 drops of water–glycerol (1:1 mixture) is used to form a film. After vigorous agitation of the solution to break down the water into minute droplets, a clean microscope slide is dipped into the liquid. When the slide is dry, the thin film is cut close to the edges of the glass slide and floated on a clean water surface. Grids are deposited on top of the film, scooped up with Whatman filter paper and left to dry.

Before coating the film with carbon, the holes are etched by applying methanol to the filter paper. After carbon coating, the Formvar may be dissolved away with a few drops of chloroform on the filter paper. The holey grids are ready to be used once they have dried.

2. The second method is described in Fukami and Adachi *(31)*. Briefly, a slide glass treated with water-repellent is cooled down in a refrigerator, and then taken out to form minute dew-droplets on the surface. Immediately, 0.5% cellulose acetobutyrate (in ethyl acetate) is poured on the surface. After drying, the slide glass is immersed in a rehydrophilic agent, rinsed in water, and then inserted in water to float the film. EM grids are placed on top of the film and scooped up with a piece of paper or Parafilm. The grids can be kept for several months. Before use, the grids are carbon-coated, and the plastic is dissolved with ethyl acetate.

3. The size of the holes usually used for imaging MTs is 1–2 μm. A regular array of holes, as found on commercially available grids (**Fig. 7B**), is more or less essential for automated image collection, which is gradually becoming popular *(32)*.

4. Unless the carbon film to be used is very fresh, it tends to be hydrophobic but the degree of hydro-philicity can be increased before the specimen is applied, by glow-discharging. The commercial grids should always be glow-discharged just before use.

## 3.2.3. Rapid Freezing

1. For electron cryo-microscopy, a solution containing MTs is placed on a grid coated with a holey carbon film (**Fig. 7A**), and the grid is rapidly frozen. Freezing is achieved, as for most specimens, by plunging the grid plus sample into a small pot of liquid ethane that is cold enough to be close to solidification, being kept cold within a larger container of liquid nitrogen. Besides fixing the structure of the MTs, rapid freezing to liquid nitrogen temperature prevents the formation of cubic ice crystals, which would distort the protein structure and also diffract electrons strongly.

2. The freezing apparatus may be driven by gravity or by some form of electrical motor that can be turned on and off quickly, for example, one that is driven by a solenoid. The apparatus has tended to be home-made. A robot for automated specimen preparation (www.Vitrobot.com) has become available recently and allows users to perform cryo-fixation under constant conditions (temperature, relative humidity, blotting conditions, and freezing velocity) more easily, but is expensive. Instructions are provided with the apparatus. The following paragraph describes the manual version of the process.

3. Ethane gas is condensed to form a slush in a small container surrounded by liquid nitrogen. A holey grid is held in forceps clamped to the plunger and the sample is applied to one side. The concentration of protein required for cryo-electron microscopy depends on the method of blotting and the hydrophilicity of the carbon (see **Subheading 3.2.2., item 4**). In our case, 0.01–0.05 mg/mL MTs is used, and the excess solution is blotted from the backside of the grid. The appropriate con-

centration of the MTs can be estimated by applying the same solution on a grid, blotting, air-drying, and checking in an EM with a high defocus. MTs tend to depolymerise at a low concentration, even in the presence of Taxol. Thus, it is better to freeze the MT solution within 1 h after dilution. Also, to avoid cold-induced depolymerization of the MTs, the specimen must be kept in a warm humid atmosphere until it is rapidly frozen. The relative humidity can be kept high by a flow of warm humid air or with an environmental chamber.

4. Four to five microliters of the diluted MT solution is pipetted onto the grid. Excess liquid is blotted away with Whatman filter paper for 2–3 s, leaving a very thin layer of solution containing the specimen across the grid. Blotting time determines how thin the layer of vitreous ice will be and consequently influences the quality of the images. With experience, rather than timing the blotting, the expanding circle of absorbed liquid on the paper may be used as a guide, before the grid is rapidly plunged vertically into the ethane slush. We use the color of the circle. First, it looks transparent, because there is liquid between the grid and filter paper. Then, it becomes white, because a layer of air forms between the grid and the filter paper. After plunge-freezing, the tip of the forceps and grid are transferred into the liquid nitrogen surrounding the pot of ethane and the frozen grid is then placed in a pre-frozen grid holder and stored in liquid nitrogen, ready for viewing by electron cryo-microscopy.

## 3.3. Electron Cryo-Microscopy

1. Electron microscopy of grids held in a cryo-stage, cooled by liquid nitrogen, can be performed on electron microscopes operating at accelerating voltages of 120, 200, or 300 kV, under low-dose conditions. The specimen holder is attached to a reservoir containing liquid nitrogen, which keeps the specimen cold enough for the ice to remain in a glassy amorphous state.

2. Images are best taken at magnifications of ↔35,000–50,000, with a defocus range between –1 and –2 µm. This amount of defocusing enhances small features in the image out to a resolution of approx 20Å (2 nm) without reversing their contrast. In other words, the contrast transfer function in the diffraction plane (close to the objective lens aperture in the microscope) is essentially a circle of 20 $Å^{-1}$ radius (**Fig. 8** shows examples of circles with smaller radii, for clarity) and it is not essential to make any corrections to the contrast. This resolution range is usually the most that can be achieved for MT specimens. In particular cases, highly ordered complexes may allow one to reach 15Å resolution or better (*[11,21,33, 55,56]* Hirose et al., in preparation) and then it becomes essential to correct for the contrast transfer function, which consists of rings of alternating contrast outside the central circle. Although the tendency for MTs to bend into curves presents difficulties, the limit to the resolution for decorated MTs that can be reached by adding more data seems to depend on how firm is the binding of the accessory proteins. Fully decorated MTs tend to be better-ordered than plain ones and our own best results to date have been obtained decorating MTs with a Kar3 motor domain, which binds more strongly than most members of the kinesin family *(35)*.

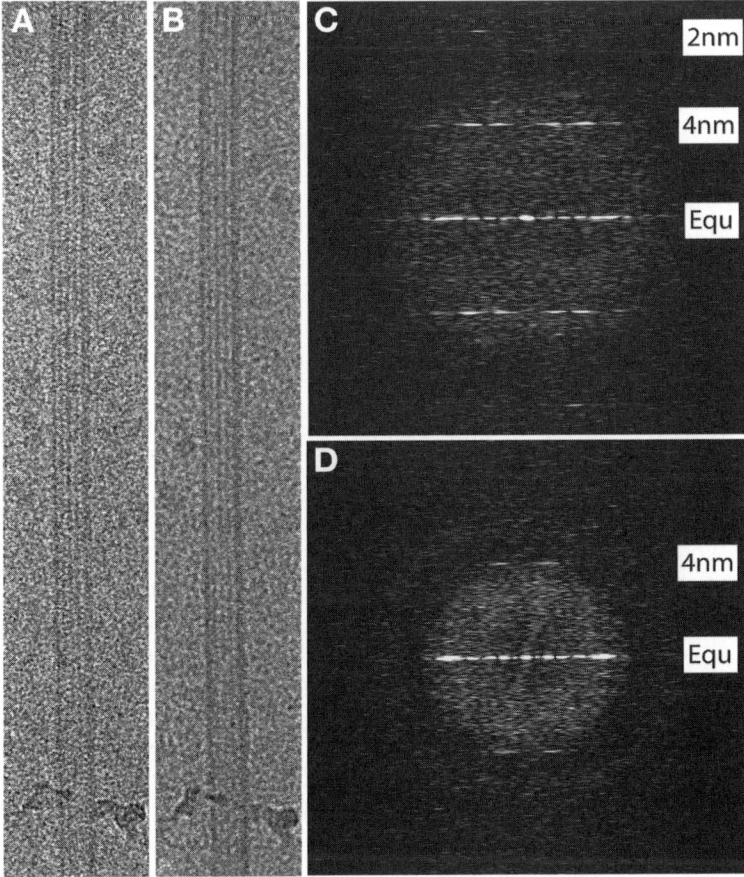

Fig. 8. The effect of underfocusing. **A,B** are images of the same specimen taken with different levels of underfocus (2 and 5 μm). In **C** (the diffraction pattern of **A**), the central circle of enhanced contrast has a radius just beyond 1/4/nm; a 2-nm reflection is picked up by one of the rings of intensity that ripple out beyond the central circle. In **D** (the diffraction pattern of **B**), the approx 5-nm$^{-1}$ radius of the central circle includes the equatorial peaks from the protofilaments, which are thus emphasized in **B**, but not the 4-nm$^{-1}$ peaks from the tubulin subunits.

3. In the past, the images have been recorded on film and scanned with a densitometer but improvements in CCD cameras will probably mean that these are more widely used in future. A pixel size of 28 μm (equivalent to ~6Å on a ×50,000 micrograph, ~8Å on a ×35,000 micrograph) is suitable when one is aiming for 20–30Å resolution reconstructed images.

4. Low-dose imaging: the grids are searched at low magnification (×3000–5000) and with a minimum spot size to minimize the dose of electrons. The contrast can

be enhanced by using a high degree of defocus (*see* **Fig. 8B**). Even at low magnification (*see* **Fig. 7A**) it is possible to select fatter-looking MTs and thus increase the proportion of images of MTs with 15 PFs. Using a low-dose attachment on the microscope, focusing at higher magnification is done on an area of carbon film close to the ice-filled hole that has been selected. Then a high magnification image of the hole is immediately recorded without any unnecessary exposure of this area to electrons.

### 3.4. Image Analysis (see Notes 1 and 2)

### 3.4.1. Initial Analysis of the Images

1. The MRC system of programs *(35,36)* has been used for most of the computer analysis mentioned in the following description. (For high-resolution data [up to 55] we have used a set of programs provided by Dr. C. Toyoshima *[37,38]* [*see also* **Note 1**]). MT images are chosen for analysis if they are long and reasonably straight; also 15-PF specimens can usually be identified from their larger diameters and shorter longitudinal repeat distances (see **Figs. 1** and **3A,B**). The selected images are digitized as a 2D array with stepsize $d$ (= 6-8Å: *see* **Subheading 3.3., item 3**) within a rectangular box and, if a MT is curved (**Fig. 5**), its axis is marked out by eye at a fairly regular series of points for spline-fitting *(39)*. The digital image is then smoothly distorted to make the fitted axial line perfectly straight. If the MT is still not straight and parallel to the sides of the box after this operation, the positions of the points along the axis are refined until the corrected image appears satisfactory. Unless there is no choice, we ignore highly curved specimens to avoid artefacts at a molecular level in the reconstructed images. But it is worthwhile to correct even slight bends to sharpen the diffraction patterns.

2. Now a box can be drawn more closely around the MT. The average density along all of the edges of the new box is calculated and subtracted from every point in the boxed image, a process known as "floating" *(35)*. This means that there are no large steps in density when the image is placed inside a larger box, padded out with zeros, with dimensions suitable for the fast Fourier transform (FFT) algorithm. We typically choose a transform box with dimensions of 512 × 1024, which means that the Fourier transform is sampled at intervals of $1/(512d)$ in X, $1/(1024d)$ in Y (some other groups sample the FFT more coarsely in the lateral dimension, to reduce computation time). A longitudinal box dimension of $1024d$ means that the lengths of MT analyzed are up to approx 0.8 μm (~100 tubulin dimer lengths, i.e., ~1500 molecules), providing enough longitudinal averaging to give good sharp layerlines in the diffraction patterns.

3. The FFT calculation produces a 2D array of complex numbers, from which the amplitudes and phases of the transform are obtained. When the array of amplitudes is displayed, as in **Fig. 3C** or **D**, the diffraction pattern is assessed to see whether further processing of the data is worthwhile. Many images are discarded at this stage. For example, the layerlines in the pattern may not be sharp, indicating that the helical structure of the MT in the image is not well-ordered, or the positions of the minor layerlines may not be consistent with a fully helical 15- or 16-PF structure.

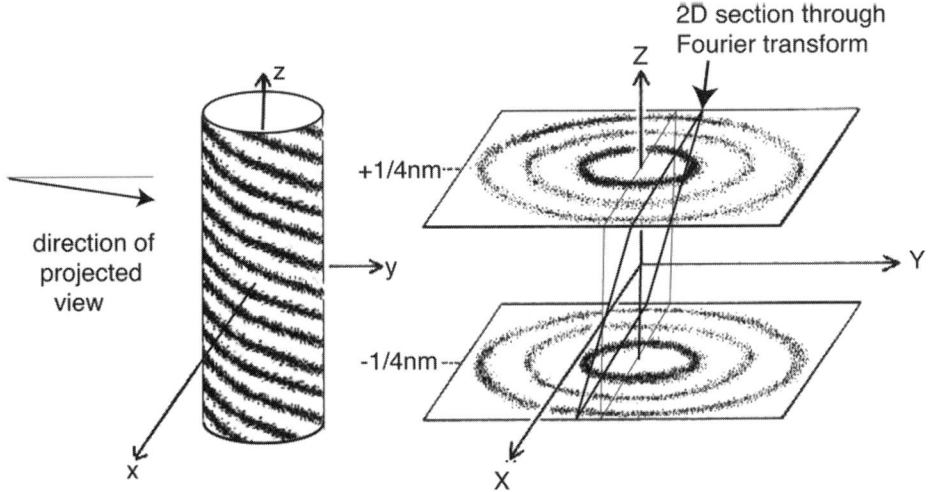

Fig. 9. The density in a MT can be represented by several sets of continuous helices that modulate each other. The Fourier transform of the family of shallow helices shown here, separated by 4 nm (each member of a three-start family, for example, would have a pitch of 12 nm) has a plane containing rings of diffracted intensity 1/4 nm$^{-1}$ above the central origin (the direct, undiffracted, beam) and a corresponding plane below. The two-dimensional (2D) Fourier transform of a projected view is a plane through the center of the three-dimensional transform. Rotating or shifting a continuous helix along its axis does not change its appearance, only its phase in proportion to the value of $n$ (as in **Fig. 4**). The intensities in the diffraction patterns are constant for different views. However, viewing the helix with a tilt changes the relationship between the front and back and the 2D transform plane cuts through the rings at points that are not directly opposite. Their relative phases are not quite correct and the points are closer than they should be. These problems can be corrected. However, a ring very close to the meridian might even be missed, so highly tilted images are avoided.

### 3.4.2. Correcting, Comparing, and Averaging Different Datasets

1. Each 2D transform corresponds, approximately, to a section through the center of the 3D Fourier transform. Provided different layerlines do not overlap (as in the case of 13-PF MTs—*see* **Fig. 4A**), the 3D transform can be generated simply by rotating each layerline around a central axis; the amplitudes remain constant, whereas the phases change by $n$ times the rotation angle (**Fig. 9**). First, some minor corrections to the extracted layerline data are necessary. When the closely fitting box was drawn around the image of the MT, it was assumed that the MT axis ran exactly along the middle of the box and the origin to be referred to in calculating the phases in the FFT was placed in the centre of the box (thus also setting a baseline in the longitudinal [$z$] direction). If the phase origin is indeed positioned on the helix axis, then equivalent peaks in the FFT, on the left and

right sides of the meridional line (arising from the near and far sides of the helix, or vice-versa in the case of a left-handed helix) should be exactly out of phase (differ by 180°) or have the same phase if the number of helices in the family is even. Another factor that disturbs this agreement is tilting of the MT axis towards or away from the observer (*see* **Fig. 9**). In practise, both the phase origin shift and a correction for the tilt angle are refined to bring all the important pairs of phases into close agreement.

2. For each image, two separate datasets (near and far-side data) are extracted from the FFT, containing lists of the amplitudes and phases along each layerline. To compare datasets from different images, they need to be brought to the same orientation. One dataset is chosen as the reference and all others are compared with it. Ranges of values for a shift along the axis and a rotation around the axis are probed, by adjusting the phases along each layerline appropriately, and selecting the shift and rotation that give best overall agreement. The extent of agreement typically obtained for the strong layerlines in a group of datasets is shown in **Fig. 10**. The data can then be averaged and used as a new reference for refined fitting of the relative orientations. The final averaged layerline data can be used to calculate a 3D-density map by Fourier-Bessel transformation *(35,40)*.

### 3.4.3. Displaying the Reconstructed Density

1. Individual 2D sections through the 3D structure are easily visualized, either as density plots or contours. Projected views can also be shown in the same ways. **Figure 11A** is a picture of a MT viewed as if its total density were projected down the PFs and was calculated from just the near-equatorial layerlines: $n = 0$ provides the cylindrical density distribution and $n = 15$ provides the 15-fold component that divides the cylinder into PFs.

2. For portraying density as a 3D structure, a single contour level is chosen, generating a 3D surface. This can be represented as a transparent net, as in **Fig. 13A,B**. Alternatively, it is possible to hide any features that lie behind the parts of the surface first encountered from the viewing direction, as in **Figs . 11** and **12**. Simulation of shadows and reflections from the surface, as if lit from a chosen direction, makes the objects appear solid.

### 3.4.4. Docking Atomic Models

Fitting near-atomic models of tubulin and kinesin obtained by X-ray crystallography or NMR into EM reconstructions of MTs has mainly been carried out by manual docking. The alternative possibilities are semi-automatic fitting of a reduced vector representation of both the model and the data *(41)* or, more directly, of density-correlation *(42,43)*. However, the shapes of tubulin and kinesin are sufficiently asymmetric to be fitted by eye and this means that some features can easily be ignored if the observer judges this to be necessary. Such flexibility might be more problematical if left to a computer program. In **Fig. 13B**, for example, the peak of density marked by an asterisk in the EM density is not

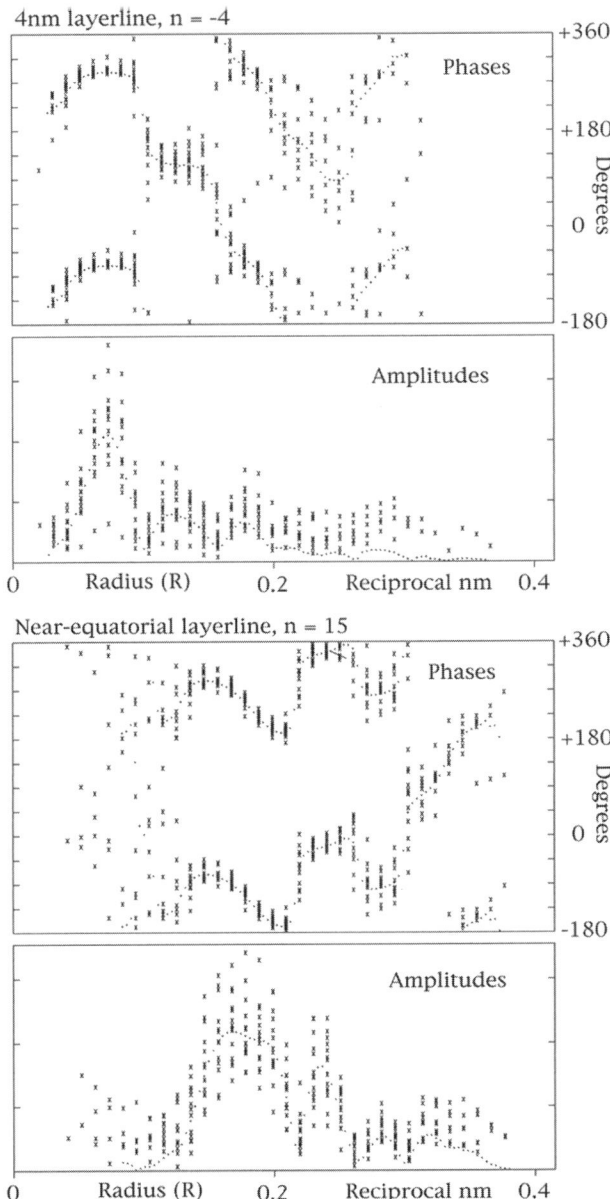

Fig. 10. Plots of the amplitudes and phases along two layerlines calculated from several different cryo-electron microscopy images of kinesin-decorated microtubules (asterisks) and their average values (dotted lines). Data from the near and far sides of each image are plotted separately. The phases of different images have been adjusted to bring them all into equivalent orientations.

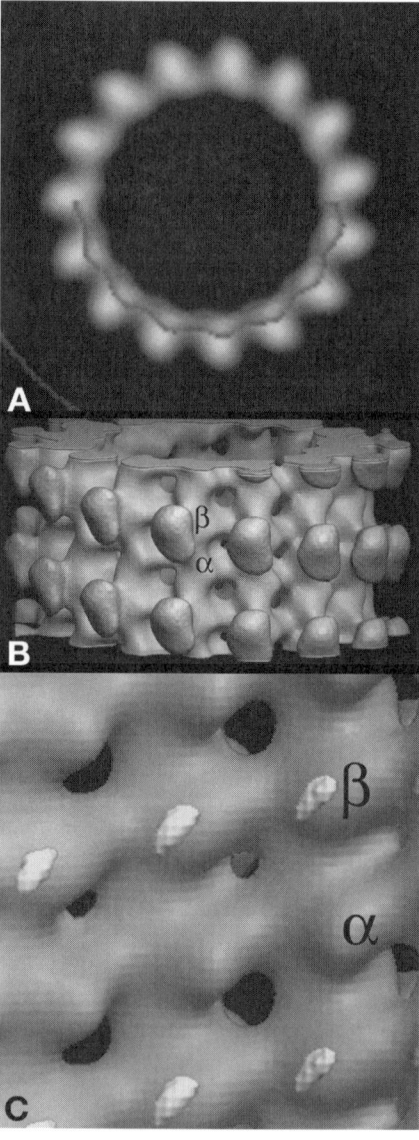

Fig. 11. Reconstructed images of microtubules (MTs) assembled with tau (*19*). (**A**)
A projected view, looking down the protofilaments from the plus end. The darker
density in the lower half of the image is difference density between a MT containing
tau labeled with nanogold in its repeat region and one containing unlabeled tau. (**B,C**)
A 3D reconstruction of a tau-containing MT, decorated with kinesin (globular subunit
on the outer surface), so as to distinguish α- and β-tubulin. Difference density representing gold label on the tau is superimposed in white on the view from inside the MI
(C) shows the label is associated with β-tubulin.

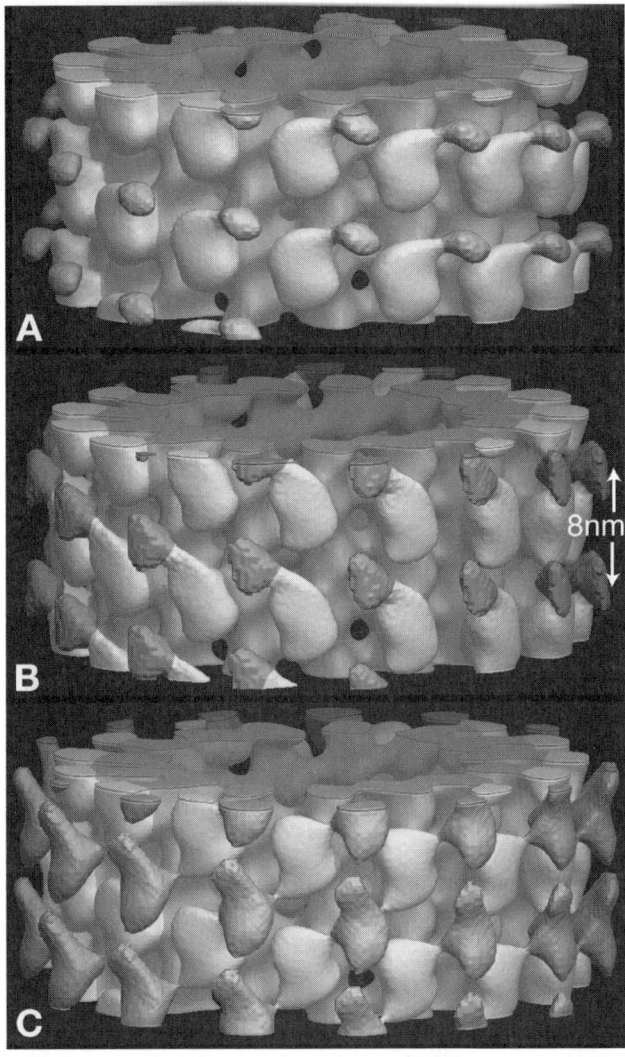

Fig. 12. Three-dimensional reconstructed images of microtubules (MTs) decorated with dimeric Neurospora kinesin (*8*). Density representing the tethered heads is shown darker than the features representing tubulin and the directly bound Kinesin heads. **A** was obtained from specimens in the presence of AMPPNP: the tethered heads appear very small, presumably because their position is not fixed and their average density is thus reduced (*see* **Note 2**). **B&C** were both obtained from specimens in the absence of nucleotides (which were removed from the solution with apyrase). Individual images were sorted into groups before averaging and show the tethered heads in very different positions. The directly bound heads also vary in appearance, which may indicate a conformational change. But note how conserved the structure of tubulin appears in **A,B,C** (best seen at the cut top surfaces of the tubes).

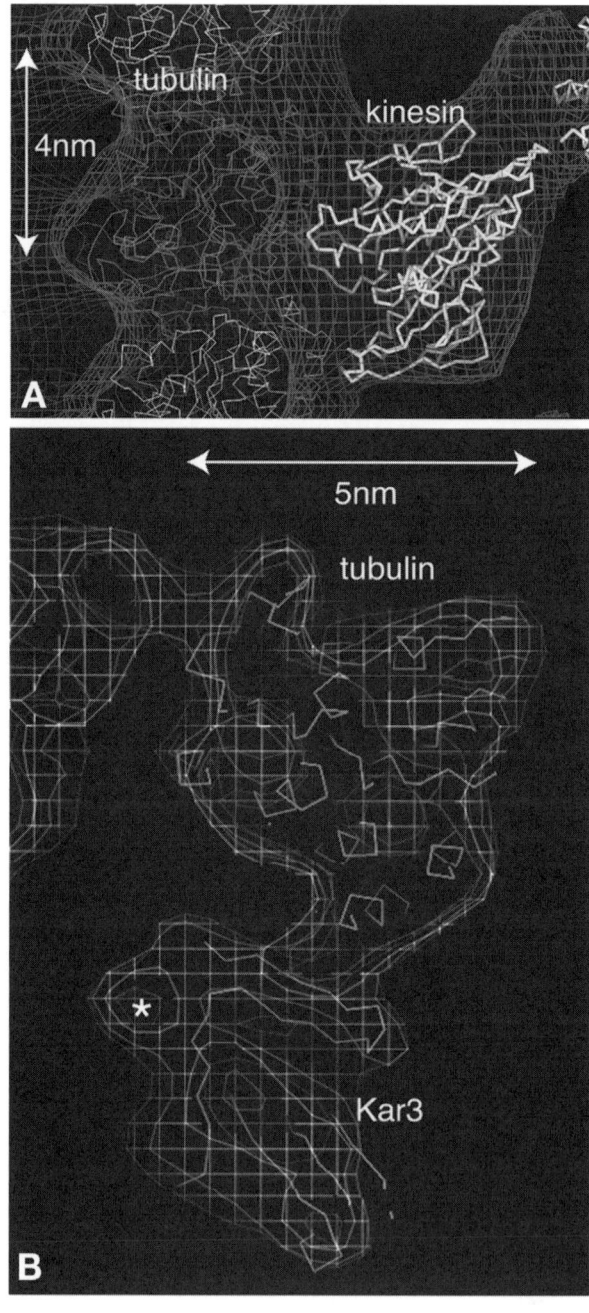

present in the crystal structure. This is understandable because parts of the sequence corresponding to certain loops in the motor domain of Kar3 are not represented in the crystal structure and are presumed to be disordered except when the motor domain is bound to tubulin.

## 4. Notes

1. ***Resolution limits:*** image processing is carried out, both to enhance details in the images and to view the structures in three dimensions, by combining pictures of many different molecules. The assumption is made that the molecules in different images are identical but that they appear different, either because they are being viewed from different directions or because random noise is superimposed. The effect of random noise is reduced more and details of the structure become clearer by averaging the images of many molecules. Furthermore, 3D images can easily be reconstructed because the different viewing directions of the 2D projections are known accurately from the symmetry of the MTs. The twisted lattice of 15-PF MTs not only increases the number of different views of tubulin dimers in a single MT image, compared with straight 13-PF MTs, but also makes it possible to combine information from different images with greater accuracy.

   The structure of tubulin PFs was determined by EM of 2D crystals to sufficient resolution (3.5 Å) to build an atomic model *(44)*. The question arises as to whether it would be possible to achieve atomic resolution for actual MTs by combining enough different images. To date, the best that has been achieved is 8 Å for plain 13-PF MT *(21)* (and 12 Å for PFs coiled to form tubes *[23]*). But in the case of the 13-PF MTs, all the monomers were averaged so the differences between α- and β-tubulin could not be resolved. MTs without a seam have been reconstructed to slightly lower resolutions; we have reconstructed Kar3-decorated MT to 10–12 Å resolution *(55)*; Kikkawa et al. *(11)* reconstructed helical MTs decorated with Kif1A 15 Å and more recently *(56)* to 10 Å; Meurer-Grob et al. *(33)* were able to distinguish α- and β-tubulin in undecorated MTs at 14–15Å. Experience suggests that near-atomic resolution will require finding a favourable

---

Fig. 13. (*opposite page*) Docking of atomic models into electron microscopy (EM) maps of microtubules (MTs) decorated with motor domains. (**A**) A trace of the backbone of the kinesin crystal structure *(51)*, superimposed on an EM map obtained for dimeric kinesin *(7)*. As in **Fig. 12A**, the tethered head of the kinesin dimer *(52)* is badly under-represented in the EM density. Because the resolution of this EM map is quite low, it is not clear whether the motor domain bound to a MT is identical to the crystal structure or locally changed. (**B**) A higher resolution (at least 12 Å) EM map of Kar3 *(55)* shows internal density peaks, some of which can be matched nicely to secondary structural features in the Kar3 crystal structure *(53)*. Others, such as the extra peak labeled *, provide evidence of conformational change. Tubulin *(54)* fits well into the MT density.

specimen. For example, we have found that MTs decorated with Kar3 monomers gave better images (diffracting to higher resolution) than undecorated MTs or MTs decorated with conventional kinesin. Evidently, some complexes assume more stable conformations than others. Eventually, it may be possible to reach the 4-Å resolution already achieved for bacterial flagella by helical analysis *(45)*.

2. Disappearing second heads: it is not always possible to reach a high resolution but often there are interesting questions to be answered at low resolution. The movement of the tethered heads of dimeric kinesins is one example. When we studied dimeric Neurospora kinesin *(8)*, 3D images reconstructed from individual micrographs were very variable with respect to density that could be ascribed to the motor domains that were expected to be tethered to their dimeric partners via the coiled coil. In fact, highly averaged images resembled MTs decorated with monomeric motors. We concluded that the tethered second heads could "park" in at least two different positions and also move around freely elsewhere. The situation was improved by sorting the individual reconstructions by eye and averaging groups of reconstructions that were most similar. The results for dimeric motors in the absence of nucleotide are shown in **Fig. 12B,C** and suggest that under these conditions there are two major positions for the tethered head, on opposite sides of the directly bound head. It is likely that the positions of the heads vary on individual MTs but the proportions are not the same in different cases. Thus, it may be impossible to reconstruct a "pure" image showing the tethered head in a particular position. However, the worst thing to do in such a case may be to indiscriminately average a large number of images.

## References

1. Amos, L. A. and Schlieper, D. (2005) Microtubules and MAPs. *Adv. Protein Chem.* **71,** 257–298.
2. Arnal, I. and Wade, R. H. (1998) Nucleotide-dependent conformations of the kinesin dimer interacting with microtubules. *Structure* **6,** 33–38.
3. Dias, D. P. and Milligan, R. A. (1999) Motor protein decoration of microtubules grown in high salt conditions reveals the presence of mixed lattices. *J. Mol. Biol.* **287,** 287–292.
4. Hirose, K., Lockhart, A., Cross, R. A., and Amos, L. A. (1995) Nucleotide-dependent angular change in kinesin motor domain bound to tubulin. *Nature* **376,** 277–279.
5. Hirose, K., Lockhart, A., Cross, R. A., and Amos, L. A. (1996) Three-dimensional cryoelectron microscopy of dimeric kinesin and ncd motor domains on microtubules. *Proc. Natl. Acad. Sci. USA* **93,** 9539–9544.
6. Hirose, K., Cross, R. A., and Amos, L. A. (1998) Nucleotide-dependent structural changes in dimeric ncd molecules complexed to microtubules. *J. Mol. Biol.* **278,** 389–400.
7. Hirose, K., Löwe J., Alonso M., Cross R. A., and Amos L. A. (1999) Congruent docking of dimeric kinesin and ncd into 3D electron cryo-microscopy maps of microtubule-motor.ADP complexes. *Mol. Biol. Cell* **10,** 2063–2074.

8. Hirose, K., Henningsen, U., Schliwa, M., et al. (2000) Structural comparison of dimeric Eg5, Neurospora kinesin (Nkin) and Ncd head-Nkin neck chimaera with conventional kinesin. *EMBO J.* **19,** 5308–5314.

9. Hoenger, A., Sablin, E. P., Vale, R. D., Fletterick, R. J., and Milligan, R. A. (1995) Three-dimensional structure of a tubulin-motor-protein complex. *Nature* **376,** 271–274.

10. Kikkawa, M., Ishikawa, T., Wakabayashi, T., and Hirokawa, N. (1995) 3-Dimensional structure of the kinesin head-microtubule complex. *Nature* **376,** 274–277.

11. Kikkawa, M., Okada Y., and Hirokawa N. (2000) 15 angstrom resolution model of the monomeric kinesin motor, KIF1A. *Cell* **100,** 241–252.

12. Kikkawa, M., Sablin, E. P., Okada, Y., Yajima, H., Fletterick, R. J., and Hirokawa, N. (2001) Switch-based mechanism of kinesin motors. *Nature* **411,** 439–445.

13. Rice, S., Lin, A. W., Safer, D., et al. (1999) A structural change in the kinesin motor protein that drives motility. *Nature* **402,** 778–784.

14. Skiniotis, G., Cochran, J. C., Muller, J., Mandelkow, E., Gilbert, S. P., and Hoenger, A. (2004) Modulation of kinesin binding by the C-termini of tubulin. *EMBO J.* **23,** 989–999.

15. Skiniotis, G., Surrey, T., Altmann, S., et al. (2003) Nucleotide-induced conformations in the neck region of dimeric kinesin. *EMBO J.* **22,** 1518–1528.

16. Song, Y. H., Marx, A., Muller, J., et al. (2001) Structure of a fast kinesin: implications for ATPase mechanism and interactions with microtubules. *EMBO J.* **20,** 6213–6125.

17. Sosa, H., Dias, D. P., Hoenger, A., et al. (1997) A model for the microtubule-Ncd motor protein complex obtained by cryo-electron microscopy and image analysis. *Cell* **90,** 217–224.

18. Wendt, T. G., Volkmann, N., Skiniotis, G., et al. (2002) Microscopic evidence for a minus-end-directed power stroke in the kinesin motor ncd. *EMBO J.* **21,** 5969–5978.

19. Kar, S., Fan, J., Smith, M. J., Goedert, M., and Amos, L. A. (2003) Repeat motifs of tau bind to the insides of microtubules in the absence of taxol. *EMBO J.* **22,** 70–77.

20. Moores, C. A., Perderiset, M., Francis, F., Chelly, J., Houdusse, A., and Milligan, R. A. (2004) Mechanism of microtubule stabilization by doublecortin. *Mol. Cell* **14,** 833–839.

21. Li, H., DeRosier, D., Nicholson, W., Nogales, E., and Downing, K. (2002) Microtubule structure at 8Å resolution. *Structure* **10,** 1317–1328.

22. Kikkawa, M. (2004) A new theory and algorithm for reconstructing helical structures with a seam. *J . Mol Biol.* **343,** 943–955.

23. Wang, H. -W., and Nogales, E. (2005) Nucleotide-dependent bending flexibility of tubulin regulates microtubule assembly. *Nature* **435,** 911–915.

24. Hyman, A. A., Salser, S., Drechsel, D. N., Unwin, N., and Mitchison, T. J. (1992) Role of GTP hydrolysis in microtubule dynamics: information from a slowly hydrolyzable analogue, GMPCPP. *Mol. Biol. Cell* **3,** 1155–1167.

25. Castoldi, M., and Popov, A. V. (2003) Purification of brain tubulin through two cycles of polymerization-depolymerization in a high-molarity buffer. *Protein Expr. Purif.* **32,** 83–88.

26. Ray, S., Meyhöfer, E., Milligan, R. A., and Howard, J. (1993) Kinesin follows the microtubule's protofilament axis. *J. Cell Biol.* **121,** 1083–1093.
27. Harrison, B. C., Marchese-Ragona, S. P., Gilbert, S. P., Cheng, N., Steven, A. C., and Johnson, K. A. (1993) Decoration of the microtubule surface by one kinesin head per tubulin heterodimer. *Nature* **362,** 73–75
28. Lockhart, A., Crevel, I. M., and Cross, R. A. (1995) Kinesin and ncd bind through a single head to microtubules and compete for a shared MT binding site. *J. Mol. Biol.* **249,** 763–771.
29. Smith, M. J., Crowther, R. A., and Goedert, M. (2000) The natural osmolyte trimethylamine N-oxide (TMAO) restores the ability of mutant tau to promote microtubule assembly. *FEBS Lett.* **484,** 265–270.
30. Tseng, H. C. and Graves, D. J. (1998) Natural methylamine osmolytes, trimethylamine N-oxide and betaine, increase tau-induced polymerization of microtubules. *Biochem. Biophys. Res. Commun.* **250,** 726–730.
31. Fukami, A., and Adachi, K. (1965) A new method of preparation of a self-perforated micro plastic grid and its application. *J. Electron Microsc.* **14,** 112–118.
32. Carragher, B., Fellmann, D., Guerra, F., et al. (2004) Rapid routine structure determination of macromolecular assemblies using electron microscopy: current progress and further challenges. *J . Synchrotron Radiat.* **11,** 83–85.
33. Meurer-Grob, P., Kasparian, J., and Wade, R. H. (2001) Microtubule structure at improved resolution. *Biochemistry* **40,** 8000–8008.
34. Song, H. and Endow, S. A. (1998) Decoupling of nucleotide- and microtubule-binding sites in a kinesin mutant. *Nature* **396,** 587–590.
35. DeRosier, D. J. and Moore, P. B. (1970) Reconstruction of three dimensional images from electron micrographs of structures with helical symmetry. *J. Mol. Biol.* **52,** 355–369.
36. Crowther, R. A., Henderson, R., and Smith, J. M. (1996) MRC image processing programs. *J. Struct. Biol.* **116,** 9–16.
37. Yonekura, K., Toyoshima, C., Maki-Yonekura, S., and Namba, K. (2003) GUI programs for processing individual images in early stages of helical image reconstruction—for high-resolution structure analysis. *J. Struct. Biol.* **144,** 184–194.
38. Toyoshima, C. and Unwin, N. (1990) Three-dimensional structure of the acetylcholine receptor by cryoelectron microscopy and helical image reconstruction. *J. Cell Biol.* **111,** 2623–2635.
39. Egelman, E. H. (1986) An algorithm for straightening images of curved filamentous structures. *Ultramicroscopy* **19,** 367–373.
40. Moody, M. F. (1990) Image analysis of electron micrographs, in *Biophysical Electron Microscopy,* (Hawkes, P. W. and Valdrè, U., ed.), Academic Press, New York, pp. 145–287.
41. Wriggers, W. and Birnens, S. (2001) Using situs for flexible and rigid-body fitting of multiresolution single-molecule data. *J. Struct. Biol.* **133,** 193–202.
42. Roseman, A. M. (2000) Docking structures of domains into maps from cryo-electron microscopy using local correlation. *Acta Cryst. D* **56,** 1332–1340.

43. Volkmann, N., and Hanein, D. (2003) Docking of atomic models into reconstructions from electron microscopy. *Methods Enzymol.* **374,** 204–225.
44. Nogales, E., Wolf, S., and Downing, K. H. (1998) Structure of the tubulin dimer by electron crystallography. *Nature* **391,** 199–203.
45. Yonekura, K., Maki-Yonekura, S., and Namba, K. (2003) Complete atomic model of the bacterial flagellar filament by electron cryomicroscopy. *Nature* **424,** 643–650.
46. Wade, R. H., Chrétien, D., and Job, D. (1990) Characterization of microtubule protofilament numbers. How does the surface lattice accommodate? *J. Mol. Biol.* **212,** 775–786.
47. Chrétien, D. and Wade, R. H. (1991) New data on the microtubule surface lattice. *Biol. Cell* **71,** 161–174.
48. Mandelkow, E. M., Schultheiss, R., Rapp, R., Muller, M., and Mandelkow, E. (1986) On the surface lattice of microtubules: helix starts, protofilament number, seam, and handedness. *J. Cell Biol.* **102,** 1067–1073.
49. Kikkawa, M., Ishikawa, T., Nakata, T., Wakabayashi, T., and Hirokawa, N. (1994) Direct visualization of the microtubule lattice seam both in vitro and in vivo. *J. Cell Biol.* **127,** 1965–1971.
50. Song, Y. H. and Mandelkow, E. (1995) The anatomy of flagellar microtubules: polarity, seam, junctions, and lattice. *J. Cell Biol.* **128,** 81–94.
51. Kull, F. J., Sablin, E. P., Lau, R., Fletterick, R. J., and Vale, R. D. (1996) Crystal structure of the kinesin motor domain reveals a structural similarity to myosin. *Nature* **380,** 550–555.
52. Kozielski, F., Sack, S., Marx, A., et al. (1997) The crystal structure of dimeric kinesin and implications for microtubule-dependent motility. *Cell* **91,** 985–941.
53. Gulick, A. M., Song, H., Endow, S. A., and Rayment, I. (1998) X-ray crystal structure of the yeast Kar3 motor domain complexed with Mg.ADP to 2.3A resolution. *Biochemistry* **37** , 1769–1776.
54. Löwe, J., Li, H., Downing, K. H., and Nogales, E. (2001) Refined structure of tubulin at 3.5Å resolution. *J. Mol. Biol.* **313,** 1045–1057.
55. Hirose, K., Akimaru, E., Akiba, T., Endow, S. A., and Amos, L. A. (2006) Large conformational changes in a kinesin motor catalysed by interaction with microtubules. *Mol. Cell* **23,** 913–923.
56. Kikkawa, M. and Hirokawa, N. (2006) High-resolution cryo-EM maps show the nucleotide binding pocket of KIF1A in open and closed conformations. *EMBO J.* **25,** 4187–4194.

# 6

# Fluorescence Microscopy of Microtubules in Cultured Cells

## Irina Semenova and Vladimir Rodionov

## Summary

Cytoplasmic microtubules are noncovalent polymers of the protein tubulin. In the cells, the main function of microtubules is to provide tracks for organelle transport. Two experimental approaches based on fluorescence microscopy are commonly used to examine organization of microtubules in mammalian tissue culture cells. The first experimental approach involves indirect immunofluorescence staining of chemically fixed cells with tubulin antibody. Fluorescence microscopy of immunostained specimens allows the examination of the distribution of microtubules in the cytoplasm at the moment of fixation. The second experimental approach involves introduction of tubulin subunits covalently labeled with a fluorochrome into the cytoplasm of living cells. Time-lapse fluorescence microscopy of cells containing labeled tubulin subunits allows to examine changes in the spatial organization of microtubules in the cytoplasm and also to directly observe their behavior. In this chapter, we describe preparation of samples for fluorescence microscopy of microtubules.

**Key Words:** Microtubules; tubulin; fluorescence microscopy; microtubule dynamics; immunostaining; live cell imaging.

## 1. Introduction

Cytoplasmic microtubules are noncovalent polymers of the protein tubulin. In the cells, the main function of microtubules is to provide tracks for organelle transport. Organization of microtubules in the cytoplasm therefore depends to a large extent on cellular transport needs and varies between the cell types and phases of the cell cycle. In interphase, microtubules are often organized into radial arrays or form bundles inside the cellular processes, such as axons and dendrites of neurons. During the transition to mitosis, interphase arrays disassemble and microtubules become organized into mitotic spindle. In both interphase and mitotic arrays microtubules are never static but continuously turn over tubulin subunits between the soluble and polymer pools through a mecha-

From: *Methods in Molecular Medicine, Vol. 137: Microtubule Protocols*
Edited by: Jun Zhou © Humana Press Inc., Totowa, NJ

nism known as dynamic instability—growth and shortening at the free ends (*see* **ref.** *1* for review). This remarkable dynamic behavior allows microtubules to continuously explore the intracellular space *(1)*. Two experimental approaches based on fluorescence microscopy are commonly used to examine organization of the microtubule cytoskeleton in mammalian tissue culture cells.

The first experimental approach involves indirect immunofluorescence staining of chemically fixed cells with a tubulin antibody. Chemical fixation involves treatment of cells with a fixative solution, permeabilization with a detergent solution, and successive incubation with primary antibody specific for tubulin and a secondary fluorochrome-conjugated antibody that binds to the primary tubulin antibody. Fluorescence microscopy of immunostained specimens allows the examination of the distribution of microtubules in the cytoplasm at the moment of fixation. The limitation of immunofluorescence staining approach is that it does not provide information about the dynamic properties of microtubules.

The second experimental approach involves introduction of tubulin subunits covalently labeled with a fluorochrome into the cytoplasm of living cells. Because concentration of tubulin subunits in microtubules is significantly higher than in the soluble cytoplasmic pool, incorporation of fluorescent tubulin subunits into microtubules results in a bright fluorescence signal. Time-lapse fluorescence microscopy of cells containing labeled tubulin subunits allows to examine changes in the spatial organization of microtubules in the cytoplasm and also to directly observe their growth and shortening.

In this chapter, we describe preparation of samples for fluorescence microscopy of microtubules. We have organized the chapter into two separate sections. The first section describes experimental protocols for indirect immunofluorescence staining with tubulin antibodies of microtubules in chemically fixed cells. The second section focuses on the methods that are used for the labeling of microtubules in living cells with fluorescently tagged tubulin subunits.

### 1.1. Indirect Immunofluorescence Staining of Chemically Fixed Cells With a Tubulin Antibody

Immunostaining involves fixation of cells, peremeablization with a nonionic detergent, and incubation with a tubulin antibody and a secondary antibody conjugated with fluorochrome, such as rhodamine or fluorescein. The most common fixatives are aldehydes, such as formaldehyde or glutaraldehyde, and methanol. Glutaraldehyde is a milder fixative and therefore may be best to preserve native localization of microtubules. In case of aldehyde fixation, it is necessary to permeabilize cells with a nonionic detergent, such as Triton X-100.

The detergent treatment perforates the plasma membrane and allows tubulin antibody to bind to microtubules. This step can be performed after the aldehyde fixation, but the best results are obtained when cells are permeabilized prior to fixation to remove soluble tubulin subunits and therefore reduce the background cytoplasmic staining. In this case, microtubule-stabilizing drug Taxol (Paclitaxel) is included in the extraction buffer to prevent disassembly of microtubules during extraction. Permeabilized fixed cells are incubated with a mouse monoclonal or a polyclonal tubulin antibody and a secondary affinity purified antibody raised against the species, in which the first antibodies are produced.

## 1.2. Labeling of Microtubules in Living Cells With Fluorescently Tagged Tubulin Subunits

Two experimental strategies are commonly used to introduce fluorescently tagged tubulin subunits into the cytoplasm of the living cells. The first strategy involves covalent attachment of small molecules of fluorescent dyes, such as rhodamine of Cy3, to biochemically purified tubulin subunits. Fluorescently labeled tubulin subunits are then microinjected into the living cells (*see* **ref. 2** for the detailed description of the microinjection technique). The second experimental strategy employs intracellular expression of tubulin cDNA fused to the cDNA of a fluorescent protein, such as green fluorescent protein (GFP). Significant advantage of the second approach is that it does not require tubulin purification, modification, and microinjection steps *(3)*. Stable cell lines of mammalian cells expressing GFP-tagged tubulin have been described *(4)*.

## 2. Materials
### 2.1. Indirect Immunofluorescence Staining of Chemically Fixed Cells With a Tubulin Antibody

1. Tissue culture cells grown on glass cover slips (thickness 1.5).
2. Phosphate buffered saline (PBS): 8 g NaCl, 0.2 g KCl, 0.2 KH$_2$PO$_4$, and 1.15 g Na$_2$HPO$_4$ per liter adjusted to pH 7.3 with NaOH prewarmed to 37°C.
3. Cell lysis buffer: if cells are extracted prior to fixation: 60 m$M$ PIPES, 25 m$M$ HEPES, 10 m$M$ EGTA, 2 m$M$ MgCl$_2$, pH 6.9 supplemented with 0.5% Triton X-100 (Pierce, Rockford, IL) and 10 µg/mL Paclitaxel (Sigma Chemical Co., St. Louis, MO) prewarmed to 37°C. If cells are extracted after fixation: 1% Triton X-100 in PBS.
4. One of the following solutions for fixation of cells: methanol (prechilled to –20°C in a freezer); 4% formaldehyde solution (e.g., 16% microfiltered stock solution from Ted Pella Inc., diluted four times) in PBS prewarmed to 37°C; 1% glutaraldehyde (electron microscopic grade) solution in PBS.
5. 1% Sodium borohydride solution in PBS if cells are fixed with glutaraldehyde.

6. Blocking solution: 1% bovine serum albumin.
7. Immunostaining chamber: large (140 mm) Petri dish with a layer of Parafilm on the bottom.
8. Washing containers: six-well cell culture dishes.
9. Tubulin antibody, such as mouse monoclonal antibody DM1A or DM1B (from Abcam Inc., Cambridge, MA, or other vendor; 1–10 μg/mL) or sheep polyclonal affinity purified antibody ATN02 (Cytoskeleton Inc., Denver, CO; 1–10 μg/mL).
10. Affinity purified secondary antibody (e.g., anti-mouse or anti-sheep secondary antibody if mouse monoclonal or sheep polyclonal primary tubulin antibody was used) conjugated with rhodamine, fluorescein, or other fluorochrome suitable for fluorescence microscopy (1–10 μg/mL).
11. Polyvinyl alcohol-based mounting medium, such as Aqua PolyMount (Polysciences Inc., Warrington, PA).

## 2.2. Labeling of Microtubules in Living Cells

### 2.2.1. Microinjection of Purified Tubulin Subunits Chemically Labeled With a Fluorochrome

1. Tissue culture cells plated onto cover slips with photo-etched locator grids (24 × 24 mm; Bellco Biotechnology, Vineland, NJ) mounted with silicone vacuum grease (Dow Corning Corp., Midland, MI) into 35-mm Petri dishes over 18-mm holes drilled in the bottom. This arrangement allows observation of the cells with an objective lens of high enough magnification for microinjection (Nikon, ×40 0.65 N.A. DL). Photoetched locator grids help locating microinjected cells.
2. Tubulin subunits labeled with rhodamine or Cy3 (3–5 mg/mL). Rhodamine-tubulin can be purchased from Cytoskeleton Inc. (Denver, CO). Alternatively, tubulin subunits can be purified from pig brain by two cycles of disassembly and reassembly of microtubules (5) and labeled with tetramethylrhodamine succinimide (Research Organics, Cleveland, OH) or Cy3 bisfunctional dye (Cy™ Bis-Reactive Dye, Amersham Biosciences, Piscataway, NJ) as described in **ref. 6**.
3. Glass capillaries (D = 1–1.5 mm with filament) for the preparation of micropipets for microinjection (can be purchased, e.g., from WPI Inc., Sarasota, FL). Exact diameter of glass capillaries is determined by the opening in micropipet holder of a micromanipulator.

### 2.2.2. Expression in the Cells of Tubulin Subunits Tagged With a Fluorescent Protein

1. Tissue culture cells (80–90% confluent) plated onto 22 × 22-mm cover slips mounted into 35-mm Petri dishes over 18-mm holes drilled in their bottoms as described in the **Subheading 2.2.1.**
2. Expression vector containing tubulin sequence fused with GFP or other fluorescent protein, such as pAcGFP1-Tubulin expression vector commercially available from Clontech (Mountain View, CA), which contains human α-tubulin

cDNA with GFP added to the 5'-end. About 0.2–0.5 µg of purified plasmid DNA is used for each transfection reaction.

3. Reagents for cell transfection: Lipofectamine 2000 transfection reagent (Invitrogen Corp., Carlsbad, CA) or other reagent suitable for transient transfection of mammalian cells.

## 2.3. Fluorescence Microscopy of Microtubules

1. For fluorescence microscopy of fixed specimens, upright or inverted fluorescence microscope is required, equipped with ×40, ×60, or ×100 objective lens and a filter appropriate for fluorochrome conjugated with a secondary antibody. Images of microtubules are taken with an automatic photographic camera or with a digital charge coupled device (CCD) camera.

2. Live imaging of microtubules is performed with a scientific grade inverted fluorescence microscope, such as Nikon Eclipse TE300 or 2000 series, with ×100 or ×63 1.4 N.A. objective lens and 100-W mercury arc lamp. Light passes through a narrow band rhodamine filter set, which is compatible with Cy3 and rhodamine fluorescence or a narrow band FITC or GFP-optimized filter set compatible with GFP fluorescence (or other appropriate filter set if a fluorescent protein other than GFP is tagged to tubulin). Images of microtubules are acquired with a CCD camera, such as Photometrix series 300 cooled CCD camera or other camera with a high (>65%) quantum efficiency at the emission wavelength of rhodamine (Cy3) or AcGFP or other fluorescent protein. The camera shutter is controlled with a software such as Metamorph software (Universal Imaging Corp., Downington, PA), which integrates individual images into time-lapse sequences. Because microtubule dynamics is temperature-dependent it is important to maintain the temperature at 37°C during the imaging process with an objective and microscope stage heaters such as Bioptechs Delta T controlled culture dish system (Bioptechs, Butler, PA).

# 3. Methods

## 3.1. Immunostaining of Chemically Fixed Cells With Tubulin Antibodies

1. To extract soluble tubulin prior to aldehyde fixation, cover slips with cells are rinsed three times with warm PBS and incubated for 3 min at room temperature in a warm microtubule-stabilizing lysis buffer, briefly rinsed with PBS to remove the excess of lysis buffer and fixed with formaldelyde or glutaraldehyde as described in **Subheading 3.1.2., step 2**.

2. For fixation, cells on cover slips are rinsed with PBS and transferred into one of the following solutions: methanol precooled to –20°C for 3–5 min; 4% formaldehyde in PBS for 30 min at room temperature; 1% glutaraldehyde in PBS for 30 min at room temperature. Cover slips are then rinsed several times with PBS to remove a fixative.

3. If **step 1** (extraction prior to fixation) is avoided, cover slips with aldehyde-fixed cells are incubated for 30 min at room temperature in 1% Triton X-100 solution in PBS and washed three times with PBS to remove the excess of Triton X-100.

4. If glutaraldehyde is used for fixation (*see* **step 2,c**), cover slips are incubated for 3 ↔ 5 min in 1% sodium borohydride in PBS and washed three times with PBS to remove borohydride.

5. Cover slips are blocked to prevent nonspecific binding of antibodies by incubating in blocking solution (1% bovine serum albumin in PBS) for 20 min at room temperature.

6. The excess of blocking solution is removed from cover slips by gently pressing an edge of a cover slip against filter paper. Each cover slip is then laid cells-down onto a drop (50 µL) of a tubulin antibody solution placed onto the Parafilm sheet in immunostaining chamber. Cover slips are incubated for 1 h at room temperature for the binding of tubulin antibody to microtubules.

7. To remove the unbound tubulin antibody, cover slips are transferred into wells of a six-well tissue culture dish filled with PBS and incubated at room temperature for 30 min. PBS solution in the wells is changed three times during the incubation.

8. Excess of PBS is removed with filter paper as in **step 6**, and cover slips are placed onto 50-µL drops of a fluorochrome-conjugated secondary antibody and incubated at room temperature for 1 h for the binding of secondary antibody to the primary tubulin antibody.

9. Washing **step 7** is repeated.

10. Excess of PBS is removed as in **step 6** and each cover slip is mounted cells down on a clean glass slide over a drop (20–30 µL) of mounting medium. Slides are incubated at 37°C to solidify the mounting medium and specimens are observed in a fluorescence microscope, equipped with an appropriate filter set for a fluorochrome conjugated to a secondary antibody. Images of microtubules are taken with a digital camera as described in **Subheading 3.2.3.**

    **Figure 1** shows characteristic distribution of microtubules in a human fibroblast revealed by immunostaining with a tubulin antibody.

## 3.2. Labeling of Microtubules in Living Cells

### 3.2.1. Microinjection of Purified Tubulin Subunits Chemically Labeled With a Fluorochrome

1. An aliquot of Cy3- or rhodamine-tubulin is thawed and centrifuged at 24,000$g$ for 30 min at 4°C to remove particulates.

2. Cells grown on a photoetched cover slip mounted into 35-mm Petri dish as described in **Subheading 2.** are injected with fluorescently labeled tubulin using a glass microneedle with a tip diameter of about 1 µm as described in **ref. 2**. The pressure source for microinjection is provided by a microinjector such as Pico-Injector from Harvard Apparatus (Holliston, MA). Alternatively, a 25-mL gas-tight syringe can be used to deliver pressure into a micropipet.

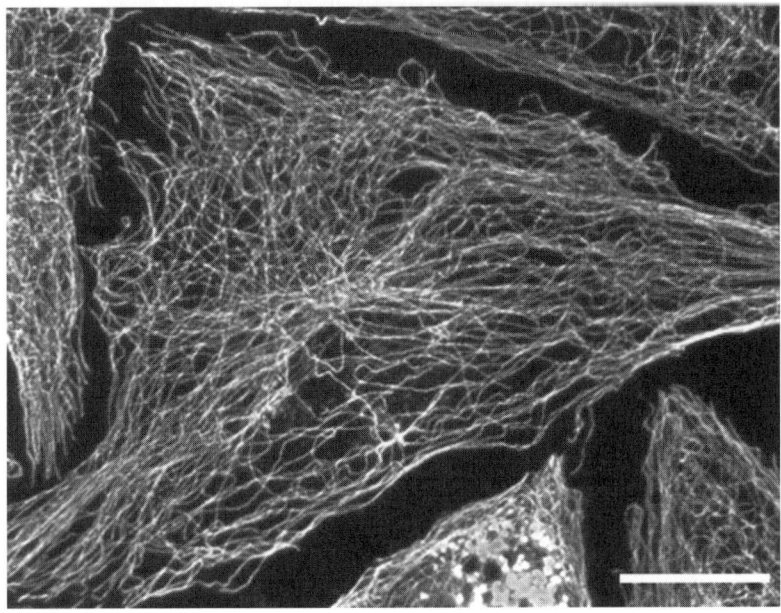

Fig. 1. Distribution of microtubules in a human 356 fibroblast revealed by immunostaining with tubulin antobody. Cell was fixed with formaldehyde, permeablized with 1% Triton X-100 in phosphate buffered saline and immunostained with monoclonal antibody DM1A against α-tubulin and a secondary antibody conjugated with rhodamine. Bar = 20 μm.

3. Injected cells are incubated for 1 h at 37°C in a $CO_2$ incubator to allow for incorporation of labeled tubulin subunits into microtubules.

### 3.2.2. Expression in the Cells of Tubulin Subunits Tagged With a Fluorescent Protein

1. Transient transfection of cells is performed using Lipofectamine 2000 (or other transfection reagent suitable for transient transfection of mammalian cells) according to experimental protocol provided by a manufacturer.
2. Cells after transfection are incubated in a $CO_2$ incubator at 37°C. Fluorescence can be usually detected within a few hours after transfection.

### 3.2.3. Live Observation of Microtubules

1. To prevent photodamage during acquisition of time-series of microtubules in living cells, tissue culture medium (3 mL) in each 35-mm tissue culture dish with labeled cells is supplied with Oxyrase (60 μL) and lactic acid (9 μL) (final dilution of Oxyrase is 2%). Mineral oil is then layered over the tissue culture medium

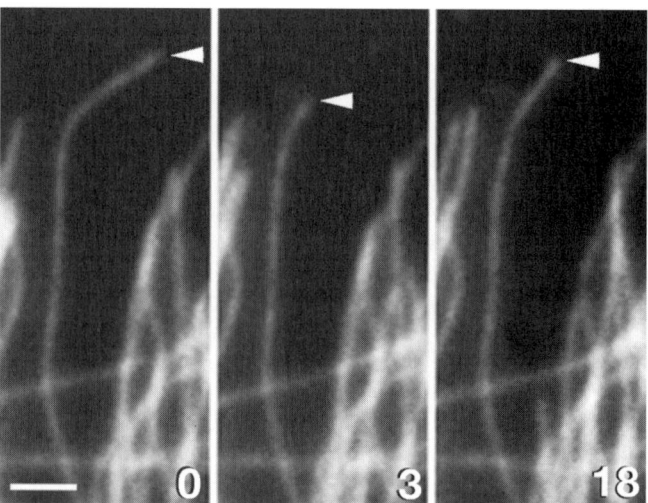

Fig. 2. Behavior at the edge of a fish melanophore of a microtubule labeled with Cy3-tagged tubulin. Cell was injected with Cy3-tubulin and time-series of images were acquired with a Photometrix series 300-cooled CCD camera with 3-s time intervals between the frames. Successive images show that microtubule grew and shortened at the free end marked with an arrow. Bar = 0.5 μm.

to prevent oxygen exchange. For efficient oxygen depletion it is essential that mineral oil layer is continuous and does not have openings.

2. A dish with labeled cells is placed on the stage of an inverted microscope, and the temperature controller is turned on to allow for the tissue culture medium to reach the desired temperature (30–37°C for the mammalian cells). Time-lapse series of images of labeled microtubules are then taken using ↔60 or ↔100 objective lens. For the observation of growth and shortening events, images of microtubules are collected with short (1–3 s) time intervals between the frames. Exposure time depends on the sensitivity of a CCD camera that is being used for acquisition of images and the brightness of the fluorescence signal, and may vary from 200 ms to 1 s.

An example of growth and shortening of microtubules in a living cell is shown in **Fig. 2**.

## 4. Notes

1. Excessive drying of cover slips during immunostaining may result in damage of cells and increased background fluorescence.
2. Some tubulin antibodies may be species-specific and therefore the information about antibody cross-reactivity should be obtained in advance from the manufacturer.
3. Immunostaining of cells with polyclonal tubulin antisera often results in excessively high background fluorescence, which can be significantly reduced by affin-

ity purification of a tubulin antibody on a column with covalently bound tubulin subunits.

4. For long-term storage, primary and secondary antibodies should be aliquoted, frozen in liquid nitrogen, and placed in a −70°C freezer. Multiple freezing-thawing cycles inactivate antibodies and therefore should be avoided.

5. Immunostaining of microtubules can be combined with immunostaining of other intracellular structures. For multiple immunofluorescence microscopy, fixation, and permeabilization of cells should be optimized to preserve not only microtubules, but also all other intracellular structures that will be examined. Each of the primary antibodies should produce clean staining when used alone. Secondary antibodies should be derived from the same host species, so that they do not recognize one another, and do not recognize other primary antibodies used in the immuofluorescence assay. Filter sets should be also selected to resolve fluorophores that will be used for multiple labeling. Secondary antibodies for multiple immunofluorescence microscopy conjugated with rhodamine, fluorescein, and Cy5 that allow triple staining of cells are commercially available from Jackson Immuoresearch (West Grove, PA).

6. Expression levels of tubulin tagged with a fluorescent protein usually vary significantly from cell to cell. At high expression levels elevated background fluorescence of the cytoplasm masks the fluorescence of the microtubules. At low expression levels, microtubule staining is too weak. To obtain live images of microtubules, it is therefore important to select cells with optimal level of expression of fluorescent tubulin subunits.

7. Confocal microscopy can be used for acquisition of images of microtubules in poorly spread cells, such as mitotic cells. Particularly clear images of microtubles have been captured with a Perkin-Elmer spinning disc confocal microscope *(4)*.

## Acknowledgments

This work was supported by National Institutes of Health grant GM-62290 to V.I.R.

## References

1. Desai, A. and Mitchison, T. J. (1997) Microtubule polymerization dynamics. *Ann. Rev. Cell Dev. Biol.* **13,** 83–117.

2. Komarova, J., Peloquin, J., and Borisy, G. (2004) Microinjection of fluorophore-labeled proteins, in *Live Cell Imaging,* (Goldman, R. D. and Spector, D. L., ed.), Cold Spring Harbor Laboratory Press, Cold Spring Harbor, NY, pp. 67–86.

3. Goodson, H. V. and Wadsworth, P. (2004) Methods for expressing and analyzing GFP-tubulin and GFP-microtubule-associated proteins, in *Live Cell Imaging,* (Goldman, R. D. and Spector, D. L., ed.), Cold Spring Harbor Laboratory Press, Cold Spring Harbor, NY, pp. 537–553.

4. Rusan, N. M., Fagerstrom, C. J., Yvon, A. M. C., and Wadsworth, P. (2001) Cell cycle-dependent changes in microtubule dynamics in living cells expressing green fluorescent protein-α tubulin. *Mol. Biol. Cell* **12,** 971–980.

5. Borisy, G. G., Marcum, J. M., Olmsted, J. B., Murphy, D. B., and Johnson, K. A. (1975) Purification of tubulin and associated high molecular weight proteins from porcine brain and characterization of microtubule assembly in vitro. *Ann. N.Y. Acad. Sci.* **253,** 107–132.
6. Rodionov, V. I., Nadezhdina, E. S., Peloquin, J., and Borisy, G. (2001) Digital fluorescence microscopy of cell cytoplasts with and without the centrosome. *Methods Cell Biol.* **67,** 43–51.

# 7

## Measurements of Stathmin–Tubulin Interaction in Solution

### Marie-France Carlier

#### Summary

Stathmin is an important phosphorylation-controlled regulator of microtubule dynamics and plays a crucial role in cell division and cell proliferation. In its non-phosphorylated form, stathmin is the protein that interacts the most tightly with tubulin, in a 2:1 tubulin–stathmin ($T_2S$) complex that does not participate in microtubule assembly. The importance of stathmin at different levels of phosphorylation in different steps of mitosis This article is a short overview of the different methods that have been or could be used to monitor the kinetic and thermodynamic parameters of tubulin–stathmin interaction and to evaluate the effects of phosphorylation. The author has tried to emphasize how hydrodynamic and spectroscopic methods measuring direct binding of stathmin to tubulin can be complemented by methods that make use of linked functions, measuring how the change in a functional property of tubulin upon binding stathmin provides information on binding parameters.

**Key Words:** Tubulin; stathmin/Op18; microtubule dynamics; tubulin sequestration; mitosis; cell division.

## 1. Introduction

Various methods are routinely used to measure the kinetic and thermodynamic parameters of complexes between two proteins in solution. Hydrodynamic methods like sedimentation-diffusion equilibrium or gel filtration, isothermal titration calorimetry, nuclear magnetic resonance, spectroscopic methods (fluorescence polarization, fluorescence correlation spectroscopy, circular dichroism), capillary electrophoresis, have proven appropriate to monitor complex formation in different situations, depending on the range of stability of the complexes (in terms of equilibrium dissociation constant). In addition, if some functional property of one of the proteins of the complex is

From: *Methods in Molecular Medicine, Vol. 137: Microtubule Protocols*
Edited by: Jun Zhou © Humana Press Inc., Totowa, NJ

known and if this functional property is affected by binding its partner, the concentration dependence of the change in function can be used to monitor complex formation. To be specific, the change in ATPase activity of myosin upon binding to F-actin provides insight into the thermodynamics of acto–myosin complex; similarly, the change in the rate of nucleotide exchange on a G-protein (resp. actin) upon binding a guanine nucleotide exchange factor (resp. profilin) similarly provides information on the affinity of the regulatory factor for its target. Alternatively, if one of the proteins of the complex binds a ligand with an associated spectroscopic change, and if the affinity or the binding mode of this ligand (i.e., the extent of spectroscopic change linked to ligand binding) is changed in the complex, this change can be used as well to derive the thermodynamic parameters for complex formation between the two proteins, with and without bound ligand, from analysis of the associated isoenergetic square, along the lines defined by Wyman in the general linked function theory (1). Hence, the physical-chemical knowledge of the specific functions of the proteins that interact greatly helps define the assays that will most adequately yield the needed information.

Stathmin (oncoprotein op18) is a major phosphorylation-regulated protein that plays a crucial role in cell division and proliferation (2–9). It was early noticed as one of the "red light" proteins that become highly phosphorylated in response to stimuli in cultured cells, and for that reason was considered as a relay in signalling (10). However, stathmin was much more abundant than a conventional signaling molecule in mammalian cells, where it is present at $10^{-6}$–$10^{-7}$ $M$. Consistently, it was shown to play a unique role in the regulation of the amount of assembled microtubules by binding tubulin, an abundant cytoskeleton protein that organizes the architecture of the cell, and preventing its self-assembly in microtubules (11). It was understood that this property of stathmin, which appears regulated by phosphorylation, is at the origin of its function in cell division (for reviews see refs. 12–14). Actually, via cycles of phosphorylation–dephosphorylation, stathmin is important both for the formation of the mitotic spindle at the entry into mitosis, and for the timely exit from mitosis.

Understanding the mechanism of stathmin therefore requires a quantitative description of the interaction of tubulin with stathmin in different states of phosphorylation of its four serines, and of the effects of stathmin in these different states on microtubule dynamics. In addition, stathmin is one member of a family of related proteins that also interact with tubulin in more neural specific context, and may bind tubulin with different affinities (15). The goal of this article is restricted to the different methods that have been or that could have been used to measure the thermodynamic and kinetic parameters of the interaction of stathmin with tubulin.

## 2. Analytical Ultracentrifugation Analysis of the Interaction of Stathmin With Tubulin

Sedimentation velocity is routinely used to analyze protein–protein interactions. Often analysis of the changes in velocity prove a simpler and faster method than sedimentation-diffusion equilibrium to obtain reliable information on the stoichiometry and stability of complexes formed between two proteins. The case of tubulin–stathmin is such an example. First, stathmin has no aromatic residues, hence only tubulin, either in free or liganded form, is spectroscopically detectable in ultraviolet absorbance in the ultracentrifuge. Second it turns out that the interaction between stathmin and tubulin is strong enough for stoichiometric interaction to be detected in the routine $10^{-6}$ to $10^{-5}$ $M$ range of investigation of the method, allowing a titration of tubulin by stathmin *(16)*. The sedimentation coefficient of heterotubulin is 6 S. When mixtures of tubulin and stathmin are analyzed in the ultracentrifuge, tubulin sedimented in two distinct species of sedimentation coefficients 7.8 S and 6 S. The relative fractions of tubulin in the two species showed a titration of tubulin by stathmin consistent with the formation of a $T_2S$ complex, i.e., all tubulin sedimented as a single species at 7.8 S as soon as the molar ratio of stathmin to tubulin was 0.5 in the mixture. This behavior was observed in the useful range of tubulin concentration of 5–20 $\mu M$. These results demonstrate that in solution the tubulin:stathmin molar ratio is 2:1 in the complex, and it adopts a nonglobular shape. The $T_2S$ complex was shown later to be an elongated kinked rod *(17)* and the crystal structure *(18)* shows that in the $T_2S$ complex the two $\alpha\beta$ tubulin subunits are organized in a $\alpha\beta$-$\alpha\beta$ polarized fashion aligned along the stahmin helix (**Fig. 1**), but also the fact that tubulin and $T_2S$ sediment as two distinct species suggests that the affinity of stathmin for tubulin is very high and the complex is likely to exhibit slow association–dissociation equilibrium. Hence, other methods should be elaborated to derive the values of the thermodynamic and kinetic parameters of tubulin–stathmin interaction, which at this point can be described by a minimum scheme as follows :

$$T + S \Longleftrightarrow TS$$
$$TS + T \Longleftrightarrow T_2S, \qquad Ks = [T]^2.[S]/[T_2S]$$

## 3. Measurement of Tubulin–Stathmin Interaction Using Fluorescence Correlation Spectroscopy

Fluorescence correlation spectroscopy is a fluorescence microscopy technique that enables to derive the dynamic parameters of protein–protein interaction in a very small volume, at nanomolar concentrations. The method therefore is suitable for the analysis of very tight complexes such as tubulin–stathmin.

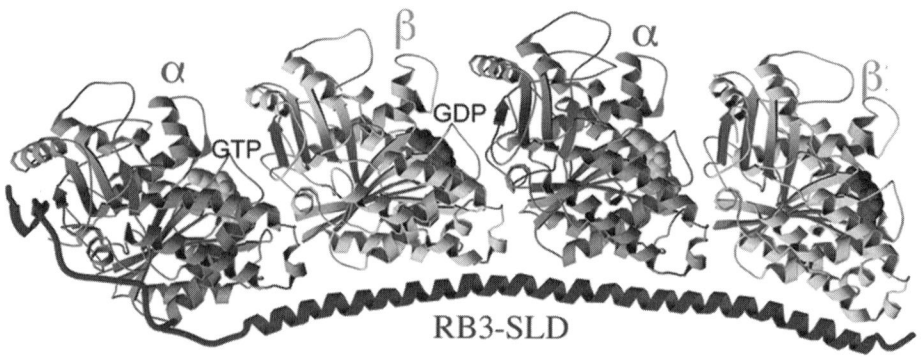

Fig. 1. Crystal structure of the complex of tubulin with RB3-stathmin like domain (courtesy of Marcel Knossow).

The method is based on the measurement of the change in diffusion coefficient of a fluorescently labeled protein upon binding a partner. Stathmin is a small protein of 149 amino acids (18 kDa), whereas tubulin is much larger (100 kDa). Hence, the stathmin core RB3 protein, which has been cristallized in complex with tubulin *(18)* was rhodamine-labeled on the N-terminal NH2 with no change in its ability to inhibit tubulin polymerization. The value of the diffusion coefficient of stathmin was $7.3 \pm 0.2 .10^{-11}$ m$^2$/s and decreased to $3.8 \pm 0.2 .10^{-11}$ m$^2$/s upon formation of the T$_2$(RB3) complex *(19)*. A time-dependent change in the fraction of RB3 in complex with tubulin was monitored. Kinetic analysis of the quadratic dependence of the apparent association rate constant on tubulin concentration, and evaluation of the dissociation rate constant of the complex by displacement of tubulin-bound labeled RB3 by unlabeled protein allowed to propose a possible kinetic mechanism for the slow association–dissociation equilibrium of stathmin-tubulin T$_2$S complex. In this scheme, formation of a TS complex in rapid equilibrium ($K_1 = 7 \mu M$) is followed by association of a second tubulin molecule followed by a slow intramolecular isomerization of T$_2$S. The on and off isomerization rates dominate the slow dynamics observed. The overall equilibrium dissociation constant of the T$_2$S complex was $0.5 \pm 0.2 \mu M^2$. Note that this value is obtained in solution, under conditions close to physiological. Other methods like surface plasmon resonance are adequate to measure the dynamics of interaction for high affinity protein complexes, but the protein is immobilized on a solid surface, hence the values of the interaction parameters may differ from the ones derived from solution studies.

## 4. Measurement of Stathmin–Tubulin Interaction Using Gel Filtration

Gel filtration is a choice hydrodynamic method that easily provides quantitative information on the Stokes radius of proteins and protein assemblies, as well as on the thermodynamic parameters of the interaction *(20)*. A method inspired from the Hummel-Dryer gel filtration method for analysis of the binding of a small ligand to a protein can be developed for stathmin–tubulin interaction and could easily be used to compare the affinities of tubulin for different wild type and mutated variants of proteins of the stathmin family. The method consists in using a gel filtration column unable to separate $T_2S$ from unliganded stathmin, equilibrated with a buffer containing tubulin at known concentrations in the $10^{-9}$ to $10^{-7} M$ range. A given amount of stathmin in a small volume is loaded on the column which is fed with tubulin-containing buffer and runs at a slow enough flow rate for the $T_2S$ complex to reach equilibrium in the early steps of the chromatographic process. Hence at each concentration of tubulin a mix of free stathmin and $T_2S$ is eluted in equilibrium with free tubulin at the imposed concentration. The concentration of tubulin present in the stathmin peak can be determined by any spectroscopic method (tryptophane fluorescence) whereas stathmin can be [35]S-labeled. Thus, the value of $Ks = [T]^2. [S]/[T_2S]$ can be derived from the measurements of the concentrations of stathmin and tubulin in the eluted peak. This method would allow a nice comparison of the relative affinities of the different stathmin related proteins, or of the differently phosphorylated forms of stathmin, for tubulin.

## 5. Measurement of the Binding of Stathmin to Tubulin Derived From Analysis of Its Tubulin Sequestering Activity

The $T_2S$ complex does not participate in microtubule assembly. In other words, stathmin is a tubulin-sequestering protein. Microtubules assembled at steady state coexist with tubulin maintained at a steady-state concentration $C_{ss}$. Addition of stathmin at a concentration $[S_0]$ causes partial depolymerization of microtubules so that at the new steady state microtubules coexist with $T_2S$, free stathmin and tubulin at the same concentration $C_{ss}$. The amount of $T_2S$ is governed by the value of Css and of the equilibrium dissociation constant Ks.

$$Ks = [T]^2.[S]/[T_2S], \text{ where } [T] = C_{ss}$$

$$\text{Hence } [T_2S] = [S_0].C_{ss}^2/(Ks + C_{ss}^2) \tag{1}$$

The value of Ks can therefore be derived from the measurement of depolymerized tubulin at different concentrations $[S_0]$ if the value of $C_{ss}$ is known. However, **Eq. 1** shows that this analysis provides a relevant measure of Ks only if Ks and Css are of the same order of magnitude. In fact, in most solution conditions routinely used for tubulin assembly in microtubules, Css is in the range of 1 to 10 $\mu M$, two to three orders of magnitude higher than $(Ks)^{1/2}$, and **Eq. 1** reduces to $[T_2S] = [S_0]$, making this method useless for the determination of Ks. To lower the value of Css, Taxotere was used as a microtubule stabilizer. The values of Css and $[T_2S]$ could be evaluated with high accuracy by sodium dodecyl sulfate-polyacrylamide gels (silver staining) of supernatants of sedimented microtubules *(21)*. The change in Ks upon phosphorylation of stathmin could be measured quantitatively using this method. Interestingly, because phosphorylation of stathmin regulates its interaction with tubulin in the cell, the measurements of the free concentration of tubulin that allows a large change to occur in the amount of $T_2S$ upon phosphorylation give an idea of the free tubulin concentration that is maintained stationary in living cells.

## 6. Measurement of the Interaction of Stathmin With Tubulin Derived From the Effect of Stathmin on Nucleotide Exchange on Tubulin

Another property of stathmin is to slow down nucleotide dissociation from tubulin. Tubulin has two bound guanine nucleotides, one is bound to $\alpha$-tubulin and nonexchangeable, the other is bound to $\beta$-tubulin and the kinetics of exchange can be monitored using the change in tryptophane fluorescence linked to replacement of bound GDP or GTP by S6-GTP *(22)*. This method was used to monitor the change in kinetics of nucleotide exchange on tubulin upon binding stathmin. Consistent with the slow association–dissociation equilibrium of tubulin and stathmin, the kinetics of nucleotide exchange were biphasic at substoichiometric molar ratios of stathmin to tubulin, the faster phase reflecting exchange on free tubulin and the slow phase exchange on $T_2S$ complex *(23)*. The exchange of nucleotide was slowed down by about 20-fold in the $T_2S$ complex. The exchange was slowed on the two $\beta$-tubulins present in the complex, although they do not have identical positions within the polar $T_2S$ complex. This result indicates that the difference in the effect of stathmin on the rate of nucleotide dissociation on each tubulin molecule in $T_2S$ cannot be detected, suggesting it must be less than fivefold. The method allowed the quantitative evaluation of the effect of phosphorylation of the different serines of stathmin on the stability of the $T_2S$ complex.

## References

1. Wyman, J. (1984) Linkage graphs: a study in the thermodynamics of macromolecules. *Q. Rev. Biophys.* **17,** 453–488.
2. Beretta, L., Dobransky, T., and Sobel, A. (1993) Multiple phosphorylation of stathmin. Identification of four sites phosphorylated in intact cells and in vitro by cyclic AMP-dependent protein kinase and p34cdc2. *J. Biol. Chem.* **268,** 20,076– 20,084.

3. Brattsand, G., Marklund, U., Nylander, K., Roos, G., and Gullberg, M. (1994) Cell-cycle-regulated phosphorylation of oncoprotein 18 on Ser16, Ser25 and Ser38. *Eur. J. Biochem.* **220,** 359–368.

4. Larsson, N., Melander, H., Marklund, U., Osterman, O., and Gullberg, M. (1995) G2/M transition requires multisite phosphorylation of oncoprotein 18 by two distinct protein kinase systems. *J. Biol. Chem.* **270,** 14,175–14,183.

5. Marklund, U., Brattsand, G., Osterman, O., Ohlsson, P. I., and Gullberg, M. (1993) Multiple signal transduction pathways induce phosphorylation of serines 16, 25, and 38 of oncoprotein 18 in T lymphocytes. *J. Biol. Chem.* **268,** 25,671–25,680.

6. Marklund, U., Larsson, N., Gradin, H. M., Brattsand, G., and Gullberg, M. (1996) Oncoprotein 18 is a phosphorylation-responsive regulator of microtubule dynamics. *EMBO J.* **15,** 5290–5298.

7. Andersen, S. S., Ashford, A. J., Tournebize, R., et al. (1997) Mitotic chromatin regulates phosphorylation of Stathmin/Op18. *Nature* **389,** 640–643.

8. Laird, A. D. and Shalloway, D. (1997) Oncoprotein signalling and mitosis. *Cell Signal.* **9,** 249–255.

9. Moreno, F. J. and Avila, J. (1999) Phosphorylation of stathmin modulates its function as a microtubule depolymerizing factor. *Mol. Cell. Biochem.* **183,** 201–209.

10. Sobel, A. (1991) Stathmin: a relay phosphoprotein for multiple signal transduction? *Trends Biochem. Sci.* **16,** 301–305.

11. Belmont, L. D. and Mitchison, T. J. (1996) Identification of a protein that interacts with tubulin dimers and increases the catastrophe rate of microtubules. *Cell* **84,** 623–631.

12. Rubin, C. I. and Atweh, G. F. (2004) The role of stathmin in the regulation of the cell cycle. *J. Cell. Biochem.* **93,** 242–250.

13. Cassimeris, L. (2002) The oncoprotein 18/stathmin family of microtubule destabilizers. *Curr. Opin. Cell Biol.* **14,** 18–24.

14. Curmi, P. A., Gavet, O., Charbaut, E., et al. (1999) Stathmin and its phosphoprotein family: general properties, biochemical and functional interaction with tubulin. *Cell Struct. Funct.* **24,** 345–357.

15. Charbaut, E., Curmi, P. A., Ozon, S., Lachkar, S., Redeker, V., and Sobel, A. (2001) Stathmin family proteins display specific molecular and tubulin binding properties. *J. Biol. Chem.* **276,** 16,146–16,154.

16. Jourdain, L., Curmi, P., Sobel, A., Pantaloni, D., and Carlier, M. F. (1997) Stathmin: a tubulin-sequestering protein which forms a ternary T2S complex with two tubulin molecules. *Biochemistry* **36,** 10,817–10,821.

17. Steinmetz, M. O., Kammerer, R. A., Jahnke, W., Goldie, K. N., Lustig, A., and van Oostrum, J. (2000) Op18/stathmin caps a kinked protofilament-like tubulin tetramer. *EMBO J.* **19,** 572–580.

18. Gigant, B., Curmi, P. A., Martin-Barbey, C., et al. (2000) The 4 A X-ray structure of a tubulin:stathmin-like domain complex. *Cell* **102,** 809–816.

19. Krouglova, T., Amayed, P., Engelborghs, Y., and Carlier, M. F. (2003) Fluorescence correlation spectroscopy analysis of the dynamics of tubulin interaction with RB3, a stathmin family protein. *FEBS Lett.* **546,** 365–368.

20. Beeckmans, S. (1999) Chromatographic methods to study protein-protein interactions. *Methods* **19,** 278–305.
21. Amayed, P., Pantaloni, D., and Carlier, M. F. (2002) The effect of stathmin phosphorylation on microtubule assembly depends on tubulin critical concentration. *J. Biol. Chem.* **277,** 22,718–22,724.
22. Yarbrough, L. R. and Fishback, J. L. (1985) Kinetics of interaction of 2-amino-6-mercapto-9-beta-ribofuranosylpurine 5'-triphosphate with bovine brain tubulin. *Biochemistry* **24,** 1708–1714.
23. Amayed, P., Carlier, M. F., and Pantaloni, D. (2000) Stathmin slows down guanosine diphosphate dissociation from tubulin in a phosphorylation-controlled fashion. *Biochemistry* **39,** 12,295–12,302.

# 8

# Analysis of Microtubule Dynamics by Polarized Light

## Rudolf Oldenbourg

### Summary

In this article, the author describes the needed instrumentation and the methods to be followed for the observation and measurement of the birefringence of single and bundled microtubules and of their ordered arrays using a polarizing microscope. As instruments, the traditional polarizing microscope and the recently developed LC-PolScope are discussed. As methods, we describe qualitative and quantitative observations, including notes on specimen preparations, which optimize the sensitivity and accuracy of measuring specimen retardance.

**Key Words:** Microtubules; alignment; birefringence; retardance; polarization; polarizer; compensator; LC-PolScope.

## 1. Introduction

The polarized light microscope was instrumental in the discovery of the dynamic character of microtubules, even before their molecular identity was established. As early as 1952, Shinya Inoué *(1)* described the reversible assembly and disassembly of fibers constituting the mitotic spindle in dividing cells that he observed with the polarizing microscope. In a number of pioneering experiments Inoué and collaborators established biophysical properties of the labile spindle fibers by observing their birefringence in living cells that were exposed to cold temperatures *(2)*, to a partial exchange of $H_2O$ with $D_2O$ *(3)*, and to hydrostatic pressure *(4)*. In 1975, after the discovery of microtubules as the major structural component of the spindle, Inoué and collaborators demonstrated that the birefringence of the spindle is indeed caused by the parallel array of aligned microtubules *(5)*; also, *see* review in **ref. 6**.

These early observations of the dynamic properties of spindle microtubules were made possible by technical improvements to the polarizing microscope, including the introduction of microscope lenses optimized for polarized light observations and the restoration of polarization distortions using polarization

From: *Methods in Molecular Medicine, Vol. 137: Microtubule Protocols*
Edited by: Jun Zhou © Humana Press Inc., Totowa, NJ

rectifiers *(7,8)*. More recently, the use of liquid crystal devices for controlling the polarization, and of image algorithms for quickly generating quantitative birefringence maps have revitalized the use of polarized light microscopy for analyzing the architectural dynamics in living cells *(9–11)*. The new technology, called LC-PolScope, is commercially available from Cambridge Research and Instrumentation in Woburn, MA (http://www.cri-inc.com), and is used in many fields, including for the observation of microtubule structure and dynamics in dividing cells *(12)*, for cell enucleation *(13)*, and for assisting in vitro fertilization procedures *(14)*.

A single microtubule (MT) molecule is a thin but stiff cylinder that is only 25-nm wide and often many microns long. Even though its diameter is well below the resolution limit of the light microscope, a single MT molecule and its dynamics can be observed in fluorescence *(15)*, dark field *(16)*, differential interference contrast (DIC) *(17)*, and polarized light microscopy *(18)*. An excellent discussion of fluorescence and DIC methods for visualizing MT dynamics in living cells is published in a recent review by Clare Waterman-Storer *(19)*. The observation of single MTs in living cells by phase sensitive techniques such as phase contrast, DIC, dark field, and polarized light microscopy, is often obscured by the seemingly random optical phase variations induced by organelles and other cytoskeletal components in the vicinity of the thin microtubule filaments. However, in special circumstances, such as in thin lamellipodia of epithelial cells, it is possible to visualize single MTs using DIC and possibly other phase sensitive techniques *(20)*. In the polarizing microscope the birefringence of single and bundled microtubules can be measured quantitatively and can thus be used to determine the number of microtubules in an observed structure *(12,18)*. Currently, only the polarized light microscope can measure an intrinsic material property of the microtubule molecules, such as birefringence, that can be used to quantify the number of microtubules or the density of aligned MTs in the field of view. In addition, the birefringence of a dense array of microtubules, such as a spindle, reveals the prevailing alignment of the filaments, even though the individual molecules are not resolved. Thus, the polarized light microscope has the unique potential to reveal the dynamics of submicroscopic structural parameters that are usually only accessible through more intrusive and static methods such as electron microscopy.

There are several recent reviews of polarized light microscopy that are helpful in the context of this article. Inoué wrote a general introduction to polarized light microscopy in biology with many helpful hints and tricks *(21)*; *see also* Appendix III in the first edition of **ref. 22**. We have published review articles on observing mitotic and meiotic spindles in the polarizing microscope *(23)* and on the use of the LC-PolScope for live cell imaging *(11)*. In the current article, we describe the instrumentation needed and methods to be followed for improving the results when observing the birefringence associated with single microtubules and microtubule arrays inside living cells or in in vitro preparations.

## 2. Instrumentation

The optical components of a traditional polarizing microscope include two linear polarizers, a compensator, and a rotation stage on which the specimen is placed. Other components, such as objective and condenser lenses, a light source, an ocular or eye piece, and a camera, are common with a standard transmitted light microscope. When used in a polarizing microscope, however, these common components might have to meet specific requirements to allow the detection of the weak birefringence of microtubule arrays. In this section, the author briefly discusses the critical components and define the quantities that can be measured in each resolved area of a birefringent specimen, namely the retardance and the slow axis orientation.

### *2.1. Polarizers*

A polarizing microscope has two polarizers one of which is located in the illumination path before the condenser lens and the other in the imaging path behind the objective lens (**Fig. 1**). The first polarizer is used to linearly polarize the light supplied by a bright halogen or arc lamp. The second polarizer serves to analyze the polarization of the light after it passed through the lenses and the specimen. This second polarizer is therefore called the analyzer. For best results the linear polarizers should have high transmission and good extinction. Modern sheet polarizers are available with transmission coefficients for unpolarized light of higher than 40% and extinction coefficients of around $10^4$. An ideal polarizer has a transmission of 50% and infinite extinction. The extinction is a measure of the purity of the polarization state of light, after it has passed through the polarizer. The extinction coefficient is measured by the ratio of light transmitted through two polarizers of the same type in series, with their transmission axes aligned parallel and perpendicular to each other (extinction = $I_{||}/I_{\perp}$). Glan-Thompson and other prism polarizers perform close to the ideal values.

The extinction or purity of polarization that is achieved in a polarization optical train directly affects the sensitivity for measuring the birefringence in a specimen: the higher the extinction the higher the sensitivity can be. However, the extinction of a polarization optical train is determined by the weakest elements in the sequence of optical components that are placed between the polarizer and analyzer. The weakest elements are those that introduce the strongest polarization distortions, which, in a microscope, are often the objective and condenser lenses.

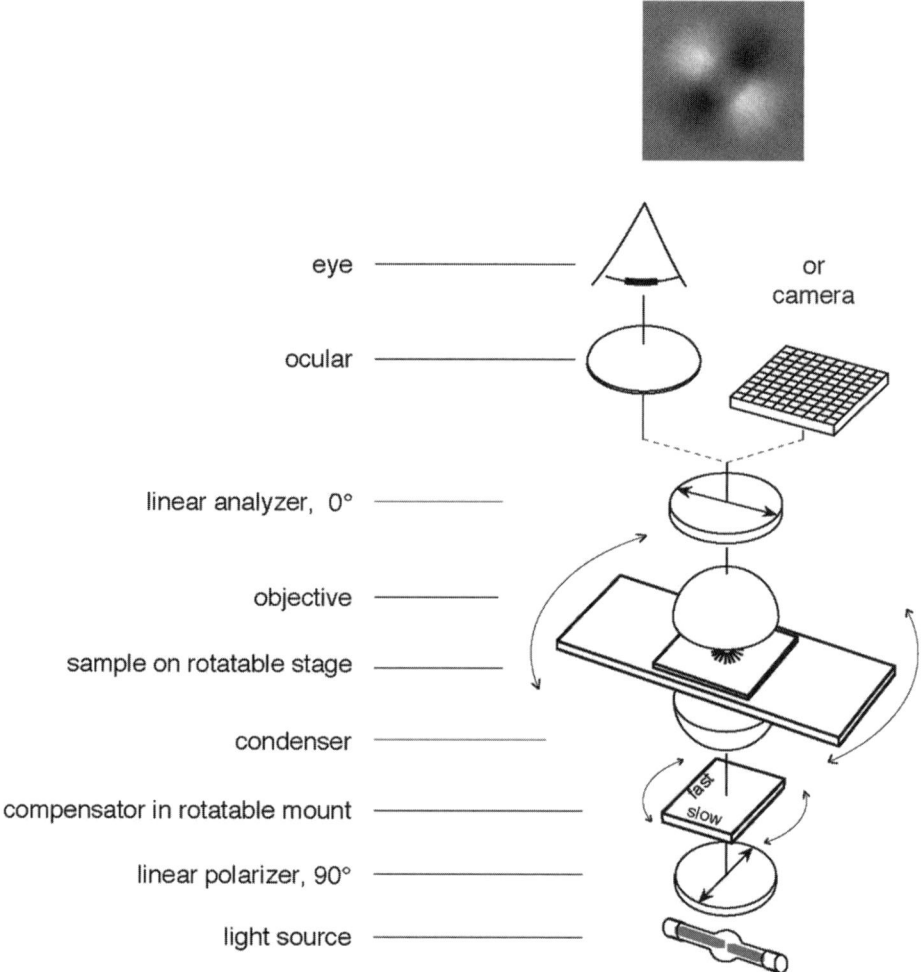

eye — or camera

ocular —

linear analyzer, 0° —

objective —

sample on rotatable stage —

condenser —

compensator in rotatable mount —

linear polarizer, 90° —

light source —

Fig. 1. Schematic of optical arrangement of the traditional polarized light microscope. The specimen image, as projected by the objective lens, is presented through an ocular to the eye and/or directly captured by a camera. The top image shows the appearance of an aster (a centrosome isolated from surf clam egg with radially arranged microtubules) between crossed polarizers and a Brace-Köhler compensator. The compensator is rotated slightly away from the extinction position in such a way that microtubule arrays that are oriented diagonally from top left to bottom right are bright because they run more parallel to the compensator slow axis. Microtubule arrays that are oriented diagonally from bottom left to top right are darker than the background because the microtubules run more perpendicular to the compensator slow axis (*see* **Subheading 3.2.**).

## 2.2. Objective and Condenser Lenses

The purity of the polarization in a light microscope is usually limited by the polarization distortions introduced by the condenser and objective lenses, especially if they have a high numerical aperture. Although a pair of sheet polarizers might have an extinction ratio of $10^4$ or higher, if a condenser and objective lens is placed between them, the extinction usually drops to around $10^3$ or below. Most microscope manufacturers offer lenses that are designated "Pol" to indicate low polarization distortions that can arise from a number of factors, including stress or crystalline inclusions in the lens glass and the type of antireflection coatings used on lens surfaces. Some lens types are available with "DIC" designation instead of "Pol." "DIC" lenses do not meet the more stringent Pol requirements but pass for use in differential interference contrast microscopy and might be the only available designation for some lens types such as Plan Apochromats that feature good image correction together with reduced polarization distortions.

Whenever possible or practical, we recommend the use of oil immersion objective and condenser lenses. The immersion medium abolishes, or at least reduces, the polarization distortions typical of so-called dry lenses that have an air space between the front lens element and the slide and cover glass of the specimen preparation. The transition of a light ray between two media of substantially different refractive index ($n_{air} = 1.00$, $n_{glass} = 1.52$) introduces polarization distortions, especially for high NA lenses. A discussion of polarization distortions introduced by microscope lenses and possible ways of rectifying the polarization distortions can be found in *(7,24)*.

## 2.3. Compensator

In addition to the polarizer and analyzer, the polarized light microscope often includes a compensator in the optical path. Although not absolutely necessary for some basic observations, especially when looking at highly birefringent objects, the compensator (1) can significantly improve the detection and visibility of weakly birefringent objects, (2) is required to determine the slow and fast axis of specimen birefringence, and (3) is an indispensable tool for the quantitative measurement of object birefringence.

There are several types of compensators, most of them are named for their original inventors. We will discuss the Brace-Köhler compensator, which is suited for the observation and measurement of weak birefringence, such as those of microtubule arrays. The Brace-Köhler compensator consists of a thin birefringent plate, often made from mica, with a retardance of a 10th to a 30th of a wavelength ($\lambda/10$ to $\lambda/30$; a definition of retardance is provided later). The birefringent plate is placed in a graduated rotatable mount and inserted either between the polarizer and condenser or between the objective lens and the analyzer. The location varies between microscope manufacturers and from one microscope type to another. In either location, the effect of the Brace-Köhler compensator on the observed image is the same and its usage does not depend on location (as long as there is not a third, strongly birefringent element between the polarizers). Further comments on the use of a Brace-Köhler compensator are found in **Subheading 3.**

Recently, liquid-crystal-based devices were introduced as compensators that can be controlled electronically *(9)*. The liquid-crystal devices can be used as traditional compensators or as part of a measurement scheme that is discussed below under the heading LC-PolScope.

### 2.4. Retardance

The retardance of a resolved specimen area is the quantity typically measured with the compensator in a polarizing microscope. In general terms, retardance characterizes the relative phase shift between two orthogonally polarized light waves that have traversed an optically anisotropic material. When linearly polarized light travels through a birefringent material, the light is split into two orthogonally polarized waves that travel with different speeds. The difference in speed between the two waves is a consequence of the difference in the refractive index $\Delta n$ of the material for the two polarization components. Retardance, also called birefringence retardation, expresses the relative phase shift between the two waves after they traversed the birefringent specimen of thickness $t$:

$$retardance = R = (n - n_\wedge) \cdot t = \Delta n \cdot t$$

Accordingly, retardance has the dimension of a distance, typically expressed in nm or as a fraction of the wavelength of light.

### 2.5. Slow and Fast Axes

The slow and fast axes describe specific orientations in a birefringent material. For a given propagation direction, light polarized parallel to the slow axis experiences the highest refractive index and, hence, travels the slowest in the material. For the same propagation direction, light that is polarized perpendicular to the slow axis experiences the lowest refractive index and, therefore, travels the fastest in the material. The polarization direction associated with the lowest refractive index is called the fast axis. The slow and the fast axes are always perpendicular to each other. A compensator is used to determine the slow and fast axis of a birefringent specimen.

### 2.6. Rotation Stage

The rotation stage is an indispensable accessory for a traditional polarizing microscope, because the contrast of a birefringent object depends on its orientation with respect to the microscope's polarizer and analyzer. For qualitative observations, the systematic variation of brightness when rotating an object between crossed polars attests to the object's optical anisotropy. For quantitative measurements, the rotation stage needs to have a graduated scale around its circular perimeter, including a vernier, so specimens can be rotated by a specific angle that usually can be measured within a tenth of a degree.

### 2.7. LC-PolScope

The LC-PolScope augments the traditional polarizing microscope by integrating a liquid-crystal (LC) compensator, electronic imaging, and digital image processing tools to build a birefringence imaging system *(9,10,25)* (**Fig. 2**). Commercial versions of the LC-PolScope are available from Cambridge Research and Instrumentation. The LC-PolScope generates high-resolution maps of the retardance and slow axis distribution in the specimen. Such maps can be generated within seconds at a sensitivity that allows the detection of single microtubules.

The two LC variable retarders and the polarizer are optically bonded together to form a universal compensator. The universal compensator can be used to produce, and rapidly switch between, circularly polarized light and elliptically polarized light of various axis orientations. In contrast to the traditional polarizing microscope, which uses linearly polarized light, the LC-PolScope uses near circularly polarized light that passes through the specimen. After the specimen the light passes through a polarization analyzer that blocks circularly polarized light. The universal compensator is switched to several predetermined settings producing circularly and elliptically polarized light. An electronic camera records images of the specimen for each of those settings and transfers the raw image data to a computer. In the computer the image data are combined using algorithms to calculate the specimen retardance in each resolved image point. The result of the computation is an image representing the retardance measured in each pixel independent of the orientation of the slow axis in that point. Hence, the retardance image is black where there is no birefringence in the sample and shows a specific shade of gray depending on the retardance of the sample at that point, regardless of the sample's orientation. In addition to the retardance, the LC-PolScope also measures the orientation of the slow axis in every image point, using the same set of raw image data. The orientation is measured as an angle between the horizontal axis and the slow axis of the specimen at each image point. The orientation values can either be displayed as a separate image or be overlaid on a retardance image using color or other means of encoding orientation. Thus, computed LC-PolScope images represent material properties rather than light intensities. As a consequence, the retardance image, for example, displays specimen retardance on a linear scale, in contrast to the quadratic relationship between image intensity and specimen retardance in a traditional polarizing microscope.

raw LC-PolScope images (intensity images)

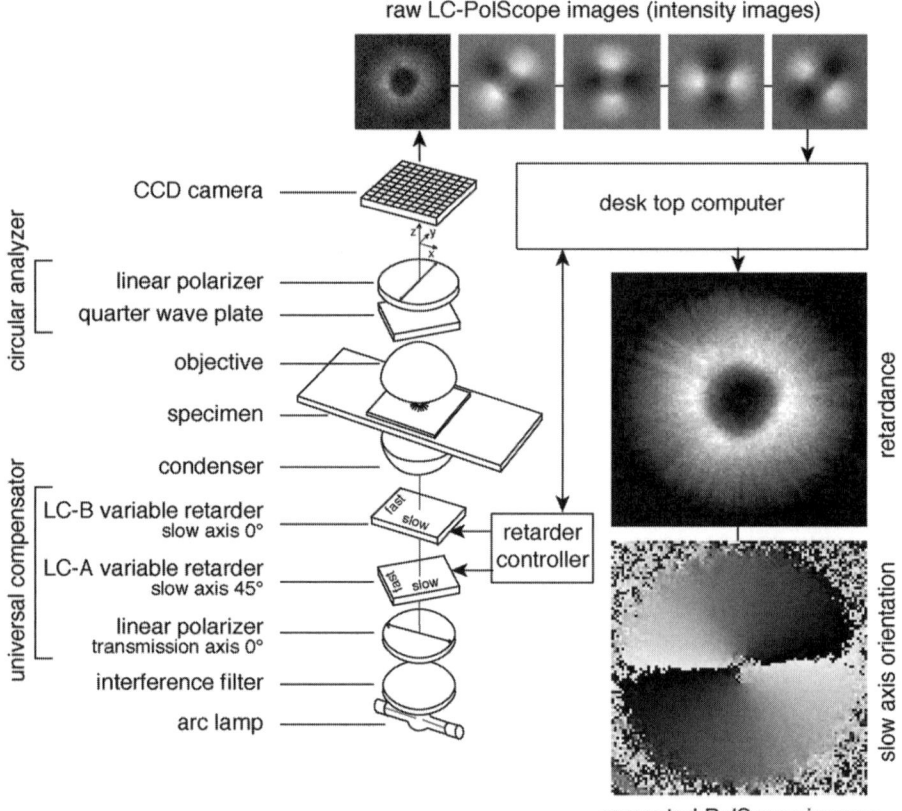

Fig. 2. Schematic of the LC-PolScope. The optical design (left) builds on the traditional polarized light microscope with the conventional compensator replaced by two variable retarders LC-A and LC-B forming the universal compensator. Images of the specimen (top row, aster isolated from surf clam egg) are captured at five predetermined retarder settings that cause the specimen to be illuminated with circularly polarized light (first, left-most image in top row) and with elliptically polarized light of different axis orientations (second to fifth images). The retardance and orientation images near the bottom right were both computed from the same five intensity images shown in the top row.

## 3. Methods

### 3.1. Observation of the Specimen Between Crossed Polarizers

Usually, the two polars (short for polarizers) are in crossed position so that the analyzer blocks (absorbs) most of the light that has passed through the polarizer. Hence, the image that is projected by the objective lens is mostly dark, except for birefringent or otherwise optically anisotropic specimen parts that appear bright. When the specimen is rotated on a revolving stage, the birefringent parts change brightness, from dark to bright and back to dark four times during a full 360° rotation. The systematic variation of brightness when rotating a specimen between crossed polars is a telltale of the presence of birefringence in the specimen.

A birefringent specimen part appears darkest when its slow and fast axis is parallel to polarizer and analyzer. This specimen orientation is also called the extinction position. Rotating it by 45° away from the extinction position makes the birefringent part appear brightest. In general, not all birefringent parts in the field of view will turn dark at the same time because each part has different axis orientations. Hence, in addition to helping recognize birefringent elements, one can also determine the axis orientations by rotating the specimen between crossed polars.

### 3.2. Observations Using a Traditional Compensator

In general, the birefringence of the compensator, when inserted into the optical path, causes the image background to become brighter, whereas birefringent parts, such as a mitotic spindle, become either brighter or darker than the background. Careful observations have established that a spindle appears brighter than the background when the spindle axis is more parallel to the slow axis of the compensator *(26)*. When the spindle axis is more perpendicular to the compensator slow axis, the spindle fibers appear darker than the background. In this latter position, the retardance of the spindle is said to be either fully or partially compensated. These results directly indicate that the slow axis of the spindle fibers are parallel to the fiber axis; or, in other words, the slow axis of microtubules is oriented parallel to the microtubule axis.

By following a number of steps in adjusting the mutual orientations of the polarizers, compensator and specimen, one can achieve best contrast and a quantitative analysis of the retardance of specimen birefringence. As an example, we consider spindle microtubules (MTs) that are nearly parallel to the spindle axis and that are observed using a Brace-Köhler compensator (*see* **Subheading 2.3.**) in the optical path of the microscope:

1. Using the rotatable mount of the compensator, rotate its slow axis parallel to polarizer or analyzer (extinction position); as a result, the background appears dark.
2. Using the rotation stage, rotate the spindle axis in the specimen to its extinction position; the spindle appears dark.
3. Now rotate the spindle axis by 45° away from the extinction position; the spindle appears brightest; during the following steps, the spindle position remains fixed at 45° to extinction.

4. Rotate the compensator either clockwise or counterclockwise to render the spindle either darker or brighter than the background; for observations by eye, a spindle that is darker than the background tends to be more easily recognized.
5. For measuring the retardance of the spindle MTs, rotate the compensator until the spindle MTs appear darkest, i.e., the light intensity in the image region of the spindle is lowest. In this postion, the spindle retardance $R_{spindle}$ is said to be compensated by the compensator retardance $R_{comp}$:

$$R_{spindle} = R_{comp} \sin(2\theta_{min})$$

$\theta_{min}$ is the angle by which the compensator was rotated away from extinction to achieve minimum light intensity in the spindle region; note that the light intensity in the spindle increases on either side of $\theta_{min}$ whereas the intensity of the surrounding background steadily increases for the same range of angles; therefore, judging the lowest spindle intensity by eye can be tricky.

It is advisable to deviate from the above protocol for generating high contrast images of small changes in orientation of the specimen's birefringence axis. For example, microtubule arrays that reorient or fluctuate by thermal or other forces are best observed in the extinction position, with the compensator rotated slightly away from extinction. With this arrangement, small changes in microtubule orientations are translated into high-contrast intensity changes against a nearly dark background. This effect was exploited for visualizing the subtle variations of DNA ordering in cave cricket sperm heads *(27)*.

A more complete discussion of the appearance of spindles in the polarizing microscope and the analytic interpretation of polarized light images of single, bundled, and arrays of microtubules can be found in **ref. 23**.

### 3.3. Observations Using the LC-PolScope

In addition to a liquid-crystal compensator, the LC-PolScope includes an electronic camera and software for image acquisition, instrument control, and data analysis. A schematic of the optical and electronic LC-PolScope components is shown in **Fig. 2**. The computer with software provides a user interface used for calibrating the instrument, for recording a sequence of raw images and for computing retardance and azimuth images. Observations with the LC-PolScope usually start with calibrating the instrument and recording a so-called background stack. The calibration, which is done without a birefringent specimen in the viewing field, determines the optimal settings of the liquid crystal devices. Subsequently, a background stack of raw PolScope images is recorded. The background stack is a record of the spurious birefringence induced, e.g., by stress in microscope lenses and optical components other than the specimen. Then the specimen is moved into the field of view and another stack of raw PolScope images is recorded. The raw specimen stack of images contains

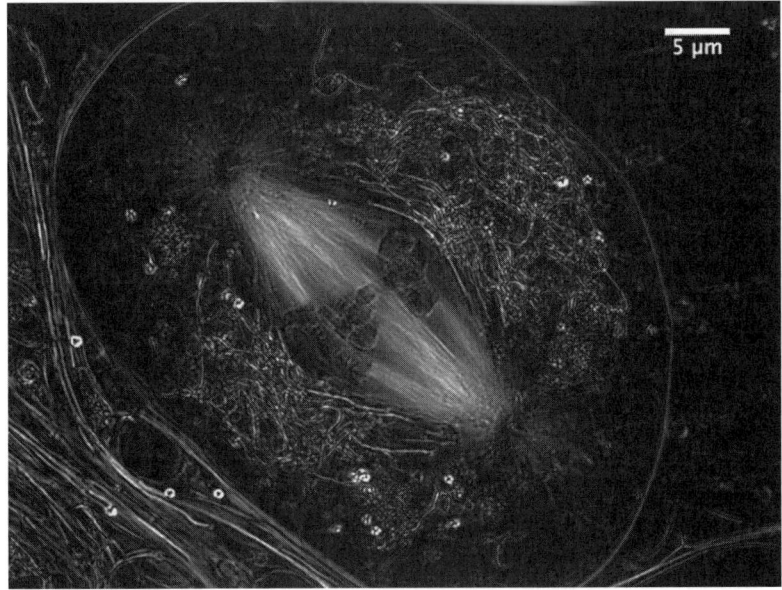

Fig. 3. Metaphase of meiosis I in a living spermatocyte from the crane fly, Nephrotoma suturalis, viewed with the LC-PolScope. The kinetochore fibers that connect the chromosomes to the spindle poles stand out with exceptional bright contrast. Each fiber is a bundle of about 60 kinetochore microtubules as inferred directly from the measured retardance *(12)*. (Image recorded with James R. LaFountain, University of Buffalo.)

the information of the specimen retardance superimposed on the background retardance. During the calculation of the specimen retardance, the influence of the spurious background retardance is removed, based on the separate background stack. As a result, the background corrected retardance image shows the birefringence of the specimen at high contrast on a dark background. The spurious background removal is especially important when observing and measuring the weak retardance of individual microtubules or small bundles of microtubules.

LC-PolScope images represent the retardance in every resolved image point measured simultaneously over the whole field of view. The recording and analysis process can take less than a second and can be repeated indefinitely, providing high-resolution time lapse records of dynamic events, such as the assembly and disassembly of microtubules and their ordered arrays.

Recently, we have published a more detailed account of the use of the LC-PolScope for imaging birefringent structures in living cells assembly and disassembly of microtubules and their ardered arrays **(Fig. 3)** *(11)*.

## 3.4. Notes on Specimen Preparation

1. Do not use plastic dishes or slides. Usually, plastic parts such as culture dishes are highly birefringent and obscure the weak birefringence of microtubule arrays. However, plastic dishes with a glass bottom, usually made of a thin cover glass, are OK.

2. Use oil immersion optics whenever possible (*see* **Subheading 2.2.**). When using high NA oil immersion optics, prepare the specimen in such a way that the structure of interest is located close to the cover slip. Otherwise, spherical aberration can noticeably reduce the resolution and sensitivity achieved in the image. If you need to focus deeper into an aequeous medium (more than about 20 μm), use water immersion optics to improve the resolution *(28)*.

3. When using the LC-PolScope, prepare the specimen so when mounted in the microscope one can easily find a clear area without birefringent parts and move it into the viewing field when required. A clear area is needed for calibrating the LC-PolScope and for recording a stack of background images. Sometimes we find it helpful to add a tiny drop of oil or other nontoxic, immiscible liquid to the preparation. The clear drop can provide an area for calibrating the instrument and take background images in a preparation that is otherwise dense with birefringent structures.

## References

1. Inoué, S. (1952) The effect of cholchicine on the microscopic and submicroscopic structure of the mitotic spindle. *Exp. Cell Res. Suppl.* **2,** 305–318.
2. Inoué, S. (1964) Organization and function of the mitotic spindle, in *Primitive Motile Systems in Cell Biology,* (Allen, R. H. and Kamiya, N., eds.), Academic Press, New York pp. 549–598.
3. Inoué, S. and Sato, H. (1967) Cell motility by labile association of molecules: the nature of mitotic spindle fibers and their role in chromosome movement. *J. Gen. Physiol.* **50,** 259–292.
4. Salmon, E. D. (1975) Pressure-induced depolymerization of spindle microtubules. I. Changes in birefringence and spindle length. *J. Cell Biol.* **65,** 603–614.
5. Sato, H., Ellis, G. W., and Inoué, S. (1975) Microtubular origin of mitotic spindle form birefringence Demonstration of the applicability of Wiener's equation. *J. Cell Biol.* **67,** 501–517.
6. Inoué, S. and Salmon, E. D. (1995) Force generation by microtubule assembly/disassembly in mitosis and related movements. *Mol. Biol. Cell* **6,** 1619–1640.
7. Inoué, S. and Hyde, W. L. (1957) Studies on depolarization of light at microscope lens surfaces II. The simultaneous realization of high resolution and high sensitivity with the polarizing microscope. *J. Biophys. Biochem. Cytol.* **3,** 831–838.
8. Inoué, S. and Kubota, H. (1958) Diffraction anomaly in polarizing microscopes. *Nature* **182,** 1725–1726.
9. Oldenbourg, R. and Mei, G. (1995) New polarized light microscope with precision universal compensator. *J. Microsc.* **180,** 140–147.
10. Shribak, M. and Oldenbourg, R. (2003) Techniques for fast and sensitive measurements of two-dimensional birefringence distributions. *Appl. Opt.* **42,** 3009–3017.

11. Oldenbourg, R. (2005) Polarization microscopy with the LC-PolScope, in *Live Cell Imaging: A Laboratory Manual,* (Goldman, R. D. and Spector, D. L., eds.), Cold Spring Harbor Laboratory Press, Cold Spring Harbor, NY, pp. 205–237.

12. LaFountain, J. R., Jr. and Oldenbourg, R. (2004) Maloriented bivalents have metaphase positions at the spindle equator with more kinetochore microtubules to one pole than to the other. *Mol. Biol. Cell* **15,** 5346–5355.

13. Liu, L., Oldenbourg, R., Trimarchi, J. R., and Keefe, D. L. (2000) A reliable, noninvasive technique for spindle imaging and enucleation of mammalian oocytes. *Nat. Biotechnol.* **18,** 223–225.

14. Wang, W. H. and Keefe, D. L. (2002) Prediction of chromosome misalignment among in vitro matured human oocytes by spindle imaging with the PolScope. *Fertil. Steril.* **78,** 1077–1081.

15. Sammak, P. J. and Borisy, G. G. (1988) Direct observation of microtubule dynamics in living cells. *Nature* **332,** 724–726.

16. Horio, T. and Hotani, H. (1986) Visualization of the dynamic instability of individual microtubules by dark-field microscopy. *Nature* **321,** 605–607.

17. Walker, R. A., O'Brien, E. T., Pryer, N. K., et al. (1988) Dynamic instability of individual microtubules analyzed by video light microscopy: rate constants and transition frequencies. *J. Cell Biol.* **107,** 1437–1448.

18. Oldenbourg, R., Salmon, E. D., and Tran, P. T. (1998) Birefringence of single and bundled microtubules. *Biophys. J.* **74,** 645–654.

19. Waterman-Storer, C. M. (1998) Microtubules and microscopes: how the development of light microscopic imaging technologies has contributed to discoveries about microtubule dynamics in living cells. *Mol. Biol. Cell* **9,** 3263–3271.

20. Cassimeris, L., Pryer, N. K., and Salmon, E. D. (1988) Real-time observations of microtubule dynamic instability in living cells. *J. Cell Biol.* **107,** 2223–2231.

21. Inoué, S. (2002) Polarization microscopy, in *Current Protocols in Cell Biology,* (Bonifacino, J. S., Dasso, M., Harford, J. B., Lippincott-Schwartz, J., and Yamada, K. M., eds.), John Wiley & Sons, New York, pp. 4.9.1–4.9.27.

22. Inoué, S. (1986) *Video Microscopy.* Plenum Press, New York.

23. Oldenbourg, R. (1999) Polarized light microscopy of spindles. *Methods Cell Biol.* **61,** 175–208.

24. Shribak, M., Inoué, S., and Oldenbourg, R. (2002) Polarization aberrations caused by differential transmission and phase shift in high NA lenses: theory, measurement and rectification. *Opt. Eng.* **41,** 943–954.

25. Oldenbourg, R. (1996) A new view on polarization microscopy. *Nature* **381,** 811–812.

26. Inoué, S. and Dan, K. (1951) Birefringence of the dividing cell. *J. Morphol.* **89,** 423–456.

27. Inoué, S. and Sato, H. (1966) Deoxyribonucleic acid arrangement in living sperm, in *Molecular Architecture in Cell Physiology,* (Hayashi, T. and Szent-Gyorgyi, A. G., eds.), Prentice Hall, Englewood Cliffs, NJ, pp. 209–248. 28. Inoué, S. and Spring, K. R. (1997) *Video Microscopy.* Plenum Press, New York.

28. Ihoué, S. and Spring, K. R. (1997) *Video Microscopy.* Plenum Press, New York.

# 9

## Visualization of Spindle Behavior Using Confocal Microscopy

### Adam I. Marcus

### Summary

Numerous comparisons have shown that confocal imaging of fluorescently labeled samples has superior image clarity compared to traditional epifluorescence microscopy, especially when imaging through thick specimens. Nevertheless, one limitation of confocal microscopy is the speed at which these systems acquire images. Newer confocal technologies, however, have overcome this drawback by rapidly scanning images while still maintaining the precision and resolution of a confocal microscope. One example is the spinning disk confocal, which can rapidly acquire images with minimal photobleaching. The following protocol provides methodology using spinning disk technology to visualize spindle microtubule behavior in live breast cancer cells expressing green fluorescent protein:$\alpha$-tubulin. It begins with sample preparation, covers equipment setup and image acquisition, and ends with image processing and archiving. Image acquisition is subdivided into two categories- imaging the complete mitotic cycle and imaging rapid mitotic events, to address the varying parameters required for the each experiment.

**Key Words:** Mitosis; microtubule; GFP; confocal; tubulin; spindle; cancer; taxol.

## 1. Introduction

Confocal imaging revolutionized the field of microscopy by providing sharply defined optical sections of specimens with minimal blurring caused by out-of-focus light. Typically, this optical system uses a laser for illumination that is guided through the objective and across the specimen, point-by-point. Emission light travels back through the objective and to a strategically placed pinhole, located conjugate (i.e., in the same focal plane) to the scanned point of interest. Thus, only emission light originating from the point-of-interest can pass and enter the detector thereby creating an optical section of the focal plane of interest.

From: *Methods in Molecular Medicine, Vol. 137: Microtubule Protocols*
Edited by: Jun Zhou © Humana Press Inc., Totowa, NJ

This point-scanning confocal setup provides researchers with a variety of capabilities and produces outstanding images of specimens; however, one drawback of this point-by-point approach is that image acquisition is relatively slow, leading to low temporal resolution that is especially obvious when imaging rapid biological events. Moreover, a slow scan speed can result in fluorophore bleaching and phototoxicity in living specimens. One alternative has been to employ widefield epifluorescence microscopy, which is faster; however confocality is sacrificed and photobleaching still remains an issue. Thus, a new confocal imaging strategy was engineered whereby multiple pinholes placed on an illuminated rotating disk, rapidly scans samples in a confocal environment. In general, these spinning disk confocal systems consist of a rapidly rotating Nipkow disc (first developed in 1884 by Paul Nipkow) with a set of pinholes that serve as the entry way for illumination light and as the confocal pinhole for emission light (**Fig. 1**). Because the pinholes themselves occupy a small area of the disc, excitation light needs to be directed through the pinholes by a microlens-studded disk placed behind the pinhole disc. Furthermore, a dichroic mirror is located between discs, redirecting emission light to the detector and preventing excitation light from entering the detector. Variations of this scanhead are available that spin at well over 1000 rpm; therefore this setup enables rapid acquisition of confocal images, resulting in reduced photobleaching and phototoxicity.

Spinning disk confocal technology is used by many cell biologists *(1–4)*, because it is well suited for five-dimensional imaging (x, y, z, time, and multifluorophore) of dynamic processes in living cells and results in only one-fifteenth the rate of photobleaching as point-scanning confocal systems *(5)*. In particular, it has great utility in imaging microtubule behavior during mitosis, which proceeds rapidly and has many key events that occur within seconds, such as centrosome separation and spindle pole formation. Although the spinning disc confocal is well suited to visualize microtubules with high spatiotemporal resolution there are technical challenges that have to be overcome to achieve optimal image acquisition. Therefore, the protocol below is to be used to visualize microtubules during mitosis in live cells using MCF-7 breast cancer cells stably expressing green fluorescent protein (GFP):α-tubulin.

## 2. Materials
### 2.1. Cell Culture and Plating
1. MCF-7 cells stably expressing GFP:α-tubulin *(6,7)*.
2. G418 (Sigma, St. Louis, MO; cat. no. G8168).
3. RPMI 1640 without phenol red (Mediatech, Herndon, VA; cat. no. 17-105-CV).
4. Fetal bovine serum (Sigma, cat. no. F6178).
5. 0.05% Trypsin (Mediatech, cat. no. 25-052-C2).

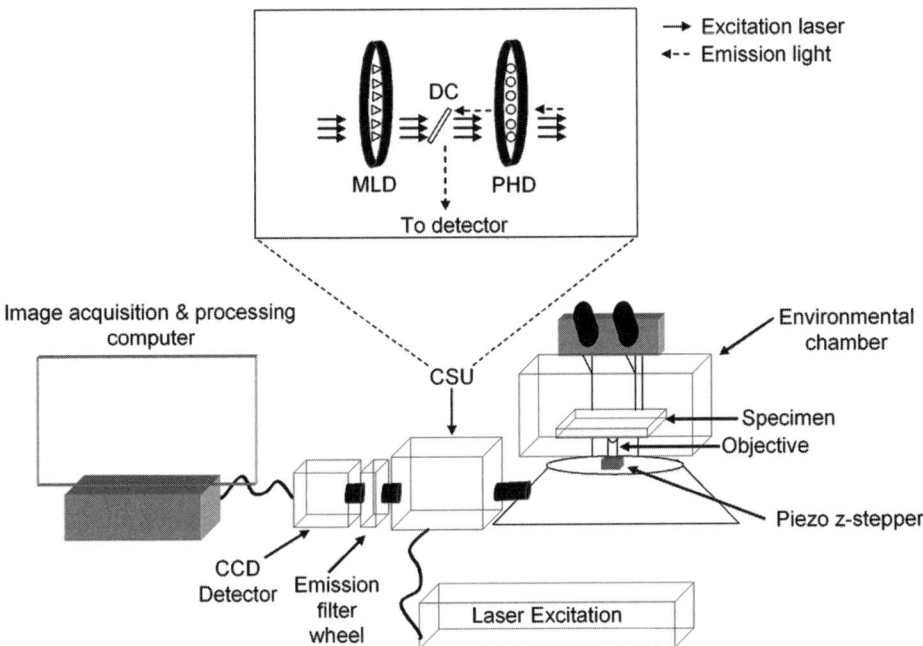

Fig. 1. Diagram of the confocal imaging setup used in our laboratory. CSU, confocal scanning unit; MLD, microlens disk; PHD, pinhole disk; DC, dichroic. Image not to scale.

6. 25-cm² Cell culture flasks (Corning, Acton, MA; cat. no. 430639).
7. 35-mm Live cell imaging dishes (World Precision Instruments, Sarasota, FL; cat. no. 500862).
8. Poly-L-lysine (Sigma, cat. no. P8920).
9. Laminin (Invitrogen, Carlsbad, CA; cat. no. 33-5300).
10. Fibronectin (Invitrogen, cat. no. 33016-015).
11. Oxyrase (Oxyrase Inc., Mansfield, OH).

## 2.2. Microscope

1. Perkin Elmer Ultraview ERS (Perkin Elmer, Boston, MA) spinning disk confocal with CSU21 confocal scan head (Yokogawa, Japan).
2. Zeiss Axiovert 200 m inverted microscope (Thornwood, NJ).
3. ORCA ER charge coupled device (CCD) camera (Hamamatsu Photonics, K.K., Japan).
4. Piezo-controlled z-stepper mounted on objective turret (Piezosystems Jena, Germany).
5. 488-nm Laser line from an Ar ion laser (488, 568, and 633 lines; Melles Griot; Carlsbad, CA).

6. HBO 100 mercury arc lamp (Osram, cat. no. 103/w2) with adjustable power supply.
7. Environmental chamber (Zeiss, cat. no. XL-3) with heat source (Zeiss, cat. no. 0421) and controller (Tempcontrol 37-2 digital; Zeiss).
8. FF Immersion oil (Cargille; Cedar Grove, NJ).
9. $CO_2$ tank for carbon dioxide perfusion and $CO_2$ controller (Zeiss, cat. no. 0506).
10. Zeiss 100X Plan-Apochromat oil objective (N.A. 1.4).
11. GFP excitation (470/40X) and emission (525/50 m) filter set with 495 LP dichroic (Chroma, Rockingham, VT; cat. no. 41017) in the microscope filter turret.
12. Rapid emission discrimination filter wheel attached to Ultraview with GFP filter set (Perkin Elmer).

### 2.3. Imaging

1. Computer with Intel 2.4 GHz Xeon processor, 2 GB RAM, and 200 GB hard drive and monitor. The computer is equipped with a NVidia GeForce Ti4800 video card with 128 MB and AGP 8X bus, as well as an 8X DVD + RW.
2. Perkin Elmer Ultraview image acquisition software.
3. Metamorph 6.1 image processing software (Universal Imaging, Downington, PA) or similar image processing software.

### 2.4. Data Archiving

1. External hard drive with at least 200 GB storage space.

### 3. Methods

The following protocol provides a step-by-step approach to imaging mitotic progression in live MCF-7 cells stably expressing GFP:α-tubulin. This protocol has also successfully been used for other cell lines expressing GFP:α-tubulin or other fluorescently labeled tubulin molecules, though minor adjustments need to be made to the exposure, gain, and other acquisition settings to compensate for differences in fluorescence intensity and photobleaching. Importantly, because some researchers want to examine the complete mitotic cycle (~2 h), whereas others are interested in observing spindle behavior over short time periods (~10 min), both methodologies have been provided (**Subheadings 3.6.** and **3.7.**, respectively). Last, it is important to note that proper maintenance of objectives, filters, and all other optical components are perhaps the most critical factors regarding image quality *(8)*.

### 3.1. Cell Culture

1. MCF-7 breast cancer cells are stably expressing GFP:α-tubulin *(6,7)*, and can be passed at least 10 times without any noticeable differences in cell integrity and GFP:α-tubulin signal intensity. On average, the cells need to be split and re-plated once every 4 d. It is recommended that G418 be removed from the solution every fifth pass to ensure optimal cell viability.

2. MCF-7 cells are maintained in RPMI 1640 media without phenol red supple
   mented with 10% FBS and 40 µg/mL G418. Cells are kept in an incubator at
   37°C with 5% $CO_2$.

## 3.2. Cell Plating

1. It is critical that the cells being plated are healthy, actively dividing, and free
   from contaminants (e.g., mycoplasma, yeast).
2. Prior to cell plating, circular glass-bottom, 35-mm imaging dishes are
   preincubated with poly-l-lysine (0.4 mg/mL) for 2 h, followed by laminin and
   fibronectin at 10 and 20 µg/mL, respectively, for an additional 2 h at 37°C, to
   promote cell adhesion and spreading.
3. Cells are trypsinized from their 25-cm² flasks with 1.5 mL of prewarmed 0.05%
   trypsin. The flask is placed at 37°C for 3–4 min, cells are shaken off, and plated
   on 35-mm imaging dishes at a concentration of $1.5 \times 10^5$ cells/mL in 2 mL of
   media.
4. Cells should be allowed to adhere for 16 h prior to imaging (*see* **Note 1**).

## 3.3. Microscope Setup

1. The spinning disk confocal imaging system used in this protocol is a Perkin Elmer
   Ultraview ERS, which uses a Yokogawa CSU21 confocal scanning head. Currently, there are other commercially available spinning disk systems and the following protocol can be adapted for them. The Ultraview ERS is attached to the
   sideport of a Zeiss Axiovert 200m (Thornwood, NY) inverted microscope
   equipped with a 100X Plan- Apochromat oil objective (**Fig. 1**). The Ultraview
   ERS, emission filter wheel, piezo z-drive, CCD camera, and laser are controlled
   by Ultraview ERS software. Alternatively, these controllers and image acquisition can be run off of Metamorph software.
2. For signal detection, a peltier-cooled ORCA-ER CCD camera is used, providing
   both high sensitivity and resolution.
3. Approximately 75% of the microscope is encased in a Plexiglass environmental
   chamber maintained at 37°C (**Fig. 1**). The plated cells are imaged in a circular
   chamber with humidified 5% $CO_2$ perfusion.
4. Turn on the environmental chamber 4–5 h prior to imaging to obtain an ambient
   temperature of 37°C (*see* **Note 2**). Confirm that the appropriate objective (in this
   case the 100X oil objective) is mounted on the piezocontrolled z-stepper (*see*
   **Note 3**).
5. Turn on the mercury arc lamp.
6. Turn on the computer and all hardware controllers (spinning disc, emission filter
   wheel, CCD).
7. Turn on the $CO_2$ controller and open the valve on the $CO_2$ tank.
8. Fire the laser in standby mode and allow it to warm for at least 10 min prior to
   imaging. The laser can be kept in standby mode prior to imaging to conserve
   laser life.

### 3.4. Imaging Software Setup

1. The various software parameters controlling the image acquisition (e.g., binning, gain, and so on) should be carefully considered for each experiment. In most cases, 2X2 binning is sufficient, and camera exposure times range from 300 to 500 ms. The protocol described below, will use imaging parameters that are optimal for observing mitotic progression at high magnification with adequate resolution. Some modifications, such as adjustments in exposure time, gain, and binning, should be made if lower N.A and/or lower magnification objectives are used.
2. Create a new experiment directory.
3. Set binning to 2X2 and gain to 0.
4. The images are acquired with a 12-bit depth (4095 gray levels). The minimum input gray level (black level) should be set to 182 and the maximum input gray level set to 498 (white level). These settings can vary between microscopy setup and cell line, and will likely require slight adjustments when imaging.
5. Adjust the laser power of the 488-nm line to 62% (*see* **Note 4**).
6. Using the software, select the 488 dichroic and if using an emission filter wheel select the GFP filter. When imaging GFP:α-tubulin alone it is not absolutely necessary to have a GFP bandpass emission filter in place, however, if the sample has multiple fluorophores then an emission filter is required.
7. Initially set the camera exposure to 400 ms.

### 3.5. Cell Preparation Prior to Imaging

1. Using a standard tissue culture microscope confirm that the cells are healthy and that mitotic cells are visible. Mitotic cells will appear as rounded cells that are still attached to the plates.
2. If these are studies assessing the effect of a pharmacologic agent(s), such as taxol or vincristine, can be added prior to imaging (*see* **Note 5**).
3. Open the environmental chamber and place a small drop of immersion oil on the 100X oil lens.
4. Put the plated cells on the stage and allow 15 min before imaging to all the cells to acclimate from their movement to the stage, and for the ambient temperature of the environmental chamber to again reach 37°C.
5. Confirm that a GFP compatible filter set is in place in the microscope filter turret and that the emission light is shuttered to the binoculars.
6. Reduce the intensity of the fluorescent light to 50% with either an adjustable mercury power supply or neutral density filter. This will reduce photobleaching of the sample when searching for cells but still provide enough fluorescent emission to view microtubules through the binoculars.
7. Open the fluorescent light shutter on the microscope and confirm that the cells are expressing GFP:α-tubulin (*see* **Note 6**). Also, confirm that cells are undergoing mitosis by quickly scanning the plate for cells with microtubule spindles. The cells that are undergoing mitosis will be in a focal plane 10–15 μm higher than interphase cells.
8. Immediately close the fluorescent light shutter.

## 3.6. Imaging the Complete Mitotic Cycle

1. This section describes a protocol for visualizing the complete mitotic cycle, beginning from prophase and ending at the completion of cytokinesis (**Fig. 2**). This protocol uses time acquisition intervals that can be used for observing the effect of an antimitotic agent on spindle microtubules but not provide enough temporal resolution for imaging extremely dynamic events such as centrosome separation or spindle pole formation.

2. The protocol for these rapid events is described in **Subheading 3.7.**

3. Set the acquisition time interval to 30 s. Intervals less than 30 s can result in photobleaching of the fluorophore and mitotic arrest.

4. Do not enter a maximum acquisition time because cell division cycles can vary between cells.

5. Take the laser off standby and put it into "on" mode.

6. Open the fluorescent light shutter and view the cells using the binoculars.

7. Search for a cell in prophase (*see* **Note 7**; **Fig. 2**). Once this cell is found immediately turn off the fluorescent light.

8. Work quickly and efficiently. Shutter the emission light to the camera.

9. Activate the laser and spinning disc to view a live image on the screen.

10. Bring the cell into focus using the on-screen controls for the z-stepper.

11. Adjust the exposure to obtain maximum contrast. The ideal exposure setting will allow one to still see relatively dim structures (e.g., the remaining interphase microtubules), with minimal saturation of the brighter structures (e.g., the inner core of the centrosome). In our experience an exposure setting of 400 ms, with no gain, and $2 \times 2$ binning is sufficient.

12. Set the top and bottom focal planes with the software controlling the z-stepper. Importantly, the top focal plane needs to be higher than the current top focal plane of the cell because the cell will continue to round up and move into a higher focal plane as mitosis progresses. Typically, the top focal plane is 10–15 µm higher than the current bottom focal plane. Use the live bottom focal plane as the actual bottom focal plane, to ensure that the remnants of the interphase microtubules are imaged.

13. Set the number of slices to six, resulting in a z-slice distance ranging from 1.5 to 2.5 µm.

14. Deactivate the spinning disc and laser.

15. Start image acquisition (the laser will automatically be activated).

16. Closely monitor the first few frames of image acquisition on the video screen to confirm that the cells remain in place and in focus.

17. Image acquisition can be stopped after completion of cytokinesis (**Fig. 2**).

## 3.7. Imaging of Rapid Mitotic Events

1. This section describes the protocol for acquiring rapid images of dynamic mitotic events that occur over the course of 5–10 min (**Fig. 3**). The example shown here, describes the protocol for imaging early spindle assembly, however, it can be

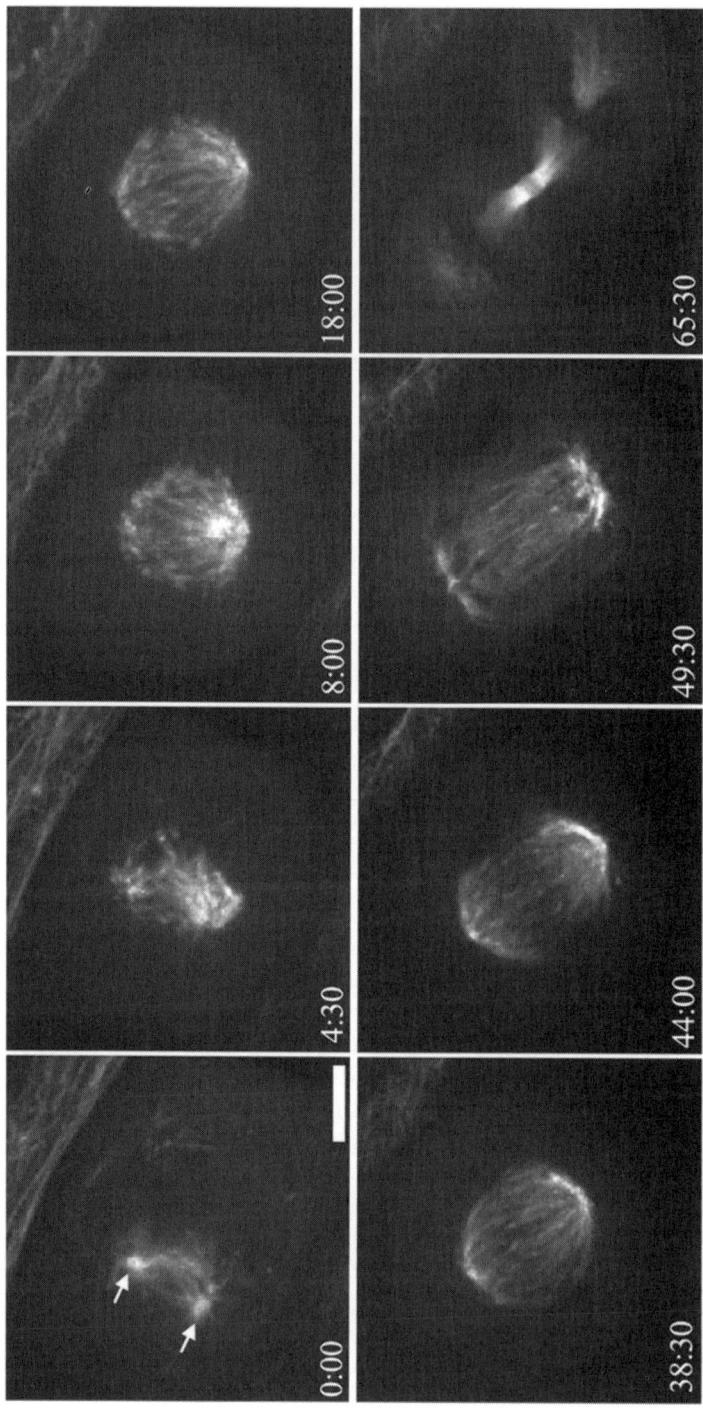

Fig. 2. Imaging of the mitotic cycle. Time lapse images where acquired every 30 s and representative images are shown here. Cell is first imaged in late prophase and proceeds through metaphase (38:30) and anaphase (44:00), and eventually ends in cytokinesis (65:30). Arrows designate the two centrosomes. Time is in minutes:seconds. Bar = 5 μm.

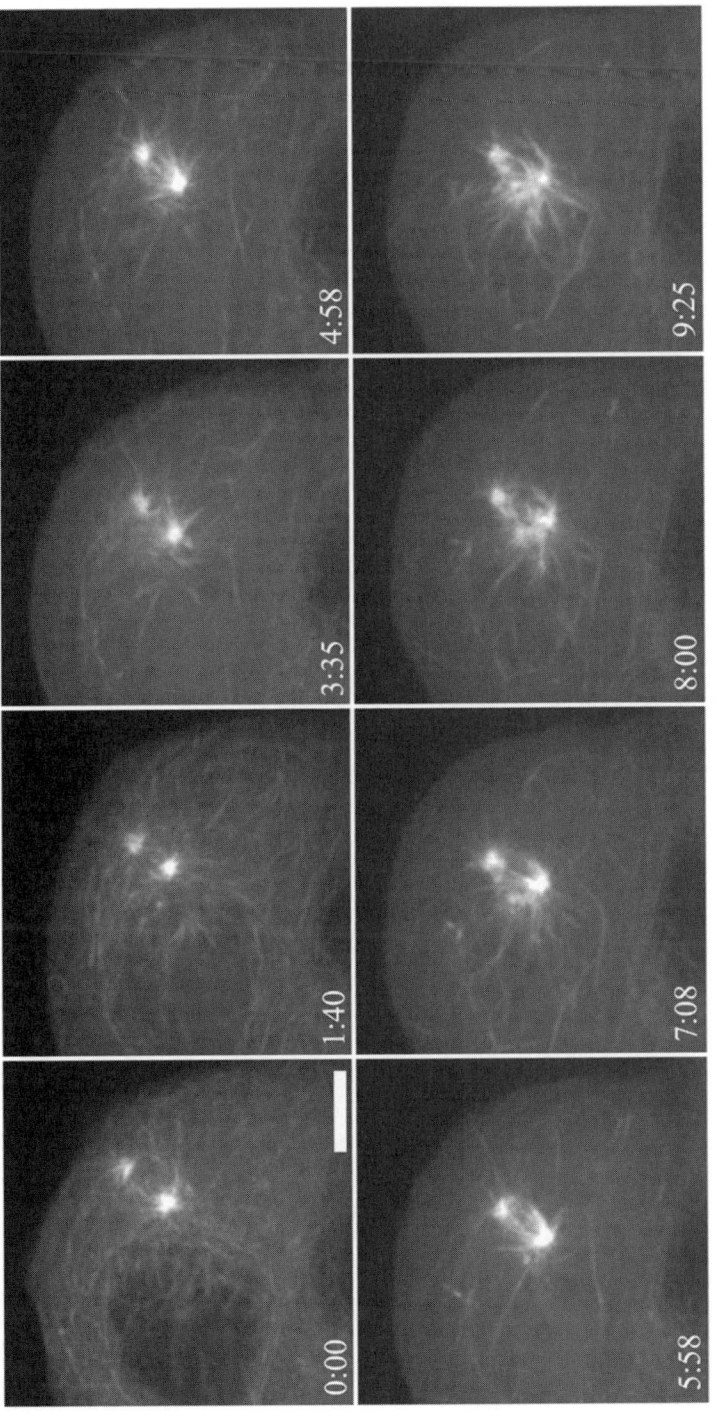

Fig. 3. Imaging of spindle pole assembly. Time lapse images were acquired every 5 s Interphase microtubule breakdown and movement of noncentrosomal microtubules to the spindle pole can be observed as previously described (1). Time is in minutes:seconds. Bar = 5 μm.

applied to any rapid mitotic event. It is important to note that once an early mitotic cell is found, the researcher needs to work in a quick and efficient manner.

2. In these short term experiments, the free radical scavenger, oxyrase, is added to the sample to reduce photobleaching at 20 µL of oxyrase per milliliter of media; however, use caution if performing long term analyses of mitotic behavior because toxicity is observed with longer exposures to oxyrase. Once oxyrase is added wait 30 min before acquiring images to allow for oxyrase to enter the cells.

3. Set the number of z-slices to acquire to four.

4. Set the time interval to 5 s.

5. Shutter the fluorescent light to the binoculars and open the fluorescent light shutter.

6. Search for a cell in prophase. Remember, once the cell is found the fluorescence shutter should be immediately closed to reduce photobleaching. Work extremely quickly and efficiently, because these events proceed rapidly.

7. Shutter the emission light to the camera.

8. Activate the laser to view a live image on the screen.

9. Bring the cell into focus using the on-screen controls for the z-stepper

10. Adjust the exposure to obtain maximum contrast. In our experience an exposure setting of 400 ms with no gain (2 × 2 binning) is sufficient for image analysis.

11. Set the top focal plane and bottom focal plane (*see* **Subheading 3.6.**, **step 11** for full details on how to set the focal planes).

12. Deactivate the laser.

13. Start image acquisition (the laser will automatically be activated).

14. Usually after 10 min of image acquisition photobleaching is observed and at this timepoint image acquisition can be ended.

### *3.8. Image Processing*

1. Once mitosis is complete the images are exported to image analysis software such as Metamorph (Universal Imaging). To export images, convert each single Z-plane and time point to an 8 bit TIF format.

2. Import the images to Metamorph or a similar image-processing program.

3. Calibrate the pixel to micrometer ratio (e.g., X: 0.58 micrometers/pixel Y:0.67 micrometers/pixel) in the image analysis software.

4. If desired, construct a maximum projection of the z-stack, which merges the z-slices into one optical plane by only displaying the brightest pixel among all slices.

5. If desired, crop the image to show the area of interest. In doing so, this will also increase CPU processing speed and decrease file size. Always keep the original, uncropped version.

6. Expand the contrast using the image histogram (*see* **Note 8**), which reassigns pixel brightness levels (**Fig. 4**). A linear expansion assigns the darkest pixel to black and the brightest to white. Although there are also other types of transfer functions (e.g., logarithmic, exponential) these are not recommended for basic users because these techniques can be subject to user bias, and are thoroughly described in **ref. 9**.

7. If desired apply a low pass filter to the image to reduce random noise in the image (**Fig. 4**).

8. A montage of the time-lapse can be made, creating a frame-by-frame depiction of the acquired data and/or a Quicktime (not PowerPoint compatible) or AVI (PowerPoint compatible) movie for presentation.

### 3.9. Image Archiving

1. Acquisition of five-dimensional confocal images requires a large amount of hard drive space. Each optical section for a single time point is about 673 kb; thus a 3-h experiment of one fluorophore with 6 z-slices acquired per acquisition, requires nearly 1 GB of hard drive space.

2. The raw data files should be archived onto an external hard drive or server. For maximum security, the external hard drive should be stored in a different location from the desktop computer.

3. For easy restoration of archived files, maintain the same directory and folder structure on the external hard drive as what was shown on the original internal hard drive. Files can then be copied and pasted from the external hard drive to the computer's hard drive accessed through the software.

## 4. Notes

1. Although these cells adhere to the plate within 4–6 h, the GFP emission signal at this time is dimmer than the signal observed at 16 h, therefore it is recommended waiting 16 h before imaging.

2. It is essential that cells are maintained at 37°C with $CO_2$ perfusion. If the system is not fully warmed to 37°C prior to image acquisition the stage and/or cells may drift in the $x$-, $y$-, and $z$-axes. Although our system uses a large environmental chamber a more cost-effective approach is to use just a stage warmer and objective warmer. It is important to warm the objective because a room temperature objective will serve as a heat sink and remove heat from the sample.

3. Choosing the proper objective depends upon the analysis one wants to perform. For example, a100X, high numerical aperture oil objective is necessary for observing individual microtubule behavior, where a 40X oil objective is well suited for observing a 10- to 15-cell population.

4. Although higher laser power will result in a brighter GFP emission signal, it will also increase photobleaching. We have found that 62% laser intensity provides an optimal balance between signal intensity and photobleaching rate.

5. In our experience, removing media from the dish and adding fresh media perturbs the cells and causes them to drift in the $z$-axis during imaging. Therefore a 2X solution of the drug diluted in prewarmed media can be made and this solution can be added to an equal volume of media already in the dish containing the cells, thus making a 1:2 dilution of the drug.

6. It is normal to observe varying intensities of the GFP:α-tubulin signal. It is always important to choose a representative cell or group of cells and to perform the experiments multiple times to control for any artifacts that could occur with overexpression of GFP chimeras.

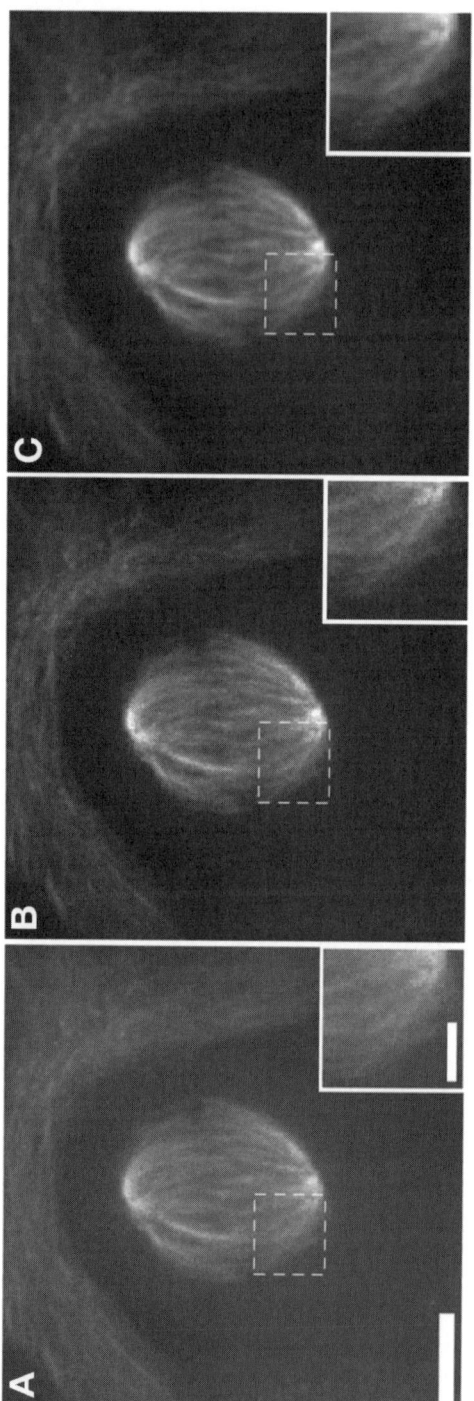

Fig. 4. Image processing. (**A**) Unprocessed image of a spindle microtubules captured with the spinning disk confocal. (**B**) Same image as in **A** after contrast expansion. (**C**) Same image as in **A** after contrast expansion and a low pass filter. Note how individual spindle microtubules can be observed after image enhancement. Insets show magnified view of the area shown with the dashed box. Bar = 5 μm.

7. A cell in early prophase will have two bright centrosomes that are still relatively close to each other (within 5 μm) and an early bipolar spindle may be observed. Furthermore, some remnants of the interphase microtubule array are likely to still be present. Typically, the prophase cell has not fully rounded up and is therefore in a similar (possibly ~5 μm higher) focal plane than an interphase cell.

8. Although, the histogram can potentially display the full dynamic range of the camera, the actual image does not have to. In many cases, adjusting the histogram to completely fill the full range from white to black can lead to an image with too much contrast that appears grainy. Remember, expanding the histogram can result in more noise and gaps in the histogram, and is not a good substitute for acquiring the image with optimal illumination and exposure. Furthermore, care needs to be taken to ensure that dim objects are not deleted from the image when enhancing the contrast and that images which are compared, are processed identically.

## Acknowledgments

The author would like to thank Drs. Mary Ann Jordan and Kathy Kamath for supplying the MCF-7 GFP:α-tubulin cell lines.

## References

1. Tulu, U. S., Rusan, N. M., and Wadsworth, P. (2003) Peripheral, non-centrosomeassociated microtubules contribute to spindle formation in centrosome-containing cells. *Curr. Biol.* **13,** 1894–1899.
2. Lampson, M. A. and Kapoor, T. M. (2005) The human mitotic checkpoint protein BubR1 regulates chromosome-spindle attachments. *Nat. Cell Biol.* **7,** 93–98.
3. Marcus, A. I., Peters, U., Thomas, S. L., et al. (2005)Mitotic kinesin inhibitors induce mitotic arrest and cell death in Taxol-resistant and -sensitive cancer cells. *J. Biol. Chem.* **280,** 11,569–11,577.
4. Rusan, N. M. and Wadsworth, P. (2005) Centrosome fragments and microtubules are transported asymmetrically away from division plane in anaphase. *J. Cell Biol.* **168,** 21–28.
5. Wang, E., Babbey, C. M., and Dunn, K. W. (2005) Performance comparison between the high-speed Yokogawa spinning disc confocal system and single-point scanning confocal systems. *J. Microsc.* **218,** 148–159.
6. Kamath, K. and Jordan, M. A. (2003) Suppression of microtubule dynamics by epothilone B is associated with mitotic arrest. *Cancer Res.* **63,** 6026–6031.
7. Jordan, M. A., Kamath, K., Manna, T., et al. (2005) The primary antimitotic mechanism of action of the synthetic halichondrin E7389 is suppression of microtubule growth. *Mol. Cancer Ther.* **4,** 1086–1095.
8. Inouâe, S. and Spring, K. R. (1997) *Video Microscopy: The Fundamentals.* Plenum Press, New York.
9. Russ, J. C. (2002) *The Image Processing Handbook.* CRC Press, Boca Raton, FL.

# 10

## Live Cell Approaches for Studying Kinetochore–Microtubule Interactions in Drosophila

### Daniel W. Buster and David J. Sharp

#### Summary

Kinetochores are essential for the proper positioning, movement and segregation of chromosomes on spindle microtubules. Live cell analyses of kinetochore movements on the spindle provide an important tool for dissecting the molecular machinery underlying kinetochore-based chromosome motility. Here, we describe contemporary techniques for studying and manipulating kinetochore function in live Drosophila syncytial blastoderm-stage embryos and S2 cells.

**Key Words:** Drosophila melanogaster; fluorescence speckle microscopy; flux; kinetochores; mitosis; RNA interference; S2 cells; syncytial blastoderm embryos; spindle microtubules.

## 1. Introduction

Kinetochores are multiprotein complexes that assemble onto centromeric regions of chromosomes and serve as a primary interface with microtubules of the mitotic spindle. These structures perform multiple mitotic functions among the most important of which is to precisely position and move chromosomes on the spindle. Kinetochores house numerous molecular motors, which can generate force along the microtubule surface lattice, and proteins that regulate the assembly dynamics of associated microtubule plus-ends. Understanding how these components work, both individually and cooperatively, to drive efficient directional chromosome motility is a fundamental issue in cell biology.

Studies of kinetochore and chromosome motility have been aided substantially by the refinement of live cell microscopy along with the development of green fluorescent protein (GFP) and related fluorescent probes to label spindles, kinetochores, and chromosomes, and the use of antibody microinjection and double-stranded RNA-interference to inhibit the activity kinetochore proteins in vivo. In this chapter, we outline some contemporary techniques for studying

From: *Methods in Molecular Medicine, Vol. 137: Microtubule Protocols*
Edited by: Jun Zhou © Humana Press Inc., Totowa, NJ

kinetochore-based chromosome motility in live Drosophila syncytial blasto-derm-stage embryos and S2 tissue culture cells. Although each cell type on its own has distinct benefits and drawbacks, their complementary use provides a powerful means of describing and dissecting the molecular machinery under-lying kinetochore-based chromosome motility.

Syncytial blastoderm-stage embryos: during the initial approx 2.5 h of Drosophila embryogenesis, embryonic nuclei proceed through 13 very rapid and nearly synchronous mitoses, which occur without intervening cytokinesis. Just before the 10th division, nuclei migrate to the surface of the embryo form-ing an interconnected monolayer and undergo divisions 10–13 in this orienta-tion. This stage of embryogenesis, termed syncytial blastoderm, is visually striking and has proven very useful in studies of spindle morphogenesis and chromosome motility. The morphology of embryonic spindles and the move-ment of chromosomes on them are remarkably consistent among nuclei in the same embryo and from embryo to embryo (*see* **Note 1**). Thus, any defects are relatively easy to spot. The injection of function blocking antibodies and/or fluorescent probes into embryos is relatively easy and, given that embryonic nuclei share a common cytoplasm, a single injection will impact multiple adja-cent nuclei. Embryos are also fairly resistant to phototoxicity and can be con-tinuously illuminated with a laser for long periods without noticeable damage. This allows very high-resolution analyses to be performed. Finally, embryos lack a strong mitotic checkpoint, which normally arrests cells in metaphase in response to certain defects. Although this may not be ideal for studying how kinetochore proteins influence the checkpoint, it makes it possible to study the anaphase functions of proteins that are also utilized earlier in the cell cycle (potentially inducing a checkpoint arrest when inhibited). There are also notable drawbacks to the syncytial blastoderm system that should be considered. In par-ticular, mutant analysis in syncytial blastoderms is often difficult or impos-sible owing to maternal loading of proteins.

S2 cells: although syncytial blastoderm embryos are unparalleled as an experi-mental system for studying mitosis because they are, essentially, an extended plane of robust and synchronous mitotic domains, they do require some practice and hardware to successfully use. On the other hand, Drosophila S2 cells can be used by anyone with a basic tissue culture facility and some bench space-mak-ing S2 cells a very useful complementary system to embryos. S2 cells are easy to culture: the culture medium ingredients are available commercially and the cells can be grown without an incubator. Most importantly, proteins of interest can be readily knocked-down in S2 cells using RNA interference (RNAi). Two reasons make S2 cells the system of choice for studies employing RNAi. First, the dsRNA oligos are easily transcribed in vitro and, second, the dsRNA oligos need only be added to the culture medium because S2 cells will take up the RNA, even without transfection reagents. The spindles of substrate-attached S2 cells are sufficiently large (averaging 10–14 ms at metaphase) to be easily visualized with standard wide-field epifluorescence and confocal microscopy. The mitotic index of S2 cells grown under standard conditions is variable but generally about 2–3%, which is expected for a cell type with a doubling time of approx 24 h and a length of mitosis of about 50 min. Therefore, the mitotic index of S2 cells is on par with that of some commonly used mammalian cultured cell lines, like HeLa, A549, and U2OS cells.

## 2. Materials

### 2.1. Syncytial Blastoderm-Stage Embryo Collection and Preparation

#### 2.1.1. Embryo Collection Cage (*Fig. 1A*)

1. Label tape.
2. Embryo lay tray: mix 11 g bactoagar, 29 g dextrose (glucose), 14.3 g sucrose, 90 mL grapejuice (unsweetened concentrate), 409 mL ddH$_2$O, and 1.25 mL 10 *N* NaOH. Microwave to boil. Add 5.6 mL of acid solution (20.9 mL proprionic acid, 2.1 mL phosphoric acid, 27 mL ddH$_2$O) and mix. Pour into 10 × 35-mm Petri dishes. This volume of molten solution should make 200–300 dishes if they are roughly half filled. Allow to solidify (at room temperature). Store at 4°C. Just before use, warm to room temperature and apply a small amount (roughly 200 µL) of yeast paste (*see* **Subheading 2.1.1.4.** below) to the center of the lay tray.
3. Polypropylene round-bottomed stock bottle (Fisher, Pittsburgh, PA). Cut a small hole in the side.
4. Yeast paste: mix type II bakers yeast (Sigma-Aldrich) with ddH$_2$O in a 100 × 20-mm Petri dish using a spatula until the paste has the consistency of frosting (add yeast or water as needed for desired consistency).
5. Small sponge or cotton cube.

Assemble collection cage as shown in **Fig. 1B** and described in legend.

#### 2.1.2. Dechorionation Slide (*Fig. 1C*)

1. Standard microscope slide (25 × 75 × 1 mm; Fisher).
2. Cover slip (22 × 50 mm-1; Fisher).
3. Double-stick tape (3M, St. Paul, MN).
4. Heptane glue: pull the double-stick tape from its roll and stuff it into a 150-mL bottle. Fill the bottle half way with 100% heptane. Rock overnight. For use, pour 1–2 mL into a 15-mL conical tube. Dip cotton applicator into tube, withdraw, and remove excess solution. In one motion, streak onto cover slip near the edge. This will be used to adhere embryos to cover slip.

Assemble as shown in **Fig. 1C** and described in legend.

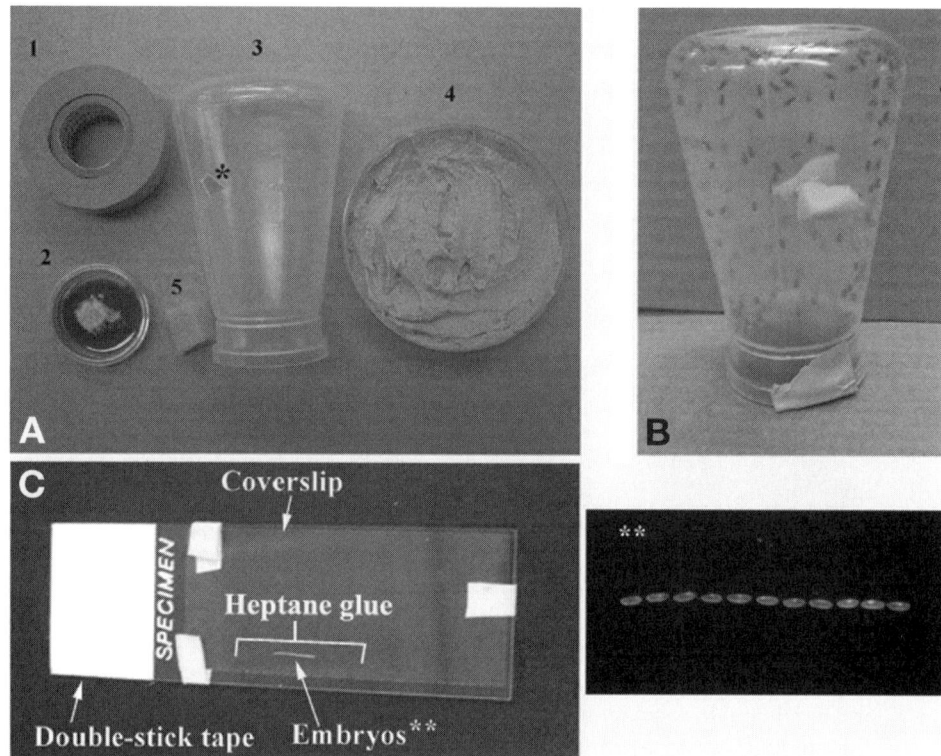

Fig. 1. The collection and preparation of Drosophila syncytial blastoderm embryos for microinjection. (A) The components of an embryo collection cage. (1) Tape. (2) An embryo lay tray constructed from the bottom half of a 35-mm Petri dish about half filled with solidified grape medium (reddish material) and dabbed with a small amount of yeast paste (brownish material in center of dish). (3) Bottle that will contain gravid adult flies. An asterisk (*) marks the ventilation hole plugged with a piece of styrofoam (5). To assemble the collection cage, gravid flies are transferred to the bottle (3) after first plugging its hole (*). Then the lay tray (without its top and with the yeast paste facing the flies) is inserted into the mouth of the bottle and held in place with tape (1). (B) An assembled collection cage with flies inside and with lay tray taped to the mouth of the bottle. The collection cage should be oriented as shown, with lay tray at bottom. Gravid flies will be attracted to lay eggs on the surface of the solid grape medium. (C) An assembled dechorionation slide. Initially, appropriately aged embryos are transferred from the lay tray to the double-stick tape, gently rolled to tear away their chorions, and then positioned in a line (**) at the edge of the cover slip. The dechorionated embryos are held in place by a film of heptane glue previously applied to the cover slip's edge (region indicated by bracket). For convenience of handling, the cover slip is taped to a microscope slide. The last panel is a higher magnification image of dechorionated embryos ready for transfer to the desiccation chamber.

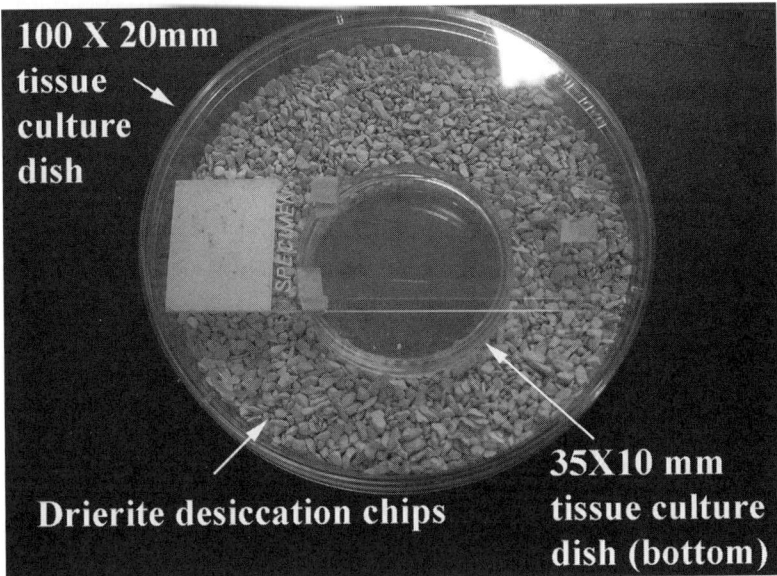

Fig. 2. The embryo desiccation chamber. In this photo of a desiccation chamber, the dechorionation slide can be seen inside the chamber, resting on the bottom of a 35-mm Petri dish so that the dechorionation slide is positioned just above the bed of desiccation chips within the chamber. After the lid of the 100-mm dish is placed to close the chamber, the humidity will decrease sufficiently to partially dehydrate the embryos.

### 2.1.3. Desiccation Chamber (*Fig. 2*)

1. Anhydrous $CaSO_4$ chips (Drierite, cat. no. 24005).
2. $100 \times 20$-mm Tissue culture dish.
3. $35 \times 10$-mm Tissue culture dish (bottom only).

To assemble, invert the bottom of the small dish onto the center of the large dish. Add drierite chips and cover.

### 2.1.4. General

1. Red wax pencil (PCG Scientifics, Gaithersburg, MA).
2. Half no. 5 forceps (Ted Pella Inc., Redding, CA). Rip forceps in two; use one arm.
3. Fine-tipped paint brush (PCG Scientifics).
4. Halocarbon oil 700 (Sigma-Aldrich).
5. BRB80 buffer: 80 m$M$ PIPES, pH 6.8, 1 m$M$ MgCl$_2$, 1 m$M$ EGTA.

## 2.2. Embryo Microinjection

### 2.2.1. Hardware

1. For injection setup (**Fig. 3**), we use a Nikon Eclipse TS-100 inverted microscope, manual micromanipulator (model KITE-R, World Precision Instruments, Sarasota, FL), magnetic holder and baseplate (Kanetech, cat. no. MB-K), and 60-mL syringe with stopcock.
2. Other items include 1 mm × 4-in. microinjection capillary tubes (World Precision Instruments), micropipet storage jar (World Precision Instruments), needle/pipet puller (Knopf, cat. no. 720), and microloader tips (Eppendorf).

### 2.2.2. Visualizing Chromosome and Spindle Microtubule Dynamics in Embryos

1. For simultaneous imaging of spindles and chromosomes or kinetochores, we inject rhodamine labeled probes into embryos from fly strains expressing other GFP-tagged proteins. Next, we list those reagents that are most commonly used in our studies. Also next is a list of references of selected studies from our lab and others in which the function of mitotic proteins has been manipulated via injection of antibodies into syncytial blastoderm-stage embryos. We also provide a protocol for fluorescence speckle microscopy in embryos that has proven useful for studying spindle microtubule dynamics and the relative movement of kinetochores along the lattice of kinetochore microtubules.
2. Fly strains expressing GFP-tagged proteins: a number of fly strains expressing GFP-tagged spindle and chromosome-associated proteins have been generated. Several useful lines are listed next. (Note that embryos from these strains should be collected, prepared and injected like wild-type embryos.) Spindle proteins: Tubulin (α-Tubulin 84B); Flybase ID: FBtp0013497 *(1,2)*. D-TACC (transforming acidic coiled-coil protein; labels centrosomes); Flybase ID: FBgn0026620 *(3)*. Chromosome/kinetochore proteins: Histone 2A variant (labels whole chromosomes); Flybase ID: FBgn0001197 *(4,5)*. Cid (centromere identifier; labels inner kinetochore); Flybase ID: FBgn0040477 *(6,7)*. Rod (rough deal; labels kinetochores throughout mitosis and kinetochore microtubules during metaphase); Flybase ID: FBgn0003268 *(8)*.

### 2.2.3. Rhodamine-Labeled Probes

We purchase rhodamine-tubulin from Cytoskeleton (Denver, CO) as lyophilized powder. To visualize whole microtubules, resuspend labeled tubulin in BRB80 buffer to 8 mg/mL. For fluorescence speckle microscopy (next) dilute to 0.8–0.4 mg/mL. Centrifuge ($20Kg$, 15 min, 4°C) to remove aggregates which can clog the needle.

Rhodamine-histones were prepared as described in **ref. 9**.

1. Forty milligrams of calf thymus histones (unfractionated, Worthington, Lakewood, NJ) dissolved in buffer A (0.2 $M$ NaCl, 20 m$M$ HEPES, pH 7.4, 1 m$M$

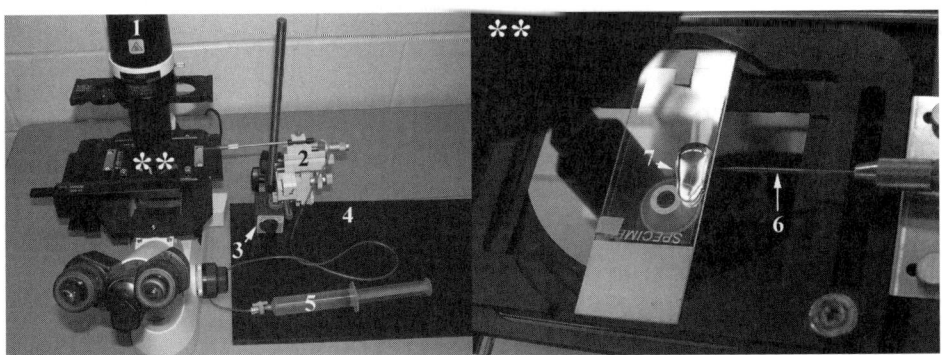

Fig. 3. The embryo microinjection apparatus. The microscope slide holding the dechorionated, partially dehydrated embryos (**) is shown on the microinjection stereomicroscope (**1**). A micromanipulator (**2**) holds the microinjection needle (already loaded with a small volume of antibody solution) in a needle holder which can be seen in the photo as the slender rod that extends from the micromanipulator to the microscope stage. A magnetic locking clamp (**3**) attaches to a metal baseplate (**4**) and holds the micromanipulator. Pressure to eject antibody solution from the needle into the embryo is generated by carefully depressing the plunger of an air-filled syringe (**5**) and is transmitted through the plastic tubing (seen coiled on the baseplate) that connects the syringe to the end of the needle holder. The second panel shows the dechorionation slide on the microscope stage, oriented such that the embryos (submerged underneath a line of halocarbon oil [**7**]) are broadside to the microinjection needle (**6**) attached to the needle holder. After the tip of the needle has been chipped open by carefully manipulating it to press against the edge of the cover slip, the needle is slid into the halocarbon oil and positioned close to the first embryo's vitelline membrane. Then the stage control is manually manipulated to impale the embryo on the needle in a quick and smooth motion, making a shallow puncture. The positive pressure within the needle should force a small bolus of solution into the embryo. After the needle is quickly withdrawn, the remaining embryos are similarly injected.

EDTA) and 1 m$M$ DTT are mixed with 20 mg ssDNA bound to cellulose (Worthington).
2. Gently rock for 24 h, 4°C, to allow histones to complex with DNA.
3. Remove unbound histones by washing four times with 30 mL buffer A + 0.2 $M$ NaCl.
4. Add a volume of buffer A (+ 0.2 $M$ NaCl) equal to packed cellulose and warm to room temperature.
5. Dissolve 9 mg of 5-(6-)carboxytetramethylrhodamine succinimidyl ester (Molecular Probes, Eugene, OR) in 1 mL dimethyl formamide.
6. Add activated rhodamine solution to histone/DNA/cellulose slurry and rock 30 min, room temperature.

7. Remove unbound rhodamine by washing six times with 40 mL buffer A (+ 0.2 $M$ NaCl, 1 m$M$ DTT).
8. Elute the rhodamine-histones by rocking slurry with 10 mL buffer A (+1.8 $M$ NaCl, 1 m$M$ DTT), 4°C, 10 min. Save eluate, then repeat this step. Pool eluates.
9. Exchange the rhodamine-histones into PBS by several cycles of dilution with PBS followed by concentration using a Centriprep filter unit (Millipore, Billerica, MS).
10. Determine the protein concentration, then aliquot and store at –80°C.

## *2.2.4. Microinjection of Antibodies to Inhibit Mitotic Proteins*

Antibodies are commonly used to inhibit protein function in syncytial blastoderm-stage embryos. Each antibody and the protein it targets are different and thus varying antibody concentrations have been used in different studies. A reference list of some studies using antibody microinjection in Drosophila embryos follows: **refs. 5** and *10–17*.

## *2.3. S2 Cell Culture and dsRNA Interference*

1. Heat-inactivated fetal bovine serum (FBS) (Invitrogen, Carlsbad, CA) (*see* **Note 2**).
2. Schneider Drosophila medium (Invitrogen).
3. Penicillin/streptomycin stock solution (Invitrogen).
4. Normal culture medium: 10% (by volume) heat-inactivated FBS in Schneider Drosophila medium. Optionally, add 50 U/mL penicillin, 50 µg/mL streptomycin (Invitrogen).
5. 20% (by volume) heat-inactivated FBS in Schneider Drosophila medium.
6. Sterile aliquots of control or target-specific dsRNA (preparation described next).
7. Standard sterile plastic ware for cell culture (Fisher): T25 flasks with vented lids, six-well plates, 35-mm culture dishes.
8. Glass-bottom microwell dishes (MatTek, Ashland, MA).
9. Concanavalin-A (Con-A) stock solution: 0.5 mg/mL concanavalin-A (Sigma-Aldrich) in H2O; sterile filter and store at –20°C.
10. Phosphate buffered saline (PBS): 154 m$M$ NaCl, 14 m$M$ $Na_2HPO_4$, 6 m$M$ $NaH_2PO_4$, pH 7.4.
11. In vitro transcription kit: T7 Megascript kit (Ambion, Austin, TX, cat. no. 1334), or RiboMax T7 kit (Promega, Madison, WI, cat. no. P1300).
12. Phenol/chloroform/isoamyl alcohol (25:24:1, by volume) (Fisher): use pH 8.0 or 5.2 solutions for extraction of DNA or RNA, respectively.
13. Chloroform/isoamyl alcohol (24:1, by volume) (Fisher).
14. Aldehyde fixation solution: 0.5% (by volume) glutaraldehyde (EM grade, Polysciences, Warrington, PA), 3% (by volume) formaldehyde (EM grade, EMS, Hatfield, PA), 0.1% Triton X-100 (Fisher), 1 µ$M$ taxol (Sigma-Aldrich), in BRB80 buffer; make fresh. **Caution:** avoid contact with solution or inhalation of fumes.

## 3. Methods

### 3.1. Preparation of Syncytial Blastoderm Embryos

The syncytial blastoderm stage of Drosophila embryogenesis begins roughly 90 min after fertilization and persists for another approx 60 min. Thus, embryo collection must be timed very carefully. Moreover, embryos must be removed from their protective shell (chorion) and slightly dehydrated prior to microinjection. Several very useful chapters describing general protocols for the collection, preparation, and visualization of syncytial blastoderm-stage embryos have been published. The protocol outlined next has been derived from these and optimized for our purposes.

1. Assemble the embryo collection cage several days before intended use. Flies need time to acclimate to the new environment and females "hold" their embryos under stressful circumstances. Thus, collecting properly staged embryos during the first day or two after transfer is difficult. Make sure to change lay trays periodically.
2. On day of use, allow flies in collection cage to lay embryos for 1–1.5 h. Time will vary depending on yield. Choose a time point that allows at least 20–30 embryos to accumulate on the tray as embryos will invariably be lost during dechorionation.
3. Change collection plate and mark the time on collected plate.
4. Allow collected embryos to age for another hour at room temperature (for syncytial blastoderm stage embryos).
5. After ageing, use a moistened, fine-tipped paintbrush to pick-up embryos from the tray and move them to double-sticky tape on the dechorionation slide.
6. Under a dissecting microscope, gently roll embryos around on double-stick tape with the tip of a half forceps. One should observe the white shell or chorion of the embryo peel away, exposing a slick and waxy vitelline layer. At this point, the embryos become sensitive to dehydration and subsequent steps should proceed quickly. Embryos also become very fragile and can burst if not handled carefully.
7. After dechorionation, embryos will usually adhere to the forceps tip. Use this to transfer embryos to the streak of heptane glue near the edge of the cover slip. On occasion, embryos will not adhere well enough to the forceps tip to be removed from the tape. In this case, roll the embryo onto its chorion (which should remain stuck to the tape) and pick it up from there. Make sure to align embryos end to end with their long axes parallel with the edge of the cover slip (**Fig. 1\*\***). After the appropriate number of embryos (10–15) has been moved to the cover slip, mark the edge of the cover slip nearest the embryos with a red wax pencil (this will help you orient during microinjection).
8. Place the slide with embryos in the desiccation chamber. Embryos must be slightly dehydrated to allow for the injection of new fluid. The time required for appropriate dehydration of embryos will vary depending on ambient humidity

(from 3 min in very dry environments to as much as 10 min in humid environments). This should be determined empirically. Take care not to over-dehydrate embryos because this will cause damage to the spindles and make injection difficult. Signs of over-dehydration include folding of the vitelline membrane and deformation of the embryo itself.

9. Remove the slide from the desiccation chamber and cover embryos with a line of halocarbon oil (*see* **Fig. 3*** 7). Try to prevent halocarbon oil from spilling over the side of the cover slip.
10. Proceed to microinjection.

### 3.2. Microinjection of Syncytial Blastoderm Embryos

1. Pull capillary tubes into microinjection needles. Generally, we pull 10–20 needles at a time as needed. These can be stored indefinitely in a micropipet storage jar. Make sure to store with needle pointing down.
2. Backload needles with injection reagent (e.g., rhodamine-tubulin, function blocking antibody, and so on) using microloader tips. Normally, we load 1–5 μL injection solution per needle. After loading, needles can be stored at 4°C until use. Generally, we do not store loaded needles for more than 1 d.
3. After the microinjection slide is loaded onto microscope (*see* **Fig. 3***), the needle tip must be broken to allow the free flow of the injection reagent. Focus on the marked edge of the slide with the needle positioned slightly above this point and out of focus. Slowly lower the needle tip into the plane of focus, close to but offset from the slide. When both the needle tip and edge of the slide are in focus, gently move the slide toward the needle (or vice versa) until they touch. A small portion of the needle tip will break away. If too much of the needle is broken, the caliber of the needle at the injection point becomes too large and will damage the embryo. If it is too small, it will become easily clogged.
4. Once the needle is broken, move it back up away from the slide and focus on the embryos under halocarbon oil. Again, move the needle tip into the focal plane and move it to be just offset from the middle of the first embryo's long axis. In a quick motion, touch the embryo to the needle tip. The tip should puncture the vitelline membrane. Make sure not to push it too far into the interior of the embryo. In many circumstances, the injection solution will immediately be drawn into the embryo and will appear as a clear bolus under the microscope. Slight pressure can also be applied to the syringe to control flow through the microneedle. Be sure not to place too much pressure on the syringe because a sudden burst of injected solution can damage the embryo.
5. Quickly remove the needle tip from embryo and proceed to the next.
6. For double injections (e.g., fluorescent tubulin and antibody), finish the first injection for all the embryos, wait at least 5 min, and then perform the second injection in exactly the same order. When injecting fluorescently labeled histones for chromosome labeling, wait at least 20 min before moving on to the second injection. Try to position the point of entry for the second injection near to the initial injection site.

Now the embryos are ready for the microscope. Imaging must be optimized for each individual microscope.

### 3.3. Microscopy of Syncytial Blastoderm Embryos

#### 3.3.1. Four-Dimensional Imaging of Spindles and Chromosomes/ Kinetochores

Spindles and chromosomes are three-dimensional structures that can change their orientation during mitosis. Thus, general tracking of chromosome and kinetochore motility for prolonged periods is often benefited by the use of four-dimensional (4D) microscopy (xyz/time) in which multiple adjacent z-planes are imaged at each time point. Most confocal microscopes are equipped with z-controllers and software necessary for 4D imaging.

1. Collect and prepare embryos as previously described.
2. Under the confocal microscope, identify embryos that are in the appropriate stage of mitosis (e.g., nuclei in prophase prior to nuclear envelope breakdown).
3. Use z-controller to set the top and bottom sections of the z-stack (this will differ from microscope to microscope) and step size. Spindles generally occupy a depth of only 5–7 μm but keep in mind that the spindles will move up and down in relation to the surface of the embryo through mitosis and a few extra z-planes will keep structures of interest in the data set. If possible, use centrosomes as landmarks and try to capture at least 1 z-plane above and below both centrosomes. It is usually not necessary to use a step size smaller than 1 μm.
4. Set camera exposure time (for CCD-based systems, e.g., spinning disk confocals) or scan time (for PMT-based systems, e.g., laser scanning confocals) so that the signal for all channels imaged are bright but not saturated.
5. Begin recording.
6. Terminate and view experimental results. Each z-stack can now be viewed as a maximum intensity projection (lose depth) or as a 3-D image. Because the latter requires a great deal of time and computing power (particularly for longer time series) we generally analyze our data sets as projections. This should be avoided if the long axis of the spindle becomes tilted in the *x*- to *y*-plane, however, because there are so many spindles in embryos it is not usually a problem to find more than a few in the appropriate orientation.

#### 3.3.2. Fluorescence Speckle Microscopy

4D imaging will not answer questions about the dynamics, movement, and turnover of kinetochore microtubules, nor will it discern the relative movement of kinetochores or chromosomes along the microtubule lattice. To examine these issues, one can employ fluorescence speckle microscopy (FSM) to place and visualize fluorescent marks in the lattice of spindle microtubules *(18)*. Spindle microtubules are marked by introducing very low concentrations of fluorescently labeled tubulin into the cell. At the appropriate concentration (which is generally determined empirically) the fluorescent tubulin incorporates into the MT lattice in small interspersed aggregates causing the microtubule to appear "speckled." This approach has been used to discern the impact of experimental manipulations on fundamental spindle behaviors such as poleward tubulin flux *(19)*. Moreover, microtubules "speckled" with rhodamine tubulin have been visualized in embryos containing chromosomes/ kinetochores marked with GFP, allowing the relative movement of chromosomes along the lattice of microtubules to be determined.

1. While embryos are aging, dilute lyophilized rhodamine tubulin to a concentration of 0.4–0.8 mg/mL. Centrifuge (20K$g$, 10 min, 4°C) to remove any aggregates. Load supernatant into needles.
2. Collect, prepare, and inject embryos as previously described.
3. Identify embryos in the appropriate stage of mitosis. Specifically, look for cases in which nuclear envelope breakdown has not yet occurred or has just begun.
4. Focus on centrosomes and begin recording. Capture images as a series of optical sections (i.e., a z-stack). Spindles will move as they form and some manual adjustment of z-position may be required—continue to focus on the centrosomes. Try to avoid frequent adjustments of the focal plane because it will make tracking of speckle movements difficult.
5. During image capture, camera exposure times on microscopes will have to be increased—on our microscope a threefold increase in exposure relative to normal rhodamine-tubulin injection generally suffices. Speckle tracking requires high signal to noise in captured images; therefore, the gain on cameras should not be maximized.

### *3.4. S2 Cells*

#### *3.4.1. S2 Cell Culture and Plating for Microscopy*

S2 cells are easy to culture. Wild-type cells grow in suspension or loosely attached to a substrate in a simple medium at room temperature (16–25°C) with normal air (i.e., without 5% $CO_2$). They are also readily susceptible to RNAi treatment by simply adding an in vitro transcribed dsRNA oligo to the cell culture medium *(20)*. But one disadvantage of S2 cells is the relatively high frequency of spindle abnormalities occurring in wild-type cells. These abnormalities include supernumerary centrosomes at the beginning of mitosis (though cells usually correct this problem by coalescing centrosomes and eventually forming bipolar spindles) (G. Rogers, UNC, personal communication), multipolar spindles (also frequently corrected during prometaphase by coalescing poles), and a variable number of chromosomes (i.e., aneuploidy) (unpublished data). In addition, wild-type S2 cells do not attach well to untreated plastic or glass substrates. Therefore, to promote the cell attachment and spreading needed for microscopy, cultured cells are transferred to dishes precoated with a reagent that induces tight adhesion. However, because attachment to a reagent-coated substrate interferes with the completion of cytokinesis (and therefore inhibits normal culture growth), S2 cells are replated onto coated dishes only a few hours before microscopy.

1. S2 cells grow as loosely adherent or nonadherent cultures in tissue culture-grade flasks or dishes. Initially, freshly passaged cells are generally weakly adherent but as the culture becomes increasingly confluent, cells will aggregate as small, free-floating clusters. Different S2 cell lines display different levels of adherence.
2. For normal culture maintenance, S2 cells are passaged to new T25 flasks at a 1:10 dilution into normal medium. With most S2 cell lines, treatment with trypsin/ EDTA is not necessary to release adherent cells. Instead, simply dislodge cells by agitating or pipetting a stream of medium onto the attached cells. Typically, cells require passaging every 5–7 d (*see* **Note 3**).
3. Healthy S2 cells are usually spherical (if unattached) and of roughly uniform diameter; very little amorphous debris should be present in the medium. The appearance of variably sized, misshapen cells (or cell remnants), or debris is a sign of a declining culture. If a pronounced amount of debris or cell remnants is seen in RNAi treated cultures, the RNAi treatment schedule may need to be shortened to have cells for analysis (*see* **Subheading 3.4.3., item 6** below).
4. For optimal microscopy, S2 cells usually require flattening. S2 cells will attach and spread on substrates coated with the lectin, concanavalin-A (con-A) *(14)*. Con-A coated substrates are prepared by applying 5–10 µgs of con-A to a cover slip (in a 35-mm tissue culture dish) or a glass-bottomed MatTek dish, then dried and sterilized by a 30-min exposure to the ultraviolet light in a tissue culture hood. After adding normal medium to the dish containing the con-A coated cover slip or glass bottom (MatTek dish), S2 cells are added in a quantity to give the desired density of attached cells (*see* **Note 4**).
5. Allow 1–2 h for S2 cells to completely spread on con-A coated substrates. These substrates are not appropriate for long-term culture because con-A coating interferes with cytokinesis to some degree. Cover slips with attached cells can then be processed for fixation and immunostaining; dishes with cells attached to glass bottoms can be used for live cell imaging or immunostaining.
6. Following RNAi treatment, we use two fixation methods for S2 cells destined for immunostaining. Methanol fixation is rapid and easy, but may interfere with immunostaining of some antigens and does not preserve cytoskeletal structures as well as aldehyde fixation. A variety of methods should be tested to empirically determine the optimal fixation method for a particular antigen.

### 3.4.2. Preparation of dsRNA for RNA Interference

The following protocol is used for the production of DNA template.

1. A suitable region of the target protein must be selected. dsRNA oligomers of about 500 bp work well for RNAi, but we have successfully used RNA oligomers ranging from about 250 to 1200 bp. S2 cells efficiently take up dsRNA present in

medium by an unknown mechanism and presumably cleave these long oligomers to generate short (21- to 23-mer) bioactive fragments *(21)*. BLASTN search the *Drosophila melanogaster* genome (http://flybase.bio.indiana.edu/) to identify regions of target protein sequence that contain a minimal amount of nucleotide sequence identity with nontarget proteins. Web sites are available to assist selection of templates (e.g., http://e-rnai.dkfz.de/). dsRNA derived from either noncoding (5' or 3') or coding transcript sequence can be used to knock-down target proteins, but avoid selecting sequence (e.g., introns) that is not present in the target mRNA. dsRNA to noncoding transcript sequence will knock-down endogenous protein in cells which can then be rescued by transfection with plasmid containing only coding sequence for the target protein. Expression of exogenous protein in a background of knocked-down endogenous protein can be used to demonstrate specificity of RNAi or to better evaluate tagged or mutant target constructs in vivo. However, the amount of noncoding transcript sequence available in the genome databases may be limited for a specific target, making it difficult to produce a 500-bp dsRNA against noncoding sequence.

2. For each template desired, set up five 100-µL PCR reactions with inexpensive Taq polymerase. Each PCR primer should have the T7 RNA polymerase promoter sequence (TAATACGACTCACTATAGGG) at its 5'-end, followed by target-specific sequence. Template can be (1) Drosophila genomic DNA (1 µg per 100 µL PCR reaction), (2) an EST clone containing the target sequence (100 ng per 100 µL reaction), or least desirably (3) a small aliquot from a previous reaction to PCR amplify the target (*see* **Note 7**).

3. After PCR is complete, pool the five reactions into an autoclaved 1.5-mL polypropylene tube. Add an equal volume of phenol/chloroform/isoamyl alcohol (25:24:1, pH 8.0). Mix vigorously to emulsify and then centrifuge 3–5 min, $20Kg$, room temperature, to separate phases. The aqueous phase will be on top; if it has a cloudy appearance, centrifuge for an additional 10 min. Carefully collect the aqueous phase while avoiding the lower organic phase and the interface (which might be white with precipitated protein). Transfer the aqueous phase to a new autoclaved 1.5-mL polypropylene tube.

4. Add a volume of chloroform/isoamyl alcohol (24:1) equal to the collected aqueous phase. Mix to emulsify and then centrifuge as above to separate phases. At this point, both phases and the interface are usually clear. Collect the upper aqueous phase and transfer to a new autoclaved 1.5-mL tube. Note: phenol/chloroform extraction kills RNases, so the DNA should be free of RNase at this point. Use caution to avoid contaminating the sample.

5. Add one-ninth volume of 3 *M* sodium acetate (to make 0.3 *M* final) and mix. Add 2 vol of cold (–20°C) 100% ethanol and mix well. Hold at –80°C, 30 min. Then centrifuge 20 min, $20Kg$, 4°C to pellet the precipitated DNA (usually visible as a small white pellet). Carefully discard the supernatant.

6. Wash the pellet with 0.5 mL 75% ethanol (prepared with DEPC-treated water) and then centrifuge for 5 min, $20Kg$, 4°C. Again, discard the supernatant carefully.

7. Air-dry the pellet until most visible liquid has evaporated, but do not completely dry the pellet. Add 50 μL sterile DEPC-treated water. The pellet should completely dissolve in about 15 min. Occasional gentle mixing helps solubilization.

8. Measure the optical density (at 260 nm) of a 100-fold dilution of the solubilized DNA. Calculate the before-dilution DNA concentration using the formula: DNA μg/mL = $OD_{260} \times 50 \times 100$. Adjust the DNA concentration to 1 mg/mL by adding DEPC-treated water and store at –20°C.

9. For the production of dsRNA, we have used two transcription kits—Ambion's T7 Megascript kit and Promega's Ribomax kit—with success. The reaction volumes given below can be scaled up or down depending on the desired RNA yield. (Avoid RNase contamination throughout.)

   a. Use Ambion's T7 Megascript kit (cat. no. 1334) and generally follow the manufacturer's instructions. After thawing the 10X reaction buffer, keep it at room temperature. Assemble the reaction at room temperature in an autoclaved 1.5-mL polypropylene tube. Increase the suggested reaction size four-fold to make the total volume 80 μL. The recommended enzyme concentration can be halved without significant loss of yield if the reaction is incubated 24 h. Order of reagent addition: nuclease-free $H_2O$ 22 (or 24) μL, rATP 8 μL, rCTP 8 μL, rGTP 8 μL, rUTP 8 μL, 10X reaction buffer 8 μL, 1 mg/mL template DNA 10 μL, enzyme mix 8 μL (or 4 μL). Total volume 80 μL. Incubate overnight (or even 24–48 h), 37°C. Alternatively, use Promega's RiboMax T7 Kit (cat. no. P1300) and generally follow the manufacturer's instructions. The rNTP mix containing 25 m*M* of each nucleotide is made by mixing equal volumes of the 100-m*M* stocks supplied with the kit. After thawing the 5X reaction buffer, keep it at room temperature. Assemble the reaction at room temperature in an autoclaved 1.5-mL polypropylene tube. The recommended enzyme concentration can be halved without significant loss of yield if the reaction is incubated 24 h. Order of reagent addition: T7 transcription 5X buffer 20 μL, rNTPs (mixture of all four nucleotides) 30 μL, nuclease-free $H_2O$ 30 (or 35) μL, 1 mg/mL template DNA 10 μL, enzyme mix (T7) 10 (or 5) μL. Total volume 100 μL. Incubate overnight (or even 24–48 h), 37°C.

   b. Remove DNA by DNase treatment (optional). This may help to reduce smearing when dsRNA is analyzed later by agarose gel electrophoresis, but has no apparent benefit regarding RNAi treatment of S2 cells. For the Ambion transcription reaction (and assuming an 80-μL reaction volume), add 4 μL of DNase 1 (supplied with Ambion kit; stock is 2 U/μL; final concentration is 0.1 U/μL); incubate at 37°C, 15 min. For the Promega transcription reaction, add 10 μL of RQ1 DNase (supplied with Promega kit; stock is 1 U/μL) to make 1 U of DNase per microgram of DNA template; incubate at 37°C, 15 min.

   c. Two procedures can be used to recover the RNA: (1) LiCl precipitation is convenient and effective; it will remove unincorporated nucleotides and most proteins; it may not efficiently precipitate RNAs smaller than 300 bp. The RNA concentration should be at least 0.1 mg/mL to ensure efficient precipitation. (2) Phenol/chloroform extraction is the most rigorous method for transcript purification and will remove all enzyme and most unincorporated nucleotides. Duplexing and quantitating RNA: (1) optional: the extent of duplexing after transcription depends on RNA length and GC content. Denature RNA secondary structures by heating the RNA solution to 65°C for 30 min, then gradually cool to room temperature to allow the ssRNA to duplex. The treatment might increase the yield and quality of duplexes and possibly decrease the amount of smearing when dsRNA is analyzed by agarose gel electrophoresis. This step is recommended for longer (>800 bp) transcripts. (2) Measure the $OD_{260}$ of solution containing 1 µL solubilized RNA mixed with 299 µL water. Calculate the before-dilution RNA concentration using the formula: RNA µg/mL = $OD_{260} \times 40 \times 300$. Adjust the RNA concentration to 2 mg/mL with DEPC-treated water.

d. Analyzing and aliquoting dsRNA: (1) analyze the RNA by agarose gel electrophoresis. Ideally, only a single band is present with a size equal to the original DNA template. The presence of a lower band could indicate ssRNA (*see* **Note 8**). (2) Aliquot the dsRNA into autoclaved 0.5-mL tubes. We normally use 20 µg (=10 µL) aliquots. Use sterile technique and avoid RNase contamination. Store aliquots at –20 or –80°C.

### 3.4.3. RNA Interference of Cultured S2 Cells

Use sterile technique for all steps.

1. Replate S2 cells into six-well culture plates. Cells should be completely dispersed and plated at a density to achieve a confluency of about 50% at the time of first dsRNA application (*see* **Note 9**). Between applications of dsRNA, store the culture plates in a covered, flat-bottomed container containing water-saturated paper towels to prevent drying of the medium.

2. Normally, S2 cells will loosely attach to the plate, so the old medium can be removed from each well with minimal loss of cells. Immediately add 0.75–1 mL of serum-free Schneider Drosophila medium to each well (*see* **Note 10**).

3. Take a small aliquot (roughly 50 µL) of medium from the well and add to the tube containing the dsRNA. Return the entire mixture to the well.

4. After adding dsRNA, swirl the six-well plate to thoroughly mix the dsRNA throughout the medium in each well.

5. Wait 1 h (*see* **Note 9**). Then add an equal volume of medium containing 20% FBS (to make 10% FBS final concentration). Apparently, the change from serum-free to 10% serum medium stimulates the uptake of dsRNA from the medium *(20)*.

6. The number and timing of dsRNA applications needed to significantly knockdown a target protein varies between proteins and is probably owing to (at least) the target protein turnover rate and ability of the selected dsRNA oligomer to yield potently bioactive dsRNA 21-mers. Therefore, the schedule of dsRNA application must be determined empirically for each target by evaluating protein knock-down by Western blotting of cell extracts or immunostaining of fixed

cells, or transcript knock-down by quantitative or semiquantitative RT-PCR. Unfortunately, no one method to evaluate knock-down is ideal; for example, we have observed that Western blotting RNAi-treated cell extracts can indicate a large decrease in target protein titer, whereas immunostaining reveals an RNAi-resistant fraction of the target persisting at functionally important sites within cells. Also, knock-down of some proteins can rapidly kill cells, so in this case the RNAi treatment schedule must be brief to have some cells available for study. As a starting point, we use an overly stringent week-long regime of dsRNA application on every second day (for a total of four dsRNA applications).

7. If cells over-grow during the course of the RNAi experiment, the cell density can be decreased by using the procedure described in **Note 10** and returning only a fraction of the resuspended cells to the well of the six-well plate.

8. On the day following the last dsRNA application, cells are gently resuspended by trituration and a small aliquot applied to a 35-mm tissue culture dish containing normal medium and a con-A coated cover slip or having a con-A coated glass bottom (MatTek dish). Cells should attach and spread on the con-A coated surfaces within 1–2 h. The quantity of cells added to the dish should be adjusted to optimize microscopy (*see* **Note 4**).

## 3.4.4. Spindle Microtubule Flux Measurement

1. Measurement of poleward microtubule flux requires a means to apply fiduciary marks to spindle microtubules to visualize the "flow" of tubulin subunits from their incorporation at microtubule plus-ends, through the polymer lattice, and ultimately to their release at minus-ends. We have used S2 cells stably expressing GFP-α-tubulin under either a metallothionen-inducible promoter (pMT/V5; Invitrogen) or a constitutively active actin promoter (pAc5.1; Invitrogen). Expression of GFP-α-tubulin under the pMT promoter is leaky without induction, so GFP-α-tubulin is present as a low ratio relative to endogenous tubulin. The rare fluorescent tubulin subunits incorporate into spindle microtubules and create a random pattern of fluorescent "speckles" on the microtubule lattice that is required for FSM. Constitutive GFP-α-tubulin expression generates uniformly labeled spindle microtubules; in this case, fiduciary marks are made by photobleaching small regions of the spindle that are visible by their dark profile against the bright spindle background.

2. Poleward flux has been reported to cease during anaphase B in Drosophila *(7)*, so the mitotic stage is an important consideration when measuring flux. We use two techniques to measure poleward flux rates in cells expressing GFP-α-tubulin. First, FSM can measure flux of individual microtubules (probably kinetochore MT bundles in our case), but only a small minority of S2 cells express GFP-α-tubulin at the level needed for clearly discernable speckling, so data are sometimes difficult to acquire. The second technique, tracking relatively large marks photobleached onto fluorescent spindles, can generate data from almost any bipolar spindle. In our experience, both techniques produce similar flux measurements. A number of rate measurements are made on each spindle because the

spindle microtubules do not necessarily flux at identical velocities. To obtain an overall flux rate for a spindle, the individual measurements can be averaged.

3. For fluorescence speckle microscopy, S2 cells expressing GFP-α-tubulin (pMT promoter) are plated on con-A coated glass-bottom MatTek dishes as previously described. Attached and spread cells are viewed by wide field epifluorescence to locate cells with spindle orientations perpendicular to the light path. Selected cells are imaged with a Ultraview spinning disk confocal system (Perkin Elmer, Boston, MA) on a Nikon TE200 inverted microscope (Plan Apo × 100 objective, 1.4 N.A.) with a Hamamatsu Orca RE digital camera. After locating an optical section with spindle poles and prominently speckled microtubules in sharp focus, the spindle is recorded at 2- or 5-s intervals for 2–3 min. Usually, images must be processed to increase the contrast between speckles and background fluorescence (probably soluble GFP-α-tubulin); this can be accomplished with commercial digital imaging software like Metamorph (Universal Imaging, Downingtown, PA). A disadvantage of Metamorph software is that it is unable to compensate for and track any large motions (e.g., swiveling) made by the entire spindle. Using Metamorph, recorded images are digitally processed by sequentially applying (1) the high-sharpen filter to make speckles more prominent, and then (2) the low-pass filter to smooth the stark reliefs generated by the high-sharpen filter. **Figure 4A** shows a single, digitally processed frame from the recording of a speckled spindle of a mitotic S2 cell. One or more kymographs are generated from each half-spindle by selecting a pole and an associated speckled microtubule (probably a kinetochore microtubule bundle, actually) and applying Metamorph's kymograph function (*see* **Note 11**). The pole of the half-spindle is included in each kymograph to serve as a landmark relative to which the rates of fluxing speckles are measured, in case the spindle as a whole moves in the direction of the spindle's long axis (**Fig. 4B**). Flux rates are calculated from the angles between the tracks of pole and speckles in a kymograph, and using the image resolution (i.e., $\mu$ms/pixel) and time interval between images. Specifically, if $\alpha$ is the angle between the pole and speckle tracks, C1 is the calibrated image resolution of the system (with units of $\mu$m/pixel), and C2 is the frequency of image capture (with units of $s^{-1}$), then the flux rate is calculated as: flux rate ($\mu$m/min) = $[1/\tan(90-\alpha)] \times$ C1($\mu$m/pixel) $\times$ C2 ($s^{-1}$) $\times$ 60 s/min.

4. For photobleaching, narrow rectangular regions are photobleached in the fluorescent spindles of S2 cells constitutively expressing GFP-α-tubulin using a TCS SP2 confocal system (Leica, Heidelberg) on a Leica DMIRE2 inverted microscope (Plan Apo ×63 objective, 1.4 N.A.). Typically, photobleached marks are most sharply defined when the bleached mark is created roughly two-thirds of the distance from pole to equator (**Fig. 4C**). Images of the photobleached spindle are recorded as single optical sections at short (3 s) intervals for the time that the photobleached mark remains clearly visible (usually 30–60 s). The position of the photobleach mark relative to the pole is measured using the calipers tool (Metamorph). Flux rate is calculated from the distance progressed and time elapsed before the bleached zone disappears due to fluorescence recovery (*see* **Note 11**).

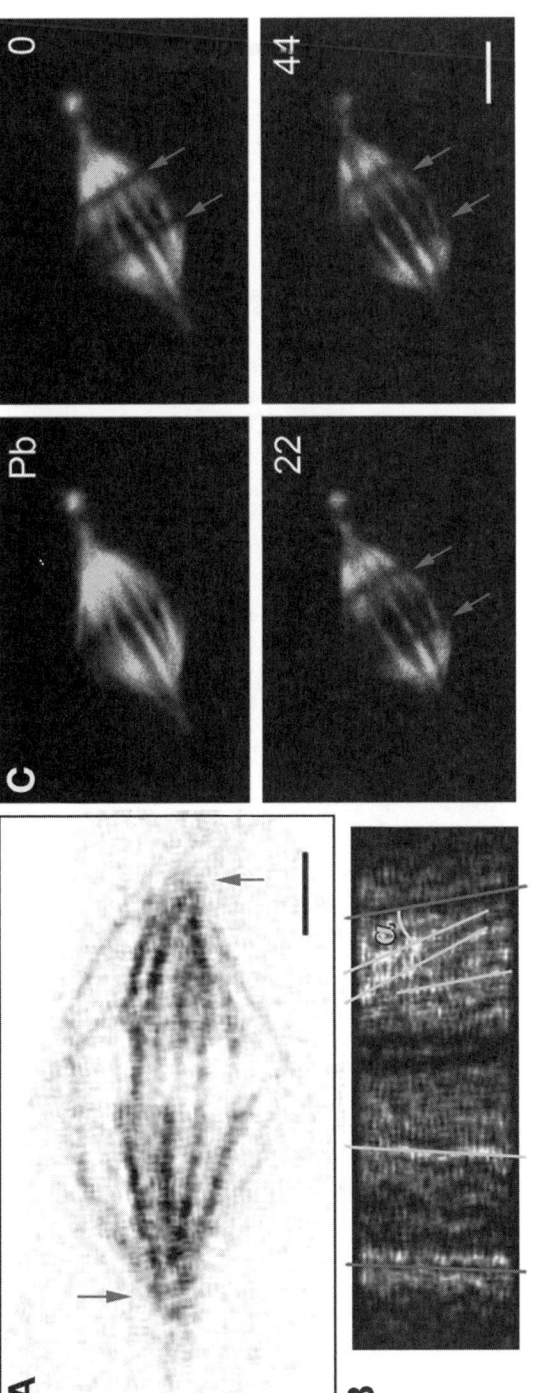

Fig. 4. Measuring poleward microtubule flux in S2 cells. (**A**) A negative image of a mitotic S2 cell expressing a low titer of GFP-α-tubulin. The uneven distribution of GFP fluorescence makes the microtubules appear "speckled"—the required state for FSM. As the spindle microtubules flux, the speckles in each half spindle can be tracked as they move toward the spindle pole. Red arrows mark the pole of each half spindle. Bar 2 = μm. (**B**) The flux velocity can be calculated by measuring the displacement of individual speckles over time, or, more commonly, kymographs are generated which show the movement of one or more speckles through space (x-axis) over time (y-axis). Red lines indicate the tracks of each spindle pole; green lines track individual speckles (only a few are indicated for clarity). Flux is measured as speckle movement relative to a pole, so the angle (α) between the tracks of a pole and a speckles can be converted to a flux velocity. Large angles indicate rapid flux, while small angles (i.e., the tracks are nearly parallel) indicate little flux. (**C**) An alternative to FSM for measuring flux is photobleaching. The uniformly fluorescent spindle of an S2 cell expressing a high titer of GFP-α-tubulin before a line is photobleached (Pb) and at time intervals (the elapsed seconds shown in upper right corners) following photobleaching. Red arrows mark the initial position of the two photobleached lines. Note the poleward movement of the bleached segments along the microtubules over time. Bar = 4 μm.

## 4. Notes

1. If making detailed analyses of spindle structure and/or function, only embryos of the same mitotic cycle (e.g., cycle 12) should be compared.
2. FBS is heat inactivated to eliminate the cell damaging affects of serum complement proteins, even though this practice probably results in some loss of serum growth factor activity. In fact, S2 cells have been cultured with normal FBS *(22,23)*. If desired, heat-inactivated FBS may be purchased or can be prepared by warming FBS to 56°C for 30 min.
3. Cell density affects the growth rate; sparse cultures grow slowly. Before beginning RNAi treatment, allow the culture to return to the high growth rate by passaging cells more frequently at a lower dilution.
4. Because con-A coating strongly promotes spreading, avoid over-seeding the dish or cells will be completely confluent and not completely spread. Inspect the dish about 30 min after seeding to evaluate the cell density. If too sparse, simply add more cells.
5. Fixation is markedly improved if the methanol is completely anhydrous. Type 3A hygroscopic molecular sieves (Sigma-Aldrich) can be added to the primary methanol container to remove traces of water. Fixation quality might be improved if a solution of 10% (by volume) formaldehyde in methanol is used instead of 100% methanol (S. Rogers, UNC, personal communication).
6. Alternatively, a more physiological buffer, HL3, can be used to wash cells (S. Rogers, UNC, personal communication). HL3 buffer: 70 m$M$ NaCl, 5 m$M$ KCl, 20 m$M$ MgCl$_2$, 10 m$M$ NaHCO$_3$, 1 m$M$ EGTA, 5 m$M$ trehalose, 115 m$M$ sucrose, 5 m$M$ HEPES, pH 7.2.
7. High fidelity amplification of the template is probably not critical because a number of different bioactive dsRNA oligomers are ultimately generated from the DNA template.
8. If high-resolution analysis is needed, a denaturing gel (containing formaldehyde, glyoxal, or 8 $M$ urea) can be used or the sample can be denatured before loading on a nondenaturing gel. To denature a sample: mix 1–2 µL of RNA with 18 µL of RNA sample buffer (buffer recipe: 10 mL formamide, 3.5 mL 37% formaldehyde, 2 mL 0.2 $M$ MOPS, pH 7.0, 10 m$M$ [final] sodium acetate, 1 m$M$ [final] EDTA). Add 5 µL of 5X gel loading buffer (contains glycerol, bromophenol blue, EDTA) and heat for 5–10 min at 65–70°C. Then load onto an agarose gel.
9. For the S2 cell lines in our hands, their tolerance of the RNAi treatment (specifically, incubation in serum-free medium) depends on their density. Cells at approx 50% confluency are not noticeably affected by the serum-free incubation. However, sparsely plated cells (e.g., only 10$^6$ cells per well of a six-well plate) often die following the serum deprivation step (even without RNA application). In this case, shorter incubation times in serum-free medium can be used (we have successfully used 10-min incubation times for some RNAi treatments). A potential alternative to serum deprivation is the stimulation of dsRNA uptake by addition of bovine insulin to the medium *(24)*, but we have not tested this protocol.

10. Some S2 cells lines primarily grow without adherence to a substrate, making it difficult to remove old medium by simply vacuuming or pipetting out the medium without significant loss of cells. In this case, resuspend the cells in a well with gentle trituration, transfer the resuspended cells to a sterile snap-cap tube, and pellet them by centrifuging at 130*g*, 3 min, room temperature. Carefully remove the supernatant and immediately add 0.75–1 mL of serum-free medium. Gently resuspend the pelleted cells and return them to a six-well plate.

11. Within one spindle, not all spindle microtubules necessarily flux at the same rate. We have not investigated this observation further, but we routinely measure as many flux rates as possible from a spindle and then calculate an average rate for that spindle.

## Acknowledgments

The GFP-α-tubulin S2 cell lines are a gift from the lab of R. Vale (Univ. of California, San Francisco). We also thank S. Rogers and G. Rogers (Univ. of North Carolina) for their large contribution in developing the S2 cell methods.

## References

1. Grieder, N. C., de Cuevas, M., and Spradling, A. C. (2000) The fusome organizes the microtubule network during oocyte differentiation in Drosophila. *Development* **127,** 4253–4264.
2. Trieselmann, N. and Wilde, A. (2002) Ran localizes around the microtubule spindle in vivo during mitosis in Drosophila embryos. *Curr. Biol.* **12,** 1124–1129.
3. Gergely, F., Kidd, D., Jeffers, K., Wakefield, J. G., and Raff, J. W. (2000) D-TACC: a novel centrosomal protein required for normal spindle function in the early Drosophila embryo. *Embo J.* **19,** 241–252.
4. Clarkson, M. and Saint, R. (1999) A His2AvDGFP fusion gene complements a lethal His2AvD mutant allele and provides an in vivo marker for Drosophila chromosome behavior. *DNA Cell Biol.* **18,** 457–462.
5. Sharp, D. J., Rogers, G. C., and Scholey, J. M. (2000) Cytoplasmic dynein is required for poleward chromosome movement during mitosis in Drosophila embryos. *Nat. Cell Biol.* **2,** 922–930.
6. Henikoff, S., Ahmad, K., Platero, J. S., and van Steensel, B. (2000) Heterochromatic deposition of centromeric histone H3-like proteins. *Proc. Natl. Acad. Sci. USA* **97,** 716–721.
7. Brust-Mascher, I. and Scholey, J. M. (2002) Microtubule flux and sliding in mitotic spindles of Drosophila embryos. *Mol. Biol. Cell* **13,** 3967–3975.
8. Basto, R., Scaerou, F., Mische, S., et al. (2004) In vivo dynamics of the rough deal checkpoint protein during Drosophila mitosis. *Curr. Biol.* **14,** 56–61.
9. Valdes-Perez, R. E. and Minden, J. S. (1995) Drosophila melanogaster syncytial nuclear divisions are patterned: time-lapse images, hypothesis and computational evidence. *J. Theor. Biol.* **175,** 525–532.

10. Sharp, D. J., Brown, H. M., Kwon, M., Rogers, G. C., Holland, G., and Scholey, J. M. (2000) Functional coordination of three mitotic motors in Drosophila embryos. *Mol. Biol. Cell* **11**, 241–253.

11. Sharp, D. J., Yu, K. R., Sisson, J. C., Sullivan, W., and Scholey, J. M. (1999) Antagonistic microtubule-sliding motors position mitotic centrosomes in Drosophila early embryos. *Nat. Cell Biol.* **1**, 51–54.

12. Kwon, M., Morales-Mulia, S., Brust-Mascher, I., Rogers, G. C., Sharp, D. J., and Scholey, J. M. (2004) The chromokinesin, KLP3A, dives mitotic spindle pole separation during prometaphase and anaphase and facilitates chromatid motility. *Mol. Biol. Cell* **15**, 219–233.

13. Rogers, G. C., Rogers, S. L., Schwimmer, T. A., et al. (2004) Two mitotic kinesins cooperate to drive sister chromatid separation during anaphase. *Nature* **427**, 364–370.

14. Rogers, S. L., Rogers, G. C., Sharp, D. J., and Vale, R. D. (2002) Drosophila EB1 is important for proper assembly, dynamics, and positioning of the mitotic spindle. *J. Cell Biol.* **158**, 873–884.

15. Ruden, D. M., Cui, W., Sollars, V., and Alterman, M. (1997) A Drosophila kinesin-like protein, Klp38B, functions during meiosis, mitosis, and segmentation. *Dev. Biol.* **191**, 284–296.

16. Mermall, V. and Miller, K. G. (1995) The 95F unconventional myosin is required for proper organization of the Drosophila syncytial blastoderm. *J. Cell Biol.* **129**, 1575–1588.

17. Savoian, M. S., Goldberg, M. L., and Rieder, C. L. (2000) The rate of poleward chromosome motion is attenuated in Drosophila zw10 and rod mutants. *Nat. Cell Biol.* **2**, 948–952.

18. Maddox, P. S., Moree, B., Canman, J. C., and Salmon, E. D. (2003) Spinning disk confocal microscope system for rapid high-resolution, multimode, fluorescence speckle microscopy and green fluorescent protein imaging in living cells. *Methods Enzymol.* **360**, 597–617.

19. Rogers, G. C., Rogers, S. L., and Sharp, D. J. (2005) Spindle microtubules in flux. *J. Cell. Sci.* **118**, 1105–1116.

20. Clemens, J. C., Worby, C. A., Simonson-Leff, N., et al. (2000) Use of double-stranded RNA interference in Drosophila cell lines to dissect signal transduction pathways. *Proc. Natl. Acad. Sci. USA* **97**, 6499–6503.

21. Elbashir, S. M., Lendeckel, W., and Tuschl, T. (2001) RNA interference is mediated by 21- and 22-nucleotide RNAs. *Genes Dev.* **15**, 188–200.

22. Caplen, N. J., Fleenor, J., Fire, A., and Morgan, R. A. (2000) dsRNA-mediated gene silencing in cultured Drosophila cells: a tissue culture model for the analysis of RNA interference. *Gene* **252**, 95–105.

23. Worby, C. A., Simonson-Leff, N., and Dixon, J. E. (2001) RNA interference of gene expression (RNAi) in cultured Drosophila cells. *Sci. STKE* **2001**, PL1.

24. March, J. C. and Bentley, W. E. (2004) Insulin stimulates double-stranded RNA uptake in Drosophila S2 cells. *Biotechniques* **37**, 898–900.

# 11

# Analysis of Tubulin Transport in Nerve Processes

## Andrey Tsvetkov and Sergey Popov

## Summary

In neurons, the molecular machinery for axonal growth and navigation is localized to the growth cone region, whereas tubulin is synthesized primarily in the cell body. Because diffusion serves as an efficient transport mechanism only for very short distances, tubulin has to be actively transported from the cell body down the axon. Two mechanistically distinct models for tubulin transport have been proposed. "Polymer model" postulates that tubulin moves in the form of microtubules preassembled in the cell body, whereas "subunit model" assumes that axonal microtubules are stationary, and that tubulin is delivered from the cell body in unassembled form. We used three independent quantitative approaches (photobleaching, fluorescence speckle microscopy, and microtubule plus end tracking) to demonstrate that axonal microtubules are stationary in rapidly growing axons produced by Xenopus spinal cord neurons in culture. These experiments strongly support subunit model for tubulin delivery.

**Key Words:** Microtubule; tubulin; slow axonal transport; photobleaching; EB1; fluorescence speckle microscopy.

## 1. Introduction

### 1.1. Axonal Transport

There are four basic problems essential for understanding any transport phenomenon in a living cell: (1) what is the molecular composition of the transport unit, (2) what cytoskeletal structures are supporting transport, (3) what motor protein is involved, and (4) how is motor protein linked to the transport unit. In neurons, membrane-bound proteins are transported by kinesins along microtubules at the rates of about a few micrometers per second, and the molecular mechanism of this "fast axonal transport" is well established *(1,2)*. Cytosolic proteins (including cytoskeletal components) are transported at the rates, which are about two orders or magnitude lower. Surprisingly, for the vast majority of cytosolic proteins, the molecular mechanisms of their transport have not been

From: *Methods in Molecular Medicine, Vol. 137: Microtubule Protocols*
Edited by: Jun Zhou © Humana Press Inc., Totowa, NJ

investigated. One notable exception is tubulin (this is hardly surprising given the essential role of microtubules in fast axonal transport). It has been reported that tubulin is transported by conventional kinesin *(3)* in unassembled form, either as a dimer or as a small oligomer *(3,4)*. Alternatively, it has been suggested that tubulin moves down the axon in the form of microtubules assembled in the cell body *(5)*. The latter model ("polymer transport model") has been difficult to prove or to reject using direct microscopy methods, primarily because microtubules in the axonal shaft form a tight bundle, which greatly complicates detection of individual microtubules.

## 1.2. Techniques to Study Tubulin Transport

The rates of slow axonal transport were measured, for the first time, with a pulse-chase approach. In a typical experiment, radioactive amino acids are injected into the nerve ganglia of a living animal and allowed to incorporate into newly synthesized proteins. A few days/weeks after injection, the nerve is isolated and the amount of radiolabeled protein is measured at different distances from the ganglia to determine the rate of transport *(6)*. This technique has been instrumental in revealing the heterogeneous nature of slow axonal transport and in identifying its various components. This method, however, is poorly suited for determining the form in which proteins are transported.

As an alternative to pulse-chase labeling experiments, various approaches based on neuronal cell culture models have been developed. Simple estimates indicate that active transport of cytosolic components is required to support elongation of even relatively short neurites (a fraction of a mm in length). Therefore, despite obvious potential limitations of any cell culture work, the ability of neurons to extend long neurites, by itself, strongly suggests that slow axonal transport system is functional in cell culture models *(7)*.

Among various approaches for the analysis of slow axonal transport, including pharmacological or electron microscopy techniques, fluorescence microscopy methods are of particular interest. These methods require labeling of tubulin or microtubules with fluorescent markers. Traditionally, such labeling has been accomplished by covalent conjugation of purified tubulin to fluorophores *(8)*. Introduction of green fluorescent protein (GFP) significantly facilitated visualization of individual microtubules in living cells and analysis of their assembly/disassembly *(9)*. Such direct analysis of microtubule dynamics in neurons is difficult, because the distance between microtubules in axonal shaft is smaller than resolution limit of light microscopy. Therefore, in an attempt to detect microtubule translocation in growing neurites, we employed three different strategies: photobleaching, fluorescence speckle microscopy, and microtubule plus end tracking *(10–12)*.

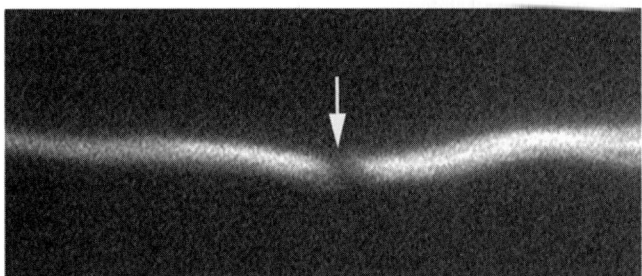

Fig. 1. Fluorescence micrograph of a Xenopus neuron loaded with Cy3-tubulin 1 min after photobleaching. The position of the center of the bleached zone is indicated by an arrow.

Photobleaching (or conceptually similar photoactivation) method is based on local labeling of the microtubule array by briefly exposing a small axonal segment to an intense laser beam (**Fig. 1**). Photobleaching typically does not affect integrity of axonal microtubules and axonal elongation *(13,14)*. While the rate of fluorescence recovery of the bleached zone reflects the rate of microtubule turnover, translocation of the zone provides information on the movement of microtubules in the axonal shaft *(13,14)*.

Fluorescence speckle microscopy technique is based on the finding that when the fraction of labeled molecules relative to the pool of endogenous unlabeled molecules is small (typically, less than 1%), fluorescently labeled structures have a characteristic "speckled" pattern (**Fig. 2**). Individual speckles arise from stochastic fluctuations in the local density of labeled molecules within the polymer *(15)* and, in the extreme case, may represent individual fluorophore molecules. In the analysis of microtubule transport, the rationale behind fluorescence speckle microscopy technique is similar to the photobleaching method: in both cases, translocation of the markers on the microtubule array (individual speckles and the bleached zone, respectively) is interpreted as the movement of underlying microtubules. In general, fluorescence speckle microscopy appears to be more sensitive and, potentially, can reveal movement of a single polymer in the array of stationary microtubules.

Finally, the movement of individual microtubules can be evaluated with microtubule plus end tracking technique. The method is based on preferential binding of certain microtubule-associated proteins (such as CLIP-170, EB1, or APC) to the plus ends of growing microtubules *(16,17)*, resulting in characteristic fluorescence "comets" (**Fig. 3**). In the case of a stationary microtubule, the rate of comet movement is equal to the rate of microtubule assembly. If,

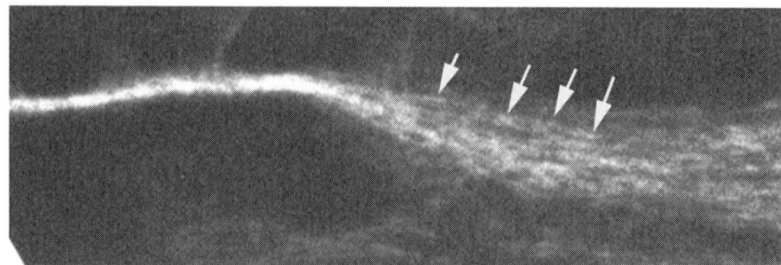

Fig. 2. Fluorescence micrograph of a Xenopus neuron loaded with Cy3-tubulin. Concentration of Cy3-tubulin in the cytoplasm is significantly lower than the concentration of unlabeled endogenous tubulin, which allows for the detection of individual speckles (marked with arrows) associated with microtubules array.

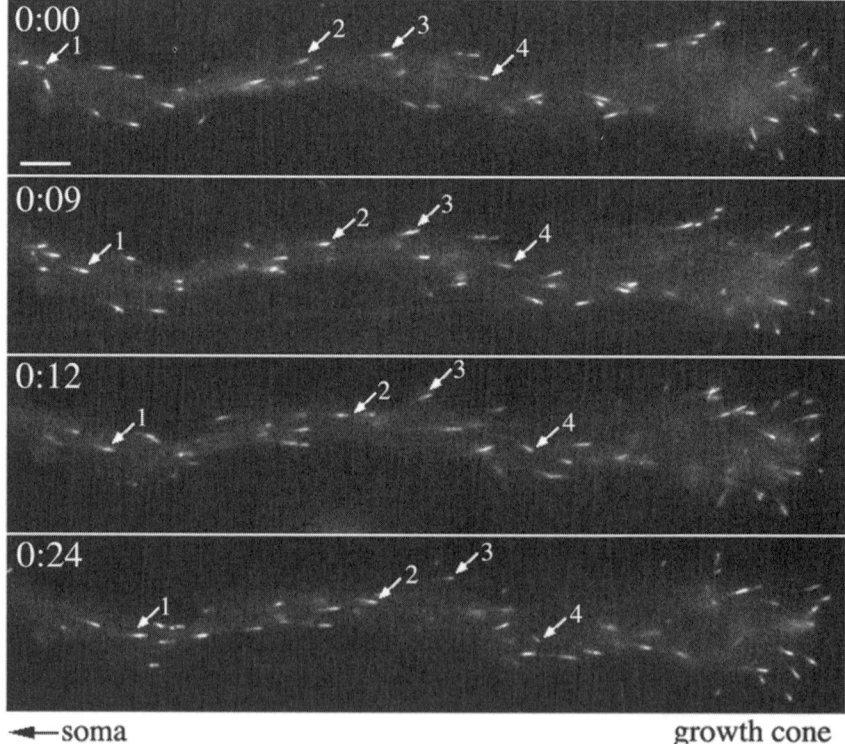

Fig. 3. A series of fluorescence micrographs of a Xenopus neuron expressing green fluorescent protein-EB1. Growth cone is on the right. Time is in minutes:seconds at the upper left corner of each panel. Individual fluorescent "comets" (marked with arrows) move towards the growth cone. Bar = 10 μm.

however, microtubule is transported, the velocity of an individual "comet" is equal to the sum of the rates of microtubule transport and growth. This technique is particularly well suited for the detection of rapid asynchronous movements of individual microtubules at the rates, which are significantly higher than the rate of microtubule assembly.

### 1.3. Tubulin Transport in Xenopus Neurons

Xenopus spinal cord neurons in culture can be loaded with fluorescently labeled tubulin by injecting Cy3-tubulin or constructs encoding GFP-tubulin into dividing Xenopus eggs *(18)*. Neuronal cultures should to be prepared on strongly adhesive substrate. This excludes a possibility of "towed axonal growth" (anterograde movement of the entire axon driven by mechanical tension produced by the growth cone), which can be mistakenly interpreted as a true microtubule transport relative to the axon *(19)*. When plated on strongly adhesive substrate coated with plant lectin Concanavalin A, axons grow rapidly (at the rates of up to 60 µm/h), survive for a few days at room temperature, and, similar to other neuronal types, follow conventional "growth from the tip" paradigm *(20)*.

The three techniques for the analysis of axonal microtubule transport are designed for the detection of a particular type of microtubule movement. Photobleaching method is expected to report synchronous *"en bloc"* movement of the entire microtubule array, fluorescence speckle microscopy can easily register relatively slow movement of individual polymers, whereas microtubule plus end tracking is ideally suited for the detection of rapid microtubule movement. Despite technical limitations of each of these individual methods, together, they provide comprehensive test for the polymer transport model. All three assays indicate that axonal microtubules are stationary in rapidly advancing axons produced by Xenopus spinal cord neurons. These experiments do not exclude a possibility of local microtubule transport in specialized neuronal segments (such as growth cone or axonal branch point). They do, however, exclude polymer transport model as the mechanism of slow axonal transport of tubulin. Interestingly, both in Xenopus spinal cord neurons *(12)* and in rat sympathetic neurons *(21)*, photobleaching experiments reveal rapid anterograde transport of short tubule-like structures labeled with GFP-tubulin. These structures are distinct from microtubules and are likely to represent membrane organelles *(12)*. It is tempting to speculate that transport of tubulin in association with these vesicles represents the mechanism of slow axonal transport of tubulin. Such "hijacking" of fast axonal transport system by cytosolic proteins might serve as a general mechanism of slow axonal transport.

## 2. Materials

### 2.1. Fluorescent Tubulin

#### 2.1.1. Recombinant Tubulin Fused With a Fluorescent Protein

1. Plasmid DNA (*see* **Note 1**) encoding recombinant tubulin fused to GFP (e.g., pAcGFP1-Tubulin, Clontech, Mountain View, CA).
2. Plasmid isolation kit (Promega, Madison, WI). Purified plasmid DNA is stored in aliquots at –20°C.

#### 2.1.2. Conjugation of Fluorophores to Tubulin

1. Purified tubulin from bovine brain. Lyophilized protein is stable up to 2 yr at 4°C.
2. Microtubule-polymerizing solution: 1 m$M$ GTP (Sigma, St. Louis, MO), 10% dimethyl sulfoxide (DMSO).
3. Cyanine-3-succinimidyl ester (Toronto Research Chemicals, North York, ON, Canada) or rhodamine-succinimidyl ester (Molecular Probes, Carlsbad, CA). Succinimidyl ester can be dissolved in DMSO and stored in single use aliquots at –80°C.
4. Resuspension buffer: 0.1 $M$ PIPES, 1 m$M$ EDTA, 0.5 m$M$ MgCl$_2$, pH 6.9.

#### 2.1.3. Recombinant EB1 Fused With Fluorescent Protein

1. Plasmid DNA encoding recombinant EB1 (or its neuronal isoform EB3) fused to GFP.
2. Plasmid isolation kit (Promega). Purified plasmid DNA is stored in aliquots at –20°C.

### 2.2. Egg Preparation and Fertilization

1. Human chorionic gonadotropin (Sigma). Dissolve 800 U in 0.5 mL of water before embryo injection.
2. Ringer's solution: 115 m$M$ NaCl, 2 m$M$ CaCl$_2$, 2.5 m$M$ KCl, and 10 m$M$ HEPES, pH 7.6 supplemented with 5% Ficoll (Sigma).
3. 20% Ringer's solution containing 5% Ficoll.
4. 2% L-cysteine (Sigma) solution in 20% Ringer's solution, pH 7.8.
5. Testes storage solution: Ringer's solution with 10% fetal calf serum (Sigma). All solutions are stored at 4°C.

### 2.3. Microinjection

1. Ringer's solution: 115 m$M$ NaCl, 2 m$M$ CaCl$_2$, 2.5 m$M$ KCl, and 10 m$M$ HEPES, pH 7.6 supplemented with 5% Ficoll (Sigma). Store at 4°C.
2. 20 µg/mL DNA plasmids or 10 mg/mL Cy3-tubulin.
3. 20% Ringer's solution containing 5% Ficoll. Store at 4°C.

## 2.4. Preparation of Neuronal Cultures

1. Microdissection scissors (DR Instruments, Palos Hills, IL).
2. 1 mg/mL Concanavalin A solution (EY Labs, San Mateo, CA). Store in aliquots at –20°C.
3. Xenopus cell culture medium: 50% (v/v) Ringer's solution, 49% L-15 Leibovitz medium (GIBCO), 1% fetal bovine serum. Store at 4°C.
4. Dissociation ($Ca^{2+}$- and $Mg^{2+}$-free) solution: 115 m$M$ NaCl, 2.6 m$M$ KCl, 0.4 m$M$ EDTA, and 10 m$M$ HEPES, pH 7.6. Store at 4°C.
5. The oxygen-depleting solution: 50% (v/v) Ringer's solution, 49% L-15 Leibovitz medium (GIBCO), 1% FBS, 10 m$M$ lactate, and 20 U/mL oxyrase (Oxyrase, Mansfield, OH). Oxyrase is added to the culture medium immediately before fluorescence microscopy recordings.

## 2.5. Videomicroscopy and Photobleaching

1. Digital images of cells are obtained at room temperature using an inverted microscope (TE200, Nikon) equipped with a ×60 PlanApo objective (NA 1.4) and a 100-W mercury arc lamp.
2. Images are acquired with a charge-coupled device (CCD) camera (CoolSnapHQ, Roper Scientific) driven by IPLab software (Scanalytics, Fairfax, VA).
3. For photobleaching, the beam of the 150-mW multimode argon laser (Spectra-Physics, Fremont, CA) is channeled into the epi-illumination port of an inverted Nikon microscope.
4. A cylindrical or spherical lens with a 700-mm focal distance and a ↔60, 1.4 NA objective lens are used to produce a focused beam in the specimen plane. Typical exposure is 50–200 ms.

# 3. Methods

## 3.1. Fluorescent Tubulin

### 3.1.1. GFP-Tubulin

1. Isolate the plasmid DNA encoding GFP fused to β-tubulin (e.g., pEGFP-tubulin or pAcGFP1-Tubulin; *see* **Note 2**) using plasmid purification kit; follow manufacture's instructions.
2. Dissolve the plasmid in water (20 μg/mL). Store at –20°C.

### 3.1.2. Purified Tubulin Covalently Conjugated With a Fluorescent Dye (see **Note 3**)

1. Thaw an aliquot of purified tubulin on ice.
2. Assemble purified tubulin into microtubules by adding 1 m$M$ GTP (Sigma) and 10% DMSO.
3. Dissolve succinimidyl-activated fluorescent dye (rhodamine or cyanine-3) in DMSO.

4. Combine assembled microtubules and activated fluorescent dye and incubate 15–20 min at 37°C. Pellet labeled microtubules at 37°C by centrifugation.

5. Remove a supernatant and add one volume of ice-cold resuspension buffer. Incubate on ice for 15 min. Remove denatured and aggregated tubulin by centrifugation at 4°C for 20–30 min.

6. Assemble labeled tubulin from supernatant into microtubules in 1 m$M$ GTP and 10% DMSO. Incubate 15–20 min at 37°C.

7. Pellet labeled microtubules at 37°C by centrifugation.

8. Repeat **steps 4–6** one or two times to eliminate unconjugated dye.

9. Resuspend pellet containing labeled microtubules in cold resuspension buffer and incubate on ice for 15 min.

10. Store fluorescently labeled tubulin in aliquots in liquid nitrogen.

### 3.1.3. Recombinant EB1 Fused With GFP

1. Isolate the plasmid encoding GFP-EB1 (or its neuronal isoform EB3) using plasmid purification kit (*see* **Note 4**).

2. Dissolve the plasmid in water (20 μg/mL).

## 3.2. Preparation of Fertilized Xenopus Eggs

1. Inject female Xenopus frog with 800 U of human chorionic gonadotropin into the hind leg below the lateral line. Leave the frog overnight in a dark place at room temperature.

2. On the day of the experiment, sacrifice the male and dissect the abdomen. The paired testes are ovoid structures approx 10 mm × 3 mm in size. Remove the testes, and rinse them with cold Ringer's solution.

3. Store the testes in Ringer's solution containing 10% fetal calf serum and 1% penicillin/ streptomycin at 4°C. The testes can be used for at least 1 wk after isolation.

4. Gently squeeze the female frog above the hind legs to lay eggs. Collect the eggs into a Petri dish.

5. Transfer one testis (or a small portion of a testis) into a tube containing 1–2 mL of 10% Ringer's solution. Crush the testis with the flat end of a glass Pasteur pipet to release spermatozoa. Add to the tube 10 mL of 10% Ringer's solution. Store on ice.

6. Cover the freshly obtained eggs with solution containing spermatozoa and incubate for 15–20 min. Wash with 10% Ringer's solution. Successful fertilization should result in the rotation of eggs bringing the dark animal pole to the top. Incubate at room temperature. The first division starts 75 min after fertilization.

7. Remove the jelly coat by incubating the eggs for 2–5 min in 2% l-cysteine solution (pH 7.9), gently shaking the dish. Removal of the jelly coat results in the dissociation of aggregates into individual eggs.

8. Rinse three times with 10% Ringer's solution to completely remove cysteine. Transfer fertilized eggs in a separate Petri dish for microinjection (*see* **Note 5**).

## 3.3. Microinjection (see Note 6)

1. Pull out microinjection needles from the capillary glass tubes using a microelectrode puller. We use borosilicate glass tubes with 1.5 mm OD and 0.75 mm ID (cat. no. BF150-75-10, Sutter Instrument, Novato, CA). Thin-walled glass tubes produce microelectrodes with very flexible tips, which may bend during penetration into the egg; such tubes should be avoided.
2. Using stereomicroscope, cut the tips of microelectrodes with the razor blade. Examine the tips under a conventional light microscope and select injection needles with diameters in 9 to 18 μm range.
3. Fill out the needles with injection solution (e.g., 20 μg/mL DNA plasmid or 10 mg/mL Cy3-tubulin).
4. Mount the needle in a holder connected to a pressure control system (e.g., Picospritzer II (General Valve, Fairfield, NJ).
5. Injections are performed in 100% Ringer's solution supplemented with 5% Ficoll. The presence of Ficoll increases viscosity of the medium and minimizes cytoplasmic leakage through the injection site.
6. Using stereomicroscope, inject fertilized eggs at the two-cell stage. The injected volume should not exceed 20–50 nl (*see* **Note 7**).
7. One hour after injection, transfer the embryos to 20% Ringer's solution containing 5% Ficoll.
8. Two hours later, transfer the embryos to 20% Ringer's solution.
9. Allow the embryos to develop to stage 19–24 (20–24 h at room temperature or about 48 h at 15°C, *see* **Note 8**).

## 3.4. Preparation of Neuronal Cultures

1. Pick up individual 22-mm cover slips with forceps, immerse them in 96% ethanol, flame in a gas burner, and transfer in a Petri dish (Falcon, cat. no. 1029). Heat the tip of the Pasteur pipet in a gas burner, and gently touch the bottom of the dish to melt the plastic and to "glue" the cover slip to the Petri dish
2. Thaw a tube with an aliquot of 1 mg/mL Concanavalin A and dilute it in water to 0.1–0.2 mg/mL. Add 100 μL of Concanavalin A solution to each cover slip. Allow it to air-dry in the laminar hood under ultraviolet illumination. Add 3 mL of a culture medium (50% Ringer's solution, 49% L-15 Leibovitz medium, and 1% fetal bovine serum) to each dish (*see* **Notes 9** and **10**).
3. Hold the embryos with the tweezers and cut the neural tube with microsurgery scissors. Transfer neural tubes into $Ca^{2+}$- and $Mg^{2+}$-free solution and incubate for 10–20 min.
4. Dissociate neural tubes into individual cells by aspiration into a glass tube and plate on the glass cover slips precoated with Concanavalin A. Leave the dishes undisturbed for 1–2 h to allow the cells to attach to the cover slips.
5. Incubate cells at 20°C. Neurites emerge from cell bodies within 4–8 h after plating and initially grow at a rate of about 40–60 μm/h.

6. Remove the cover slip from the Petri dish and attach it to the bottom of a chamber for light microscopy analysis.

## 3.5. Photobleaching and Time-Lapse Microscopy

1. Inject the embryos with Cy3-tubulin, GFP-tubulin, or GFP-EB1 and allow them to develop to stages 19–24. Prepare the cell cultures and use them 14–20 h after plating.
2. Supplement culture medium with Oxyrase to reduce photodamage.
3. Photobleaching of GFP-tubulin-expressing cells is accomplished by briefly (50–200 ms) exposing a small axonal segment to the flash of argon laser (*see* **Note 11**).
4. To minimize photodamage, the shutter in epifluorescence optical path is synchronized to the shutter on CCD detector.
5. Exposure time depends on the expression levels, and is typically in 0.2 to 1.0 s range. Images are acquired at 1–3 s intervals. The number of acquired images is dictated by photobleaching and photodamage and usually is in the range of 100 to 600 frames.
6. The protocol for the analysis of time-lapse sequences depends on a specific problem to be addressed. In general, analysis of the movement of fluorescence objects (GFP-EB1 comets or GFP-tubulin-labeled vesicles) requires identification of the objects based on their fluorescence intensity and tracking their movement throughout the time-lapse sequence. Such tracking can be performed with most of the currently available software packages. The image series can be presented as a kymograph to evaluate the rate of transport. Computational algorithms used for the analysis of fluorescence speckle microscopy data in nonneuronal cells *(22)* are yet to be applied to the problem of slow axonal transport.

## 4. Notes

1. Traditionally, expression of exogenous proteins in Xenopus neuronal cultures is accomplished by injecting in vitro synthesized mRNA into dividing Xenopus eggs *(23)*. We find that injection of DNA plasmids leads to similar expression levels. Because mRNA is less stable and is more difficult to handle, we prefer to use DNA vectors.
2. Clontech sells a wide variety of vectors encoding fluorescent proteins such as newly developed vectors based on monomeric fluorescent proteins from reef corals. The plasmid encoding fluorescent protein fused to tubulin can be purchased or β-tubulin cDNA (OriGene Technologies, Rockville, MD) can be subcloned into a vector of interest.
3. For the detailed protocol, including troubleshooting steps, *see* **ref. 8**. Tubulin is labeled with fluorophore in polymeric state to protect residues essential for tubulin polymerization; nonfunctional tubulin is excluded by the cycles of polymerization/depolymerization. Alternatively, rhodamine-tubulin (cat. no. TL331M-A) can be purchased from Cytoskeleton (Denver, CO). If only a few experiments are planned, this alternative is more cost-efficient. However, in our hands, survival

rate of Xenopus embryos injected with rhodamine-tubulin from Cytoskeleton was rather low.

4. The plasmid encoding GFP-EB1 has been generously provided by E. Morrison *(24)*. This plasmid (as well as the plasmid encoding GFP-tubulin from Clontech) can be directly used for embryo injection. However, we found that subcloning of GFP-tubulin and GFP-EB1 into pCS2+ vector leads to higher rates of embryo survival and somewhat higher protein expression levels in neuronal cultures. A combination of two different fluorescence markers with different spectral characteristics (e.g., mRFP and GFP) allows for the simultaneous analysis of the dynamics of two proteins.

5. If the quality of eggs is good, virtually 100% of them will be fertilized. Low percentage of fertilized eggs is usually indicative of the poor quality of eggs (the problems with the spermatozoa are very rare).

6. In comparison with microinjection of individual neurons in cell culture, injection of Xenopus eggs is very straightforward. The quality of micromanipulator and optics, as well as the size and the shape of the microelectrode tip are not crucial. The most important factor is the quality of the eggs. The volume of solution injected into the egg typically should not exceed 20–50 mL, but this number varies a lot between different batches of eggs.

7. The size of the microinjection pipets is relatively large, and, if necessary, pipets can be easily calibrated. Usually we chose not to calibrate the pipets, because the exact amount of solution injected into the egg is not crucial. Typically we start microinjections at the 2-cell stage and continue until 16- or 32-cell stage. The plasmids are relatively immobile in the egg cytoplasm, and only a small fraction of cells in culture is expected to express the construct. To increase the percentage of cells expressing the construct, we perform injections in four to eight different sites of the egg.

8. Development of embryos injected with the plasmid DNA or Cy3-tubulin might be slowed down in comparison with control uninjected embryos. The injected embryos that reach stages 17–20 can be stored at 4°C for a few days to slow down their development.

9. Although initially we used Concanavalin A from Sigma, the quality of neuronal cultures appears to be much better if Concanavalin A from EY Laboratories is used.

10. We prefer to avoid using antibiotics during cell culture preparation. Although conventional penicillin/streptomycin cocktail apparently does not affect neuronal development and neurite extension, it can mask potential contamination problems.

11. Epifluorescence port of an inverted Nikon microscope was modified to allow for the simultaneous exposure of cells to the laser beam and to a conventional mercury arc lamp. Detailed description of the optical path is beyond the scope of the current protocol. Photobleaching can be also accomplished with most confocal microscopes. We did not observe any detectable effect of laser irradiation on

axonal growth or on microtubule assembly/disassembly even when exposure time exceeded a few seconds.

## References

1. Goldstein, L. S. and Yang, Z. (2000) Microtubule-based transport systems in neurons: the roles of kinesins and dyneins. *Annu. Rev. Neurosci.* **23,** 39–71.
2. Hirokawa, N. and Takemura, R. (2004) Molecular motors in neuronal development, intracellular transport and diseases. *Curr. Opin. Neurobiol.* **14,** 564–573.
3. Terada, S., Kinjo, M., and Hirokawa, N. (2000) Oligomeric tubulin in large transporting complex is transported via kinesin in squid giant axons. *Cell* **103,** 141–155.
4. Galbraith, J. A., Reese, T. S., Schlief, M. L., and Gallant, P. E. (1999) Slow transport of unpolymerized tubulin and polymerized neurofilament in the squid giant axon. *Proc. Natl. Acad. Sci. USA* **96,** 11,589–11,594.
5. Baas, P. W. (1999) Microtubules and neuronal polarity: lessons from mitosis. *Neuron* **22,** 23–31.
6. Hoffman, P. N. and Lasek, R. J. (1975) The slow component of axonal transport. Identification of major structural polypeptides of the axon and their generality among mammalian neurons. *J. Cell Biol.* **66,** 351–366.
7. Sabry, J., O'Connor, T. P., and Kirschner, M. W. (1995) Axonal transport of tubulin in Ti1 pioneer neurons in situ. *Neuron* **14,** 1247–1256.
8. Peloquin, J., Komarova, Y., and Borisy, G. (2005) Conjugation of fluorophores to tubulin. *Nat. Methods* **2,** 299–303.
9. Komarova, Y. A., Vorobjev, I. A., and Borisy, G. G. (2002) Life cycle of MTs: persistent growth in the cell interior, asymmetric transition frequencies and effects of the cell boundary. *J. Cell Sci.* **115,** 3527–3539.
10. Chang, S., Rodionov, V. I., Borisy, G. G., and Popov, S. V. (1998) Transport and turnover of microtubules in frog neurons depend on the pattern of axonal growth. *J. Neurosci.* **18,** 821–829.
11. Chang, S., Svitkina, T. M., Borisy, G. G., and Popov, S. V. (1999) Speckle microscopic evaluation of microtubule transport in growing nerve processes. *Nat. Cell Biol.* **1,** 399–403.
12. Ma, Y., Shakiryanova, D., Vardya, I., and Popov, S. V. (2004) Quantitative analysis of microtubule transport in growing nerve processes. *Curr. Biol.* **14,** 725–730.
13. Okabe, S. and Hirokawa, N. (1992) Differential behavior of photoactivated microtubules in growing axons of mouse and frog neurons. *J. Cell Biol.* **117,** 105–120.
14. Okabe, S. and Hirokawa, N. (1990) Turnover of fluorescently labelled tubulin and actin in the axon. *Nature* **343,** 479–482.
15. Waterman-Storer, C. M., Desai, A., Bulinski, J. C., and Salmon, E. D. (1998) Fluorescent speckle microscopy, a method to visualize the dynamics of protein assemblies in living cells. *Curr. Biol.* **8,** 1227–1230.
16. Akhmanova, A. and Hoogenraad, C. C. (2005) Microtubule plus-end-tracking proteins: mechanisms and functions. *Curr. Opin. Cell Biol.* **17,** 47–54.

17. Perez, F., Diamantopoulos, G. S., Stalder, R., and Kreis, T. E. (1999) CLIP-170 highlights growing microtubule ends in vivo. *Cell* **96,** 517–527.
18. Tanaka, E. M. and Kirschner, M. W. (1991) Microtubule behavior in the growth cones of living neurons during axon elongation. *J. Cell Biol.* **115,** 345–363.
19. Okabe, S. and Hirokawa, N. (1993) Do photobleached fluorescent microtubules move? Re-evaluation of fluorescence laser photobleaching both in vitro and in growing Xenopus axon. *J. Cell Biol.* **120,** 1177–1186.
20. Craig, A. M., Wyborski, R. J., and Banker, G. (1995) Preferential addition of newly synthesized membrane protein at axonal growth cones. *Nature* **375,** 592–594.
21. Wang, L. and Brown, A. (2002) Rapid movement of microtubules in axons. *Curr. Biol.* **12,** 1496–1501.
22. Danuser, G. and Waterman-Storer, C. M. (2003) Quantitative fluorescent speckle microscopy: where it came from and where it is going. *J. Microsc.* **211,** 191–207.
23. Alder, J., Kanki, H., Valtorta, F., Greengard, P., and Poo, M. M. (1995) Overexpression of synaptophysin enhances neurotransmitter secretion at Xenopus neuromuscular synapses. *J. Neurosci.* **15,** 511–519.
24. Morrison, E. E., Moncur, P. M., and Askham, J. M. (2002) EB1 identifies sites of microtubule polymerisation during neurite development. *Brain Res. Mol. Brain Res.* **98,** 145–152.

# 12

# Analysis of Microtubule-Mediated Intracellular Viral Transport

## Chunyong Liu, Min Liu, and Jun Zhou

### Summary

Host microtubules and motor proteins are crucial to the intracellular transport of a number of viruses. Disruption of microtubules or suppression of motor functions can remarkably inhibit the movement of viruses in host cells. It is now known that incoming viruses use motor proteins to travel along microtubules from the plasma membrane to the nuclear or perinuclear replication site, whereas progeny viruses depend on microtubules and motors to move from the assembly site to the cell periphery. Here, we describe several major methods for analyzing microtubule-mediated intracellular viral transport, using adenovirus as an example.

**Key Words:** Microtubule; viral transport; adenovirus; Texas red; nocodazole; dynamitin; fluorescence microscopy.

## 1. Introduction

The infectious cycle of viruses can be divided into three major steps: viral entry, replication, and egress. As obligate parasites, viruses depend on the host cell for all these steps. There is a growing body of evidence demonstrating that host microtubule transport machinery driven by motor proteins, such as dynein and kinesin, is involved in both viral entry and viral egress *(1–6)*. Understanding the interaction between viruses and microtubules will not only provide insights into the pathogenesis of viruses, but can also facilitate the development of new vectors for gene therapy as well as new antiviral drugs.

A variety of cell biological and virological methods are available to study the roles of microtubules and motor proteins in intracellular viral transport. The localization of viruses in fixed cells can be examined by electron microscopy and immunofluorescence microscopy, whereas individual viral particles in living cells can be tracked by fluorophore-tagging and green fluorescent protein (GFP)-labeling methods *(7–12)*. Microtubule-depolymerizing agents,

From: *Methods in Molecular Medicine, Vol. 137: Microtubule Protocols*
Edited by: Jun Zhou © Humana Press Inc., Totowa, NJ

such as nocodazole, have been shown to disrupt the movement of viruses to nuclear or perinuclear replication sites *(2,4,9,12)*. In addition, overexpression of dynamitin, which interferes with dynein function, has also been demonstrated to inhibit intracellular viral transport *(4,6,13)*.

Adenoviruses are nonenveloped viruses that enter host cells by endocytosis. Adenoviral particles can be chemically tagged with a fluorophore, which remains coupled to the viral capsids after infection. These tagged particles can be visualized by fluorescence microscopy either in living cells or in fixed cells. Incoming adenoviral particles have been shown to hijack the dynein motor to transport along microtubules toward the nucleus *(4,14)*. In this chapter, we describe the approach for cross-linking adenoviruses with Texas Red and the use of fluorescence microscopy to visualize adenoviral movement in host cells. In addition, we describe the nocodazole treatment and dynamitin overexpression methods for analyzing the roles of microtubules and motor proteins in viral transport.

## 2. Materials
### 2.1. Cell Culture

1. A549 and 293 cell lines (ATCC).
2. Dulbecco's modified Eagle's medium (DMEM).
3. Fetal bovine serum (FBS).
4. Glutamine.
5. Antibiotics: penicillin and streptomycin.

### 2.2. Preparation of Viruses

1. Adenovirus type 5 (Ad5; ATCC).
2. 1X TD buffer: 140 m$M$ NaCl, 5 m$M$ KCl, 25 m$M$ Tris, 0.7 m$M$ Na$_2$HPO$_4$, pH 7.4.
3. CsCl.

### 2.3. Texas Red Conjugation

1. Dimethyl sulfoxide (DMSO).
2. Texas Red (TR; Molecular Probes) was made as 5-mg/mL stock solution in DMSO.
3. Hydroxylamine.

### 2.4. Nocodazole

1. Nocodazole (Sigma) was made as 20-m$M$ stock solution in DMSO.

### 2.5. Dynamitin Overexpression

1. pEGFP-dynamitin plasmid and pEGFP vector.
2. Lipofectamine reagent (Invitrogen).

## 2.6. Immunolabeling

1. Phosphate-buffered saline (PBS).
2. MT stabilizing buffer: 80 m$M$ PIPES, 2 m$M$ MgSO$_4$, 10 µ$M$ paclitaxel, pH 6.8.
3. Goat serum.
4. Mouse monoclonal anti-β-tubulin antibody.
5. Mouse monoclonal anti-dynamitin (MAb 50.1) antibody.
6. Goat anti-mouse IgG-Cy5.
7. 4', 6'-diamidino-2-phenylindole (DAPI).

# 3. Methods

## 3.1. Cell Culture

A549 and 293 cells are grown in the DMEM medium containing 10% FBS and 2 m$M$ glutamine, 100 U of penicillin/mL, and 0.1 mg of streptomycin/mL at 37°C in a humidified incubator with 5% CO$_2$.

## 3.2. Preparation of Viruses (15,16)

1. Infect 293 cells (~70% confluence) at a multiplicity of infection (MOI) of 100 by adding serum-free DMEM medium containing Ad5 viruses to cells.
2. After incubation at 37°C for 1 h, add normal DMEM growth medium and continue incubation.
3. Harvest the media and cells 72 h after infection. Freeze (at –70°C) and thaw (at 37°C) for five times to release the viruses.
4. Spin at 1000$g$ for 10 min and save the supernatant
5. Load the supernatant onto a two-layer CsCl gradient (2.5 mL 1.2 g/mL and 2.5 mL 1.4 g/mL) prepared in 1X TD buffer and centrifuge at 80,000$g$ for 2 h at 15°C.
6. Collect the virus band through a small puncture hole at the bottom of the tube, transfer the virus band to a fresh 10-mL centrifuge tube and fill the tube with 1.33 g/mL CsCl solution, centrifuge at 120,000$g$ for 18 h at 15°C.
7. Collect the virus band, concentrate the virus in a collodion bag by vacuum dialysis against 0.01 $M$ Tris-HCl (pH 8.0), 150 m$M$ NaCl and store in aliquots at –70°C.
8. Determine the titer of viruses as described in **ref. 16**.

## 3.3. Texas Red Conjugation (17) (see Note 1)

1. Dilute Texas Red from stock solution into dialysis buffer to make 0.5 mg/mL Texas Red.
2. Add 8 µL of 0.5 mg/mL Texas Red to 400 µL of 0.8 mg/mL adenoviruses, incubate in the dark on a rocker at room temperature for 1 h.
3. Terminate the reaction with 1 $M$ hydroxylamine, and purify the virus on a CsCl gradient as described in **Subheading 3.2.**
4. Measure the amount of dye incorporated into Ad5-TR by recording absorbance at 595 nm using a spectrophotometer.

5. Calculate dye-to-capsid ratio on the basis of the surface exposure of 252 viral capsids per virion.
6. Determine the titer of the fluorophore-conjugated virus Ad5-TR.
7. Add the glycerol to Ad5-TR virus to a final concentration of 30%, and store the preparations at –20°C.

### 3.4. Nocodazole Treatment

1. Plate A549 cells onto glass cover slips to obtain 60–70% confluence after incubation for 24 h.
2. Add serum-free DMEM growth medium containing 20 µ*M* nocodazole or just the DMSO solvent (control). Incubate for 1 h at 37°C (*see* **Note 2**).
3. Replace with 0.2-mL binding buffer, pipet Ad5-TR onto the cells at MOI of 10, and incubate for 1 h at 4°C with gentle agitation (*see* **Note 3**).
4. Remove unbound viruses by three washes with binding buffer.
5. Add serum-free DMEM medium (± drugs) and shift to 37°C.
6. At different times after infection, either fix cells for immunofluorescence microscopy (*see* **Subheading 3.6.**) or maintain the cover slips on stage heater for time-lapse fluorescence microscopy of live cells (*see* **Subheading 3.7.**).

### 3.5. Overexpression of Dynamitin (see Note 4)

1. Transfect A549 cells (~70% confluence) with either pEGFP-dynamitin or the pEGFP vector.
2. Twenty-four hours after transfection, wash cells three times with the binding buffer, replace with the binding buffer.
3. Pipet Ad5-TR onto the cells at MOI of 10, incubate for 1 h at 4°C with gentle agitation.
4. Remove unbound virus by three washes with the binding buffer.
5. Add serum-free DMEM medium and shift to 37°C.
6. Fix cells for immunofluorescence microscopy (*see* **Subheading 3.6.**) or maintain the cover slips on stage heater for time-lapse fluorescence microscopy of live cells (*see* **Subheading 3.7.**).

### 3.6. Fluorescence Microscopy in Fixed Cells (2)

1. Wash cells three times with PBS.
2. Preextract cells with 0.5% Triton X-100 for 10 s at 37°C in a microtubule-stabilizing buffer.
3. Fix cells in 100% methanol at –20°C for 4 min, wash the cells with PBS buffer.
4. Transfer cells to 10% normal goat serum for 30 min at room temperature.
5. Label with the primary antibody in 10% goat serum for 20 min. For microtubule labeling, use mouse anti-α-tubulin antibody and for dynamitin labeling, use rabbit anti-dynamitin antibody.
6. Rinse the cover slips three times, and incubate with fluorescently labeled secondary antibodies in 10% goat serum for 20 min.

7. After extensive washing, mount the cover slips with AquaMount containing 2.5% (w/v) DAPI on glass slides.
8. Examine with a fluorescence microscope.

### 3.7. Fluorescence Microscopy in Living Cells (4)

1. Maintain the cover slips at 37°C by a thermostat-controlled stage heater, and observe cells on an electronically controlled inverted fluorescence microscope.
2. Acquire fluorescence images with a digital, back-illuminated charge coupled device camera.
3. Acquire the time-lapse series at 1- to 2-s intervals for 90 s.
4. Exposure time in the RT channel is between 0.2 and 0.4 s.
5. Process the images using a PCI interface card and the Metamorph software package (Universal Imaging).
6. Track the individual viral particles showing linear translocations, and calculate the velocities (distance between subsequent frames per time interval) of viral movement using the Metamorph tracking module.

## 4. Notes

1. Ad5 was conjugated with Texas Red to allow the examination of adenoviral entry by fluorescence microscopy. However, fluorophore-labeling can not be used to study adenoviral egress because newly replicated viruses cannot be specifically labeled.
2. For microtubule depolymerization, the concentration of nocodazole used varies according to different cell lines. In addition, the drug-containing medium should be used throughout the experiment because nocodazole effect is reversible.
3. For viral infection, it is important to use proper MOI such that most cells are infected but not detached from the plate.
4. Besides dynamitin overexpression, several other methods are available for the inhibition of the motor function of dynein, e.g., microinjection of anti-dynein antibodies, introduction of dominant negative proteins, and so on.

## Acknowledgments

The authors would like to thank Professors Yunqi Geng, Qimin Chen, and Wentao Qiao for discussion and collaboration on the project of virus–host interactions.

## References

1. Smith, G. A. and Enquist, L. W. (2002) Break ins and break outs: viral interactions with cytoskeleton of mammalian cells. *Annu. Rev. Cell Dev. Biol.* **18**, 135–161.
2. Sodeik, B., Ebersold, M. W., and Helenius, A. (1997) Microtubule-mediated transport of incoming herpes simplex virus 1 capsids to the nucleus. *J. Cell Biol.* **136**, 1007–1021.

3. Sodeik, B. (2002) Unchain my heart, baby let me go—the entry and intracellular transport of HIV. *J. Cell Biol.* **159,** 393–395.
4. Suomalainen, M., Nakano, M. Y., Boucke, K., Keller, S., Stidwill, R. P., and Greber, U. F. (1999) Microtubule-dependent minus and plus end-directed motilities are competing processes for nuclear targeting of adenovirus. *J. Cell Biol.* **144,** 657–672.
5. Dohner, K., Nagel, C. H., and Sodeik B. (2005) Viral stop-and-go along microtubules: taking a ride with dynein and kinesins. *Trends Microbiol.* **13,** 320–327.
6. Döhner, K., Wolfstein, A., Prank, U., et al. (2002) Function of dynein and dynactin in herpes simplex virus capsid transport. *Mol. Biol. Cell.* 13, 2795–2809.
7. Bartlett, J. S., Wilcher, R., and Samulski, R. J. (2000) Infectious entry pathway of adeno-associated virus and adeno-associated virus vectors *J. Virol.* **74,** 2777–2785.
8. Bearer E. L., Breakefield, X. O., Schuback, D., Reese, T. S., and LaVail, J. H. (2000) Retrograde axonal transport of herpes simplex virus: evidence for a single mechanism and a role for tegument. *Proc. Natl. Acad. Sci. USA* **97,** 8146–8150.
9. Ward, B. M. and Moss, B. (2001) Visualization of intracellular movement of vaccinia virus virions containing a green fluorescent protein-B5R membrane protein chimera *J. Virol.* **75,** 4802–4813.
10. Glotzer, J. B., Michou, A., Baker, A., Saltik, M., and Cotton, M. (2001) Microtubule-independent motility and nuclear targeting of adenoviruses with fluorescently labeled genomes. *J. Virol.* **75,** 2421–2434.
11. Seisenberger, G., Ried, M. U., Endreb, T., Buning, H., Hallek, M., and Brauchle, C. (2001) Real-time single-molecule imaging of the infection pathway of an adeno-associated virus. *Science* **294,** 1929–1932.
12. Mabit, H., Nakano, M. Y., Prank, U., et al. (2002) Intact microtubules support adenovirus and herpes simplex virus infections. *J. Virol.* **76,** 9962–9971.
13. Valetti, C., Wetzel, D. M., Schrader, M., et al. (1999) Role of dynactin in endocytic traffic: effects of dynamitin overexpression and colocalization with CLIP-170. *Mol. Biol. Cell* **10,** 4107–4120.
14. Kelkar, S. A., Pfister, K. K., Crystal, R. G., and Leopold, P. L. (2004) cytoplasmic dynein mediates adenovirus binding to microtubules. *J. Virol.* **78,** 10,122–10,132.
15. Lawrence, W. C. and Ginsberg, H. S. (1967) Intracellular uncoating of type 5 adenovirus deoxyribonucleic acid. *J. Virol.* **1,** 851–867.
16. Greber, U. F., Willetts, M., Webster, P., and Helenius, A. (1993) Stepwise dismantling of adenovirus 2 during entry into cells. *Cell* **57,** 477–486.
17. Greber, U. F., Suomalainen, M., Stidwill, R. P., Boucke, K., Ebersold, M. W., and Helenius, A. (1997) The role of the nuclear pore complex in adenovirus DNA entry. *EMBO J.* **16,** 5998–6007.

# 13

## Xenopus Oocyte Wound Healing as a Model System for Analysis of Microtubule–Actin Interactions

### Tong Zhang and Craig A. Mandato

### Summary

Microtubule–actin interactions are fundamental to many cellular processes such as cytokinesis and cellular locomotion. Investigating the mechanism of microtubule–actin interactions is the key to understand the cellular morphogenesis and related pathological processes. The abundance and highly dynamic nature of microtubules and F-actin raise a serious challenge when trying to distinguish between the real and fortuitous interactions within a cell. *Xenopus* oocyte wound model represents an ideal system to study microtubule–actin interactions as well as microtubule-dependent control of the actin polymerization. Here, we describe a series of cytoskeleton specific treatments in *Xenopus* oocyte wound healing experiments and use confocal fluorescence microscopy to analyze fixed oocytes to examine microtubule–actin interactions.

**Key Words:** Microtubule; actin; microtubule–actin interaction; wound healing; *Xenopus*; cytokinesis; rho GTPases.

## 1. Introduction

The animal cell cytoskeleton is composed of three distinct systems—microtubules, actomyosin, and intermediate filaments. Recently, it has become apparent that interactions among these three systems not only occur, but also are crucial for essential cellular processes (*1*). For instance, both animal cytokinesis and cellular locomotion require the cooperative interactions of actomyosin and microtubules: the actomyosin provides the force to power the contractile ring, whereas microtubules control and direct the force (*2*). In addition, microtubules regulate actomyosin-based contraction (*3*), pseudopodial extension (*4*), focal adhesion in mammalian cells (*5*), retrograde flow in *Aplysia* growth cones (*6*), and cortical flow in *Xenopus* oocytes (*7*).

Two general mechanisms have been proposed that permit microtubules to control actomyosin. First, microtubules transport actomyosin regulators via

From: *Methods in Molecular Medicine, Vol. 137: Microtubule Protocols*
Edited by: Jun Zhou © Humana Press Inc., Totowa, NJ

kinesin or dynein motor molecules because disruption of microtubule motor function perturbs cell locomotion *(8)* and cytokinesis *(9)*. Second, microtubules indirectly control actomyosin via microtubule polymerization or depolymerization because cell locomotion is inhibited under conditions that block microtubules dynamics without obvious changes in microtubule density or distribution *(10)*. Further, it has recently been found that two key regulators of the actin cytoskeleton: rac and rho GTPases are controlled by microtubule dynamics. Specifically, microtubule polymerization activates rac *(11)*, whereas microtubule depolymerization activates rho *(12)*.

We still lack a clear understanding of the cellular and molecular processes that link microtubules to the actomyosin machinery. The cytoskeleton proteins are highly dynamic and extremely abundant, making identification of bona fide vs fortuitous interactions quite difficult. The polymeric nature of the two systems frustrates many classic biochemical approaches making most of the processes that rely on microtubule–actomyosin interactions are inherently difficult to study. Conversely, cellular wound repair is a valid and probable way to study the assembly of microtubule-dependent actomyosin structures. Here, we describe a series of cytoskeleton specific treatments in *Xenopus* oocytes wound healing experiments and use confocal fluorescence microscopy to analyze fixed oocytes to examine microtubule–actin interactions. The same fixation protocol is also useful for studying microtubule–actin interactions during *Xenopus* embryo development.

## 2. Materials

### 2.1. Animal Husbandry

Adult female frogs are purchased from Nasco (Fort Atkinson, WI) and housed in 3′ × 3′ × 2′ polypropylene tanks at a density of less than 20 frogs per tank. The tanks are filled to a depth of 12 in. with water that has passed through a commercial charcoal filtration system to remove chlorine and ammonia and to ensure the pH range is around 7.0. The tanks are cleaned and the water replaced twice a week after the frogs have been fed Nasco frog brittle. The frogs are kept on a 12 h light/12 h dark cycle and remain at 18°C.

### 2.2. Frog Surgery Solution and Relative Tools

1. Benzocaine (Sigma): dissolve 2 g of Benzocaine in 20 mL of 90% ethanol. Mix 1 mL of benzocaine ethanol solution with 500 mL filtered water as the survival frog anaesthetic solution, or mix 4 mL of Benzocaine ethanol solution with 400 mL filtered water as the terminal frog anaesthetic solution.
2. Suture: coated VICRYL braided (Polyglactin 910) suture (4-0, 1.5 metric, for maximum oversize 0.017 mm) (Ethicon Inc., cat. no. J422II).

3. Watchmaker's forceps (Dumont no. 5) (Fine Science Tools).
4. Vannas Iris Scissors (7.5 cm) (Harvard Apparatus, Canada).

## 2.3. Oocytes Solution

1. Oocyte Ringer 2 Solution (OR-2): for 1 L 10X Solution I: 48.2 g NaCl, 1.85 g KCl, 1.4 g $Na_2HPO_4$, 11.9 g HEPES at pH 7.4 in water (*see* **Note 1**). For 1 L 10X solution II: 1.45 g $CaCl_2 \cdot 2H_2O$, 2.05 g $MgCl_2 \cdot 6H_2O$ with water. 10X solutions I and II are both stored at 4°C for further use. The working concentrations are 1X for both of them. For instance, when prepare 1 L of 1X working solution of OR-2, mix 100 mL of 10X solution I and 100 mL of 10X solution II with 700 mL of water, then adjust the pH to 7.4. Bring the whole volume up to 1 L. Store 1X OR-2 solution at 16°C for further use.
2. Collagenase Type II (Gibco): store at 4°C. Use as 0.2% (w/v) in OR-2 solution.

## 2.4. Cytoskeleton Specific Drugs

1. 10 m*M* Taxol (Paclitaxel) (Sigma, cat. no. T7402) in dimethyl sulfoxide (DMSO) (Fisher): store at –80°C. Light sensitive. Avoid refreezing. Use as a 1:1000 dilution (*see* **Note 2**).
2. Alexa Fluor 488 Phalloidin in Methanol (Molecular Probes: A12379): store at –20°C. Light sensitive. Avoid refreezing.

## 2.5. Capillary Glass Needle Preparation

Ten microliters Drummond Microdispenser Tubes (Drummond Scientific Company, cat. no. 3-000-210-G): a capillary tube is pulled by Flaming/Brown Micropipette Puller Model P-97 machine (Sutter Instrument Co.). The pulled capillary glass needle is cut on the tip under a dissection microscope with a blade to a diameter of approx 150 µm.

## 2.6. Fix Solutions (13)

1. 37% Paraformaldehyde (Fisher): add 3.7 g of paraformaldehyde into a 50-mL polypropylene screw tube and fill with water to a total volume of 10 mL. In a fumehood, boil the solution tube in a water bath for 5 min, vortex; boil for another 5 min, and then add 1 drop of 10 *N* NaOH, vortex; boil for another 5 min. At this point, paraformaldehyde should completely dissolve in solution. 37% paraformaldehyde should be kept in the fumehood at room temperature and good for 1 wk (*see* **Note 3**).
2. Fix buffer: 100 m*M* KCl, 3 m*M* $MgCl_2$, 40 m*M* HEPES, 10 m*M* EGTA pH 8.0, 150 m*M* sucrose in water. Store at 4°C for further use.
3. Fix solution: 3.7% paraformaldehyde, 0.1% glutaraldehyde, 0.1% Triton X-100, 1 µ*M* Taxol in fix buffer. The fix solution is stored at 4°C and good for 1 wk.
4. Phosphate buffered saline (PBS) solution: 137 m*M* NaCl, 2.7 m*M* KCl, 10 m*M* $Na_2HPO_4$, 2 m*M* $KH_2PO_4$, in water adjust pH to 7.4.

5. 100 m$M$ Sodium borohydride (NaBH$_4$) (Sigma) PBS solution: prepare fresh solution every time and use immediately after preparation (*see* **Note 4**).
6. Nonidet P 40 substitute solution (NP-40) (Sigma): 10% NP-40 is purchased from BioChemika.
7. Tris buffered saline (TBS) solution: 0.1 $M$ Tris-HCl, 0.9% (w/v) NaCl.
8. TBS with 0.1% NP-40 solution (TBS-N).
9. Bovine serum albumin fraction V (BSA) (Fisher): store at 4°C.
10. TBS-N with 2 mg/mL bovine serum albumin solution (TBS-N-BSA): store at 4°C for further use.

### *2.7. Antibodies*

1. Monoclonal anti-α-tubulin antibody (Sigma, cat. no. T9026): store at −20°C. Avoid refreezing. Clarify the solution by centrifugation before use.
2. Alexa Fluor 647 goat anti-mouse IgG (H+L) antibody (Molecular Probes, cat. no. A21235): store at 4°C. Light sensitive. Clarify the solution by centrifugation before use.

### *2.8. Confocal Fluorescence Microscopy Preparation*

1. Microscope cover slips (Fisher): Fisherfinest Premium Microscope Slides Plain 3" × 1" × 1 mm (cat. no. 12-544-1) and microscope cover glass no. 1.5 thickness 22 × 22 mm (cat. no. 12-541B).
2. Dow Corning High Vacuum Grease (Dow Corning Co.): as a spacer between slide and cover glass.

## 3. Methods

### *3.1. Xenopus laevis StageVI Oocytes Preparation*

#### *3.1.1. Surgery Removes Oocytes from Adult Female* X. laevis *(14)*

1. Submerge one adult female frog in benzocaine survival frog anesthetic solution for about 15 min until the frog is well anesthetized. Test by laying the frog in a prone position, and pull on one leg whereas squeezing one foot pad. No response indicates that she is sufficiently anesthetized to continue. Place the frog prone to operate. Make a small incision through the skin and body wall on one side of the abdomen (about 1 cm). Carefully remove a small lobe of the ovary with a pair of blunt forceps. The frog should not bleed much. Suture the wound with coated VICRYL braided suture. Place the postsurgery frog in a separate container with charcoal filtration system filtered water to recover for 6 mo before next surgery. An adult female frog can be done four surgeries maximum.
2. After removing lobes of ovary from the frog, they are cut into small pieces and incubated into 0.2% collagenase OR-2 solution with gentle rotation at room temperature for 1 h. Wash with OR-2 solution four to five times, until the brown collagenase is removed. The washed oocytes are kept in Petri dish containing OR-2 solution at 16°C to recover for at least 3 h to overnight. Healthy oocytes should retain the uniform animal and vegetal hemispheres (*see* **Note 5**).

3. The surgically removed oocytes are a mixture of all stages. A StageVI oocyte can be recognized under a dissection microscope with a clear border between animal and vegetal hemispheres and about 1 mm in diameter size.

### 3.1.2. Manual Defolliculation

Manually defolliculating oocytes requires practice. This step is critical for wound induced oocyte cortex experiments. Use Watchmaker's forceps (Dumont no. 5) to anually remove the follicle cells under a dissection microscope. Collect all defolliculated StageVI oocytes in a separate plate with OR-2 solution.

## 3.2. Cytoskeleton Drugs Treatment

Defolliculated StageVI oocytes are incubated with $10 \, \mu M$ Taxol OR-2 solution at $16°C$ for 1 h (*see* **Note 6**).

## 3.3. Glass Needle Wound Oocytes

1. Transfer the drug treated oocytes into a 0.5-mm Nylon grid plate containing the same concentration drug OR-2 solution (*see* **Note 7**).
2. Gently rotate oocytes with the animal pole (the darker surface) facing upward. They are then stabbed in the animal pole with a piece of glass needle (*see* **Note 8**). Oocytes are removed away from the light source to recover at room temperature for 5 min before being put into the fix solution (*see* **Note 9**).

## 3.4. Wounded Oocytes Fixation (see Note 10)

1. Wounded oocytes are transferred into 1.5-mL Eppendorf tubes containing fix solution with a density of 20 oocytes per tube. The tubes are placed on a rotator at $4°C$ for 12 h.
2. Gently transfer the fixed oocytes from fix solution into a Petri dish containing PBS (*see* **Note 11**). Gently wash on a rotator at room temperature for 10 min. Repeat two more times with new PBS solution.
3. Transfer washed oocytes into fresh 100 m$M$ NaBH$_4$-PBS solution, and gently incubate them on the rotator for 45 min. You will see many bubbles in the solution.
4. Wash NaBH$_4$-PBS solution treated oocytes with PBS solution as **step 2**.
5. Bisect the washed oocytes with Vannas iris mini-scissors around the wound under a dissection microscope (*see* **Note 12**).
6. Transfer the bisected oocytes into a 1.5-mL Eppendorf tube containing TBS-N. Put tubes on a rotator at $4°C$ for 12 h.
7. Block bisected oocytes with TBS-N-BSA solution in a 1.5-mL Eppendorf tube on a $4°C$ rotator for 6 h.
8. Incubate the bisected oocytes in a 0.5-mL Eppendorf tube with 1:200 Alexa Fluor 488 Phalloidin and 1:250 monoclonal anti-α-tubulin antibody in TBS-N-BSA solution on a $4°C$ rotator for 12 h.

9.  Wash the primary antibody from oocytes in a 1.5-mL Eppendorf tube with TBS-N solution on rotator at 4°C for 6 h.
10. Stain the washed bisected oocytes in a 0.5-mL Eppendorf tube with 1:200 Alexa Fluor 488 Phalloidin and 1:200 Alexa Fluor 647 goat anti-mouse antibody in TBS-N-BSA solution on a 4°C rotator for 12 h.
11. Wash the secondary antibody from bisected oocytes in a 1.5-mL Eppendorf tube with TBS-N solution on rotator at 4°C for 12 h.

### 3.5. Cover Slips Preparation

1.  Fill a 20-mL syringe with Dow Corning high vacuum grease. Push the grease from the syringe to make a small doughnut on the microscope slide.
2.  Use a transfer pipet to put a bisected oocyte with wound facing up on the microscope slide inside the doughnut, and use a piece of Kimwipes to remove the extra solution (*see* **Note 13**).
3.  Carefully cover with no. 1.5 cover slide under a dissection microscope for the confocal fluorescence microscopy (*see* **Note 14**).

### 3.6. Confocal Fluorescence Microscopy

Confocal fluorescence microscopy is done with Zeiss LSM 510 META confocal microscope. Samples are observed with Plan-Apochromat 63X/1.4 oil DIC objective lens and Plan-Apochromat 100X/1.4 oil DIC objective lens. Confocal Z-sections are stacked with Carl Zeiss LSM 510 Expert Mode v3.2 SP2 software with 3D View Projection program. Example pictures are shown in **Fig. 1**.

## 4. Notes

1.  All solutions should be prepared in water that has a resistivity of 18.2 M$\Omega$-cm. This is referred as "water" in this chapter.
2.  DMSO should be carefully handled. It can be rapidly absorbed by the skin into the bloodstream.
3.  Paraformaldehyde powder and solution steam are very toxic. Keep and use it in the fumehood.
4.  NaBH$_4$ powder is very toxic and flammable. Store and use it in the fumehood.
5.  Remember to change OR-2 solution everyday after finishing the experiment. The good quality oocytes can live in OR-2 for approx 4 d.
6.  Vortex to mix well when preparing any drug OR-2 solution because some of them are slightly soluble in water.
7.  The use of a Nylon grid plate can avoid the rotation of oocytes during glass needle wounding.
8.  The wounding of the dark animal pole of an oocyte is easier to be detected by direct observation.
9.  The heat of the light source may inhibit oocyte wound healing.

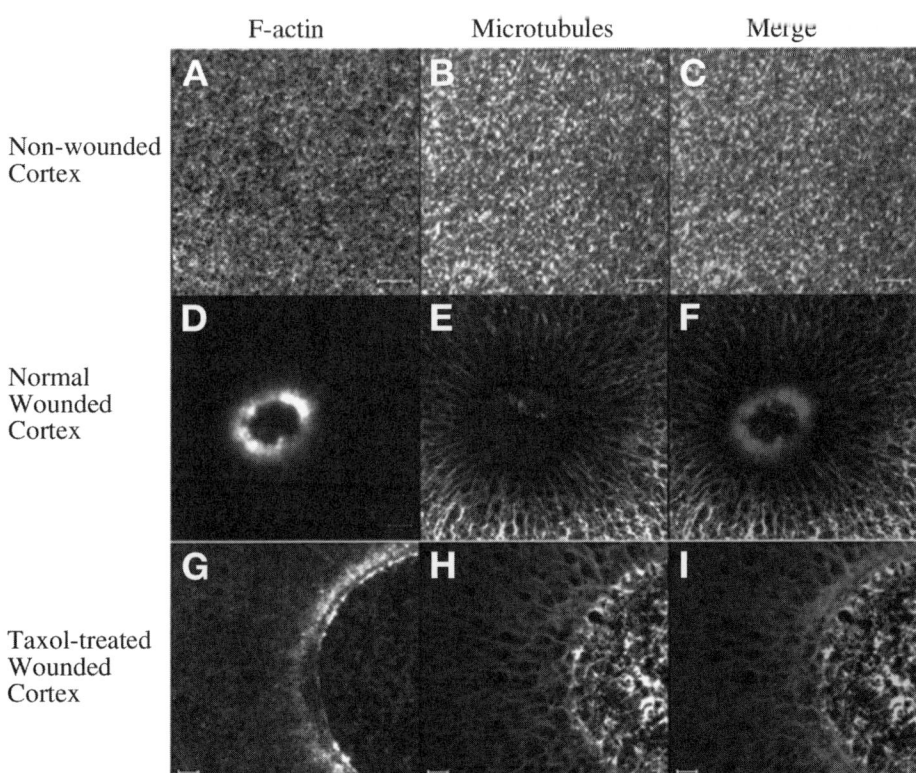

Fig. 1. Confocal fluorescence microscopy Z-sections stacked images of StageVI oocytes cortexes. **A**, **D**, and **G** are stained with Alexa Fluor 488 Phalloidin for F-actin. **B**, **E**, and **H** are stained with Alexa Fluor 647 goat anti-mouse IgG (H+L) antibody for microtubules. **C**, **F**, and **I** are merged pictures. Actins and microtubules are randomly distributed on the nonwounded cortex (**A–C**). Microtubules form a radial array around the F-actin contractile ring (**D–F**). Wound-induced Taxol-treated oocytes (**G–I**) have a double ring of F-actin (**G**). (**A–C**) Bar = 10 µm. (**D–F**) Bar = 5 µm. (**G–I**) Bar = 5 µm.

10. The same protocol is also useful for fixing the whole *X. laevis* embryos and tadpoles.
11. The fixed oocytes are very fragile.
12. Try to keep the wound in the center of the cut piece. It will be helpful for the microscopy observation because the whole oocyte is a sphere.
13. Solution can reflect light during observation and increase the background.
14. The fixed oocytes are very fragile. Too much pressure can easily break the contractile ring.

## References

1. Rodriguez, O. C., Schaefer, A. W., Mandato, C. A., Forscher, P., Bement, W. M., and Waterman-Storer, C. M. (2003) Dynamic microtubule-actin interactions: conserved mechanisms underlying directed cell movement and polarized morphogenesis. *Nat. Cell Biol.* **7,** 599–609.
2. Mandato, C. A., Benink, H. A., and Bement, W. M. (2000) Microtubule-actomyosin interactions in cortical flow and cytokinesis. *Cell Motil. Cytoskeleton* **45,** 87–92.
3. Danowski, B. A. (1989) Fibroblast contractility and actin organization are stimulated by microtubule inhibitors. *J. Cell Sci.* **93,** 255–266.
4. Rosania, G. R. and Swanson, J. A. (1996) Microtubules can modulate pseudopod activity from a distance inside macrophages. *Cell Motil. Cytoskeleton* **34,** 230–245.
5. Bershadsky, A., Chausovsky, A., Becker, E., Lyubimova, A., and Geiger, B. (1996) Involvement of microtubules in the control of adhesion-dependent signal transduction. *Curr. Biol.* **6,** 1279–1289.
6. Schaefer, A. W., Kabir, N., and Forscher, P. (2002) Filapodia and transverse retrograde F-actin arcs guide the assembly and transport of two populations of microtubules with unique dynamic parameters in neuronal growth cones. *J. Cell Biol.* **158,** 139–152.
7. Canman, J. C. and Bement, W. M. (1997) Microtubules suppress actomyosin-based cortical flow in *Xenopus* oocytes. *J. Cell Sci.* **110,** 1907–1917.
8. Rodionov, V. I., Gyoeva, F. K., Tanaka, E., Bershadsky, A. D., Vasiliev, J. M., and Gelfand, V. I. (1993) Microtubule-dependent control of cell shape and pseudopodial activity is inhibited by the antibody to kinesin motor domain. *J. Cell Biol.* **123,** 1811–1120.
9. Gonczy, P., Pichler, S., Kirkham, M., and Hyman, A. A. (1999) Cytoplasmic dynein is required for distinct aspects of MTOC positioning, including centrosome separation, in the one cell stage *Caenorhabditis elegans* embryo. *J. Cell Biol.* **147,** 135–150.
10. Mikhailov, A. and Gundersen, G. G. (1998) Relationship between microtubule dynamics and lamellipodium formation revealed by direct imaging of microtubules in cells treated with nocodazole or taxol. *Cell Motil. Cytoskeleton* **41,** 325–340.
11. Waterman-Storer, C. M., Worthylake, R. A., Liu, B. P., Burridge, K., and Salmon, E. D. (1999) Microtubule growth activates Rac1 to promote lamellipodial protrusion in fibroblasts. *Nat. Cell Biol.* **1,** 45–49.
12. Ren, X. D., Kiosses, W. B., and Schwartz, M. A. (1999) Regulation of the small GTP-binding protein Rho by cell adhesion and the cytoskeleton. *EMBO J.* **18,** 578–585.
13. Gard, D. L. (1991) Organization, nucleation and acetylation of microtubules in *Xenopus laevis* oocytes: a study by confocal immunofluorescence microscopy. *Dev. Biol.* **143,** 346–362.
14. Sive, H. L., Grainger, R. M., and Harland, R. M. (1997) *Early Development of Xenopus Laevis.* Cold Spring Harbor Press, New York, NY.

# 14

## Screening for Inhibitors of Microtubule-Associated Motor Proteins

### Frank Kozielski, Salvatore DeBonis, and Dimitrios A. Skoufias

#### Summary

The mitotic spindle is an important target for cancer chemotherapy. The main protein target for drugs in clinical use is tubulin, the building block of microtubules. In recent years, other proteins of the mitotic spindle have been identified as potential targets for the development of more specific drugs with the hope that these will have fewer side effects than known antimitotics (taxanes, vinca alkaloids). The human genome contains more than 40 members of the kinesin superfamily, with at least 12 of these involved in mitosis and cytokinesis. HsEg5 (also called KSP, kinesin spindle protein), a member of the kinesin-5 family, involved in the formation of the bipolar spindle, is a very promising target for cancer chemotherapy with specific inhibitors in Phase I and II clinical trails. Several successful approaches exist today to screen Eg5 for inhibitors, including phenotype-based assays and simple in vitro assays that explore the intrinsic enzymatic ATPase activity of Eg5. Here, we describe a robust and straightforward in vitro method to rapidly screen Eg5 for inhibitors. The assay can easily be adapted to other mitotic kinesins that may be identified in the future as potential drug targets, or simply to obtain specific kinesin inhibitors for use in "chemical genetics" to study the function of this important class of proteins.

**Key Words:** Kinesins; cancer chemotherapy; cell division; mitotic arrest; inhibitors; high-throughput screening; microtubules; apoptotic cell death.

## 1. Introduction

Anticancer drugs that target the mitotic spindle, act on tubulin *(1)*, the building blocks of microtubules that are responsible for a wide range of cellular functions *(2)*. In recent years some members of the kinesin superfamily that are important for different aspects of cell division (mitotic kinesins) have come to the fore as potential drug targets for chemotherapeutic applications *(3,4)*. A second aspect of inhibitor screening, more driven by fundamental research, is to use specific motor protein inhibitors to study the function of these proteins

From: *Methods in Molecular Medicine, Vol. 137: Microtubule Protocols*
Edited by: Jun Zhou © Humana Press Inc., Totowa, NJ

in the cell, a process called "chemical genetics" *(5)* and that requires highly specific inhibitors. The most advanced is human Eg5 (or KSP, KIF11) a protein involved in the formation of the bipolar spindle *(6)*. Several inhibitors have been described *(7–11)* and the most advanced are being used in phase I and phase II clinical trials *(12)*. These inhibitors act on Eg5 by inducing a mitotic arrest phenotype leading eventually to apoptotic cell death. Several other members of the kinesin superfamily have been discussed *(3,4)* as possible drug targets, but they have not been validated yet in practical terms.

Several procedures have been described to screen molecular motors for inhibitors, including in vitro microtubule (MT)-gliding assays to screen a limited number of mixtures of natural products from sponge, which lead to the discovery of Adociasulfate-2 *(13)* the first kinesin inhibitor discovered, phenotype-based assays *(7)*, virtual drug screening *(14)* and in vitro screening *(9)* based on the intrinsic basal or MT-stimulated ATPase activity of kinesins. We describe a simple and straightforward in vitro screening procedure for human Eg5, based on the inhibition of basal ATPase activity, which can easily be adapted for other kinesins that represent potential drug targets. Another advantage is that this test is already used in many laboratories to characterize the enzymatic activity of kinesins and thus requires only a minimal effort to automate. In our laboratory we successfully miniaturized and automated this procedure for two other human kinesins, centromere-associated protein E *(15)* and mitotic kinesin-like protein 1 *(16)*. The procedure described herein can be equally well performed manually by academic laboratories in 96-well plates using preselected small molecule libraries, or adapted on screening robots at an industrial scale, screening several hundreds of thousands molecules from industrial libraries. The in vitro identified small molecule inhibitors can then be evaluated for their ability to induce monoastral spindles, the expected phenotype resulting from Eg5 inhibition in cell-based assays as previously described *(7)*.

## 2. Materials

### 2.1. Enzyme-Coupled ATPase Assay

Assays for kinesin basal and MT-stimulated ATPase activity have been previously described in detail by Hackney and Jiang *(17)*, who provide a troubleshooting guide. The continuous enzyme-coupled assay with regeneration of adenosine-5'-triphosphate (ATP) below, is a condensed description:

1. (*N*-[2-Acetamido]-2-aminoethanesulfonic acid (ACES, Sigma, St. Louis, MO).
2. Magnesium acetate (MgAc$_2$, Sigma), prepared as a stock solution of 1 *M*.
3. 0.5 *M* Ethylene glycol-bis (2-aminoethylether)N,N,N',N'-tetraacetic acid (EGTA, Sigma), adjusted to pH 8.0 with KOH.

4. 0.5 *M* Potassium ethylenediaminetetraacetic acid (EDTA, Sigma), adjusted to pH 8.0 with KOH.

5. β-Mercaptoethanol (β-ME, Sigma).

6. Potassium hydroxide (KOH, Sigma), 2 *N* solution.

7. A25 buffer: 25 m*M* ACES/KOH, pH adjusted with 2 *N* KOH to 6.9, 2 m*M* MgAc$_2$, 2 m*M* K-EGTA, 0.1 m*M* EDTA, and 1 m*M* β-ME.

8. 3 mg/mL Lactate dehydrogenase (LDH, Sigma, type XXX-S) dialyzed in buffer A25, 290-μL aliquots, frozen in liquid nitrogen, and stored at –80°C.

9. 10 mg/mL Pyruvate kinase (PK, Sigma, type III) dissolved in buffer A25, aliquots of 170 μL, frozen in liquid nitrogen, and stored at –80°C.

10. 30 m*M* β-Nicotinamide adenine di-nucleotide, reduced form (NADH, Sigma) in distilled water, frozen in small aliquots of 416 μL in liquid nitrogen, and stored at –80°C.

11. 200 m*M* Phosphoenol pyruvate monosodium salt hydrate (PEP, Sigma) in distilled water. The pH is adjusted to 7.0 with KOH, frozen in small 500-μL aliquots in liquid nitrogen, and stored at –80°C.

12. 200 m*M* MgATP: dissolve 1.2 g Na$_2$ATP (grade I, Sigma) in 4 mL cold water and keep on ice. Adjust the pH to 7.0 with cold 2 *N* KOH. Add 2 mL of 1 *M* MgAc$_2$. Determine the ATP concentration at 259 nm using an extinction coefficient of 15.4 m*M*$^{-1}$cm$^{-1}$. Add water to reach a final MgATP concentration of 200 m*M*. Determine the exact concentration at 259 nm. Prepare 1-mL aliquots of the solution, freeze in liquid nitrogen, and store at –80°C.

13. Dimethylsulfoxide (Me$_2$SO, MERCK, Darmstadt).

14. Control inhibitor: 1 m*M* *S*-trityl-L-cysteine (STLC, Acros Organics, 93166 Noisy Le Grand) in Me$_2$SO.

15. Small molecule libraries (1 m*M* stock solution).

16. Multichannel pipet and suitable tips.

17. White, half-area 96-well μclear plates (Greiner Bio-One, Frickenhausen).

18. Eg5 stock solution at 160 μ*M* (7 mg/mL) Eg5 (*see* **Note 1**).

## 2.2. Inhibitor Screening Using the Basal Eg5 ATPase Activity

These instructions assume the use of a 96-well spectrophotometer equipped with filters at 340 and 650 nm and also 8-, 12-, or 24-channel pipets or a robot for inhibitor screening. In addition, we assume that inhibitor daughter plates have been prepared from a mother plate. The composition and the volume of the buffer given below are sufficient to screen five inhibitor plates.

1. Inhibitor screening solution: to 50 mL buffer A25 supplemented with 300 m*M* NaCl, add 250 μL MgATP (200 m*M* stock solution) to a final concentration of 1 m*M*, 500 μL PEP (200 m*M* stock solution) to a final concentration of 2 m*M*, 416 μL NADH (30 m*M* stock solution) to a final concentration of 0.25 m*M*, 290 μL LDH (3 mg/mL), to a final concentration of 17.4 g/mL, and 170 μL PK (10 mg/mL) to a final concentration of 34 μg/mL and mix gently. Prepare 10-mL single use aliquots in Falcon-15 tubes, frozen in liquid nitrogen, and stored at –80°C.

2. 500 µL Eg5 at 16 µ*M* (0.7 mg/mL): dilute 50 µL stock solution 160 µ*M* Eg5 (7 mg/mL) by a factor 10 with 450 µL buffer A (Piperazine-N,N'-bis[2-ethanesulfonic acid]) (PIPES, Sigma) pH 7.3, 200 m*M* NaCl, 1 m*M* EGTA, 2 m*M* MgCl$_2$, 1 m*M* β-ME).

## 2.3. Optional: Inhibitor Screening Based on the MT-Stimulated Eg5 ATPase Activity

The composition of the buffer volume given next is sufficient to screen five inhibitor plates.

1. MT inhibitor screening solution: to 50 mL buffer A25, add 250 µL MgATP (200 m*M* stock solution) to a final concentration of 1 m*M*, 500 µL PEP (200 m*M* stock solution) to a final concentration of 2 m*M*, 416 µL NADH (30 m*M* stock solution) to a final concentration of 0.25 m*M*, 290 µL LDH (3 mg/mL), to a final concentration of 17.4 g/mL, and 170 µL PK (10 mg/mL) to a final concentration of 34 µg/mL and mix gently. Prepare 10-mL single use aliquots in Falcon-15 tubes, frozen in liquid nitrogen and stored at –80°C.
2. Buffer A25: 25 m*M* ACES/KOH, pH 6.9, 2 m*M* MgAc$_2$, 2 m*M* EGTA, 0.1 m*M* EDTA, 1 m*M* β-ME.
3. 500 µL Eg5 at 0.8 µ*M* (0.035 mg/mL) in buffer A: dilute 2.5 µL stock solution 160 µ*M* Eg5 (7 mg/mL) by a factor 200 with 497.5 µL buffer A25.
4. Tubulin purified as described in Chapter 2 of this book or bought from commercial sources (e.g., cytoskeleton).

## 2.4. CytoPhos Phosphate Assay for Secondary Screening

The CytoPhos phosphate assay *(18)* is very sensitive to phosphate contamination (*see* **Note 2**). Care should be taken to use plastic and no glassware to avoid carry-over of phosphate. The protein used for screening should not contain ATP. The maximal ATP concentration used in this assay should not exceed 0.3 m*M* because the background level of phosphate present in the ATP stock solution starts to become significant at this concentration.

1. Prepare 500 mL buffer A25 (25 m*M* ACES/KOH, pH 6.9, 2 m*M* MgAc$_2$, 2 m*M* EGTA, 0.1 m*M* EDTA, 1 m*M* β-ME) supplemented with 300 m*M* NaCl and store at 4°C.
2. 200 m*M* MgATP (prepared as in **Subheading 2.1.**) frozen at –80°C.
3. 1 m*M* STLC in Me$_2$SO.
4. 70 mL CytoPhos phosphate assay (Cytoskeleton, Denver, CO), stored in single use aliquots of 5 mL in Falcon-15 at 4°C.
5. PS 96-well microplate, U-shaped (Greiner Bio-One).
6. Half-area 96-well plate (Greiner Bio-One).
7. Multichannel pipets and suitable tips.
8. 96-Well spectrophotometer with filter at 650 nm.
9. Purified Eg5 (*see* **Note 3**) at about 1.4 mg/mL (32 µ*M*) in buffer A, frozen in liquid nitrogen and stored at –80°C.

### 2.5. Optional: Preparation of MTs Sodium Salt From Tubulin

1. 200 mM Guanosine-5'-triphosphate (GTP, type III, Sigma).
2. PEM buffer (100 mM PIPES/KOH, pH 6.9, 1 mM EGTA, 1 mM MgCl$_2$, 1 mM GTP), aliquots of 1 mL, stored at –20°C.
3. 12 mg/mL Tubulin in PEM buffer, aliquots of 100 µL, frozen in liquid nitrogen, and stored at –80°C.
4. 10 mM Taxol (paclitaxel, Sigma) in Me$_2$SO, stored at –20°C.
5. 10% (w/v) Sodium azide (NaN$_3$), aliquots of 1 mL, frozen in liquid nitrogen, stored at –20°C

### 2.6. Small Molecule Libraries

Small molecule libraries can be obtained from different commercial and noncommercial sources. The National Cancer Institute, National Institutes of Health (NCI/NIH) provides small, preselected libraries for the scientific community, which can be obtained for cancer-related screening projects. In several countries, "National" libraries exist that have been assembled and centralized by chemists from public institutions. These libraries often contain small molecules or even natural products that are not commercially available and which might represent a rich source of unusual molecules. Other commercial libraries can simply be bought from specialized chemical companies. A small selection of addresses where to obtain small molecule libraries and natural products are listed next:

1. Synthetics and natural products from the NCI/NIH: http://www.dtp.nci.nih.gov/repositories.html.
2. Natural products from the Institut de Chimie des Substances Naturelles: http://www.icsn.cnrs-gif.fr.
3. Hans-Knöll-Institut für Naturstoff-Forschung Jena: http://www.hki-jena.de.
4. Commercial compound libraries from ChemBridge Corporation: http://chembridge.com/chembridge/compound.html.

### 2.7. Phenotype Cell-Based Assays

This protocol can be used to evaluate if an inhibitor of the Eg5 motor, identified by in vitro assays, is also able to induce monoastral spindles in cell based assays. Lack of Eg5 activity at the G2/M transition of the cell cycle leads to failure of separation of the duplicated centrosomes. As unseparated centrosomes do not loose their microtubule nucleating activity, an array of microtubules is formed, which can interact with mitotic chromosomes. Under these conditions chromosomes are not aligned properly and are distributed around a monoastral spindle because the two kinetochores of each chromosome fail to attain proper bipolar microtubule capture. This results in a loss of tension between the two sister kinetochores. Incomplete microtubule capture and lack of microtubule generated tension between sister kinetochores are two potent activators of the spindle assembly checkpoint leading to a mitotic arrest. Thus, loss of Eg5 motor activity induces a monoastral spindle assembly resulting in an accumulation of cells in mitosis owing to the activation of the spindle assembly checkpoint. A detailed method for studying spindle assembly checkpoint protein recruitment to the kinetochores owing to spindle defects induced by microtubule inhibitors is described elsewhere *(19)*.

1. Circular glass cover slips, no. 1, 12 mm diameter (Bellco, Vineland, NJ).
2. Poly-D-Lysine (Sigma).
3. Ethanol, absolute.
4. Fine-tip forceps.
5. Paraformaldehyde solution prepared as an 8% stock from powder in water and subsequent heating at 70°C (about 5 min) followed by filtration for the solution to become clear under a chemical hood. Store the stock solution at 4°C for up to 30 d.
6. Phosphate buffered saline solution (PBS): 136 m$M$ NaCl, 2 m$M$ KCl, 10.6 m$M$ Na$_2$HPO$_4$, 1.5 m$M$ KH$_2$PO$_4$, pH 7.4. Working solution should be prepared fresh from a 20X stock solution (kept at room temperature) prior to use.
7. Triton X-100 (Sigma), 10% (v/v) solution kept at 4°C.
8. Bovine serum albumin (BSA, Sigma).
9. NaN$_3$ (Sigma), 20% w/v solution kept at 4°C.
10. Tween-20 (Sigma), 10% v/v solution kept at 4°C.
11. Antibody buffer: (PBS containing 3% BSA, 0.05% Tween-20, and 0.04% NaN$_3$); kept at 4°C.
12. Secondary antibodies (Jackson Immunoresearch, West Grove, PA).
13. Aluminum foil.
14. Precleaned microscope slides (3× 1 × 1 mm) (Fisher Scientific, Pittsburgh, PA).
15. Vectashield mounting medium (Vector Laboratories, Inc., Burlingame, CA).
16. Kimwipes 1-Ply Wipers (Kimberly-Clark, Roswell, GA).
17. Nail polish, clear or with color.
18. Mouse anti-β-tubulin monoclonal antibody, clone 2.1 (Sigma) (**Note 9**).
19. Absolute methanol, HPLC-grade (Fisher Scientific, Fair Lawn, NJ).
20. Propidium iodide (Sigma), 5 mg/mL in distilled H$_2$O, kept at –20°C.

## 3. Methods

1. The use of the enzyme-coupled assay and the CytpPhos phosphate assay for inhibitor screening is very different from using these tests for enzymatic studies of motor proteins (e.g., determination of k$_{cat}$ values), where considerable efforts are undertaken to exactly control the experimental conditions and to work at the optimum of the enzymes activity, and its partner, the microtubules. For inhibitor screening, other factors such as the stability of the test assay over a longer time period (eventually days) is more important, conditions, which may be far from the enzymatic optimum. The different steps necessary for the in vitro identification of Eg5 inhibitors and subsequent characterization in cell-based assays is shown in **Fig. 1**. The following calculations are based on the preparation of one-half-area 96-well plate with a final volume of 100 µL per well.

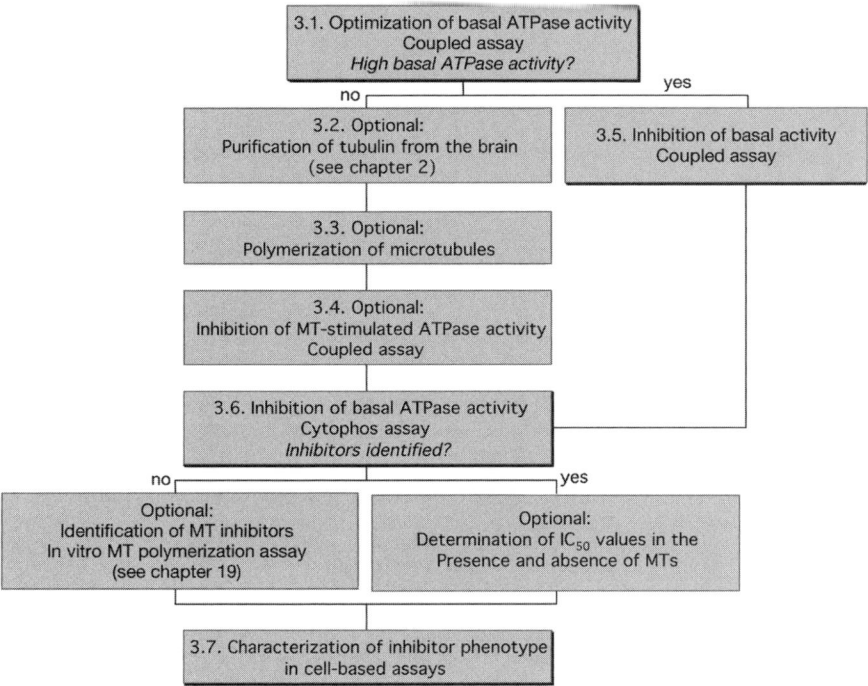

Fig. 1. Flowchart for the identification of kinesin inhibitors using the basal or microtubule-stimulated ATPase activity of kinesins. The numbers in the boxes refer to the corresponding subchapters in the **Subheading 3.**

## 3.1. Optimization of Basal ATPase Activity

MTs are expensive, when bought from commercial sources and enormous quantities will be needed, when large small molecule libraries are to be screened. A cheaper alternative is to purify tubulin from bovine brain (*see* Chapter 2), but the process has become administratively more demanding because the discovery of bovine spongiform encephalopathy (BSE). In contrast, screening without MTs is fast and direct. To determine whether the basal kinesin ATPase activity is high enough to be used for inhibitor screening, the activity should be determined and optimized for the following parameters: often but not always, the basal ATPase activity is salt-dependent and the optimum can be determined by measuring it in the presence of increasing amounts of salt (**Fig. 1A**).

1. Thaw 10 mL MT screening buffer (without salt), warm to room temperature, and pipet 80 mL into the 12 wells of the first line (A1–A12) of a half-area 96-well mclear plate.

2. Dispense 15 mL of increasing NaCl concentrations (prepared from the 4 M stock solution) covering a range from 0 mM to 600 mM NaCl.
3. Add 5 mL Eg5 (0.7 mg/mL stock solution) to all wells of the first line of the 96-well plate and mix gently.
4. Record the decrease in absorbance at 340 nm over a period of about 10 min.
5. Determine the slope ($\Delta A_{340}$/min) of the linear region for each well (often automatically performed by the software that is installed on the 96-well spectrophotometer.
6. Represent data graphically (Kaleidagraph, Excel, and so on) as shown in **Fig. 2**. The determined, optimal salt concentration can be used for all following measurements. The conditions of the test assay are a compromise between several factors, including for example the speed, with which one wants to proceed with the inhibitor screening, the amount of protein available, and the test conditions (manually or automated with robots), and the size of the compound library to be screened. In the most favourable case, where large amounts of proteins can be easily purified and screening robots with 96-channel heads are available, the actual time period for screening one 96-well plate can be reduced to as little as 3–4 min by increasing the Eg5 concentration in the test assay. Otherwise, if the test is performed manually (which usually takes 2–3 min) or only limited amounts of protein are available, one can slow down the test by decreasing the Eg5 concentration. In the later case the decrease of the absorbance at an optical density of 340 nm ($OD_{340nm}$) can be followed over a time period as long as 10–20 min (**Fig. 2B**).
7. Thaw 10 mL screening buffer (now supplemented with 300 m$M$ NaCl), warm to room temperature, and pipet 95 µL into three wells of the first column (B1–D1) of the same half-area 96-well µclear plate.
8. Add 5 µL Eg5 (160 µ$M$ stock solution) to well B1, 5 µL Eg5 (80 µ$M$ stock solution) to well C1, and 5 µL Eg5 (16 µ$M$ stock solution) to well D1 of the 96-well plate and mix gently.
9. Record the decrease in absorbance at $OD_{340\ nm}$ over a period of at least 8 min.
10. Choose the optimal amount of Eg5 (and test velocity), which best suits the test conditions. If the basal ATPase activity of the motor is sufficiently high ($k_{cat}$ = 0.06/s for $Eg5_{2–386}$), the following three sections can be ignored and one can directly proceed with **Subheading 3.5.**

---

Fig. 2. *(opposite page)* (**A**) Salt dependence of basal Eg5 ATPase activity. Eg5 is about four times more active in the presence of 300 m$M$ NaCl than without salt. (**B**) The decrease of the optical density at 340 nm in the presence of increasing Eg5 concentrations followed over a time period of 8 min. The reaction can be slowed and followed from 4 min for 8 µ$M$ Eg5 (filled circles), to 6 min for 4 µ$M$ Eg5 (filled squares), and more than 20 min, when using 0.8 µ$M$ Eg5 (filled triangles). (**C**) Determination of the optimal microtubule (MT) concentration for screening the inhibition of the MT-stimulated Eg5 ATPase activity. (**D**) The Pi standard curve using the CytoPhos phosphate assay is linear up to 0.5 $OD_{650\ nm}$. (**E**) Titration of $Eg5_{2–386}$ over a range of 0.25 µ$M$ to 1.5 µ$M$ measured in the presence of buffer A25 supplemented with 300 m$M$ NaCl.

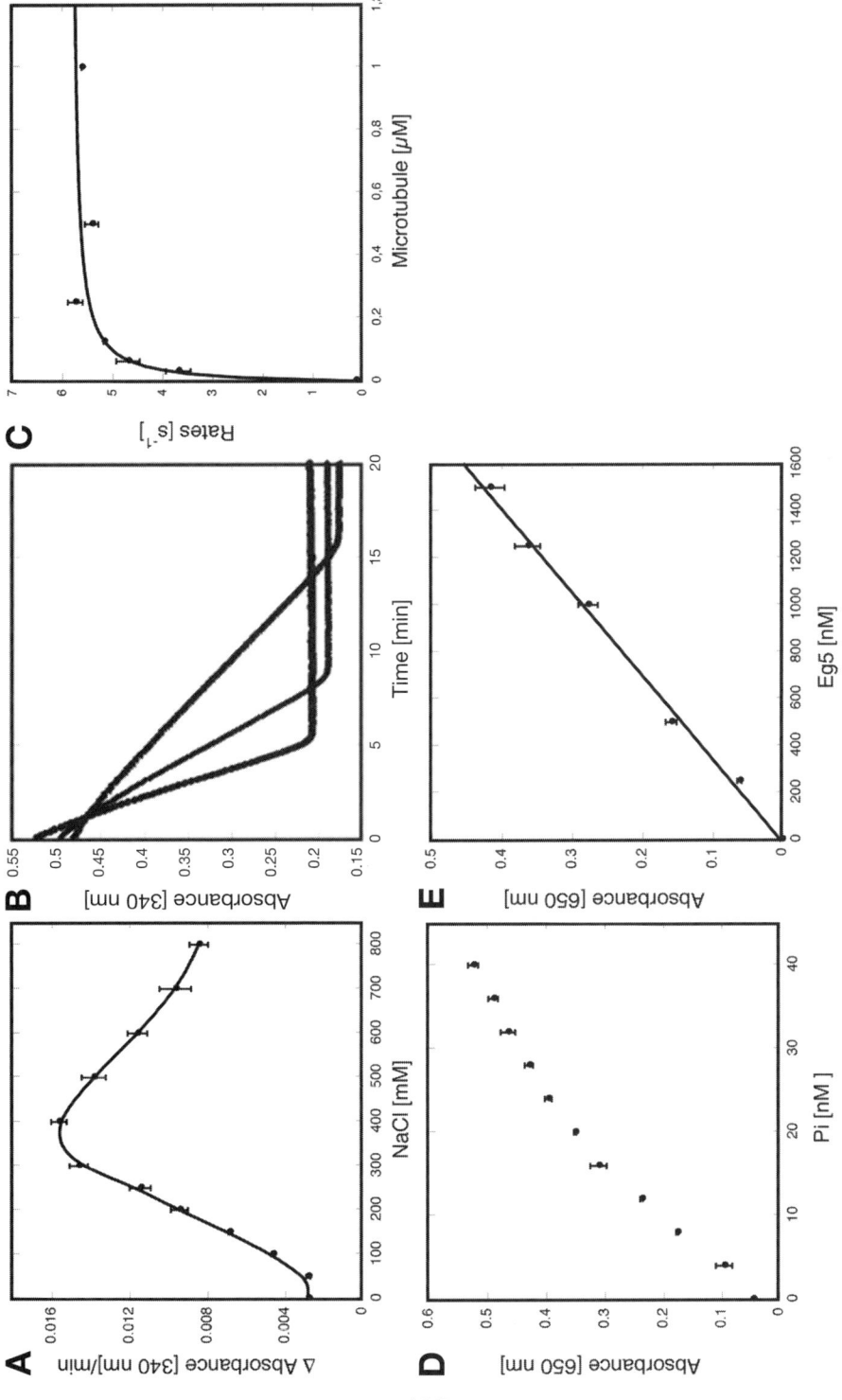

197

### 3.2. Optional: Purification of Tubulin From the Brain

In cases where sufficient amounts of the motor protein are not available to screen for the inhibition of basal ATPase activity, or when the basal ATPase activity is too low, one can use the MT-stimulated ATPase activity for inhibitor screening: the amount of motor protein is considerably reduced owing to the stimulation of the basal ATPase activity by MTs up to factor of several hundreds to thousands. The disadvantage however, is that one needs large amounts of MTs instead. Another inconvenience is that one might obtain false-positive hits, inhibitors that act on MT dynamics (polymerization or depolymerization) rather then the motor. These have to be eliminated at a later step (**Subheading 3.6.**). The purification of tubulin is described in detail in Chapter 2. Alternatively, tubulin can be bought from commercial sources (e.g., Cytoskeleton).

### 3.3. Optional: Preparation of Taxol-Stabilized Microtubules

We measure the inhibition of the MT-activity Eg5 ATPase activity at a MT concentration of $1-1.2\,\mu M$, were the stimulation of the ATPase activity usually has reached the plateau (**Fig. 2C**). In the presence of an antibacterial compound the MTs prepared as described next are stable for several days.

1. Thaw a 100-µL aliquot from the tubulin stock ($120\,\mu M$, 12 mg/mL) on ice.
2. In the meantime warm 140 µL PEM buffer to 37°C.
3. Add 1.2 µL of (Add 1.2 $M$ taxol (10 m$M$ stock) solution in $Me_2SO$) to the prewarmed PEM buffer to a final concentration of $50\,\mu M$.
4. Add 2 µL of 10% (w/v) $NaN_3$ solution to the prewarmed PEM buffer to avoid bacterial contamination.
5. Add the 100-µL tubulin aliquot to the prewarmed PEM buffer and mix gently.
6. After incubation over night at 37°C to allow polymerization the MTs are ready to be used for inhibitor screening.

### 3.4. Optional: Primary Screen Using the MT-Stimulated Eg5 ATPase Activity

Small molecule libraries are available at a given concentration in 96-well plates with columns 1 (A1–H1) and 12 (A12–H12) left empty to be used for negative and positive controls, if an inhibitor is available. In the presence of increasing salt concentrations the MT-stimulated Eg5 ATPase activity decreases. Therefore, inhibitor screening based on the MT-stimulated ATPase activity should be performed in the absence of salt.

1. Thaw 10 mL MT screening buffer, warm to room temperature, and add 2 µL taxol (10 m$M$ stock solution) to a final concentration of $2\,\mu M$.
2. Add 240 µL microtubules (**Subheading 3.3.**) to a final concentration of $1.2\,\mu M$ and mix gently by inverting the Falcon 15 tube several times.
3. Dispense 92.5 µL into each well of a half-area 96-well µclear plate.

4. As a negative inhibition control (no inhibitor is added), transfer 2.5 μL Me₂SO (*see* **Note 4**) into the wells of the first plate column (A1–H1).
5. Transfer 2.5 μL from the first inhibitor plate (1 m*M* stock solution) into the equivalent wells of the 96-well μclear plate (A2–H2 to A11–H11) using for example an eight-channel pipet, to a final inhibitor concentration of 25 μ*M*.
6. As a positive control, transfer 2.5 μL STLC (1 m*M* stock solution) to a final concentration of 25 μ*M* into the wells of the last column (A12–H12).
7. Measure the optical density at 340 nm to identify wells that are saturated (OD$_{340}$ larger than 2.0). If necessary dilute the inhibitor concentration of these wells by a factor 2 using the buffer left over from **step 1**.
8. Add 5 μL Eg5 (0.85 μ*M* [35 μg/mL]) to a final concentration of 42 n*M* to all wells of the 96-well plate, and mix.
9. Record the decrease in absorbance at 340 nm over a period of about 10 min.
10. Determine the slope (Δ$A_{340}$/min) of the linear region for each well (often performed automatically using the software installed on the 96-well spectrophotometer.
11. Represent data graphically (Kaleidagraph, Excel, or others) An example of a possible outcome is shown in **Fig. 3A**.
12. Select potential Eg5 inhibitors for secondary screen (*see* **Note 5**).
13. Repeat **steps 1–12** until all inhibitor plates of the compound library have been measured. Molecules that inhibited the MT-stimulated Eg5 ATPase activity have to be validated in a secondary screen using the CytoPhos phosphate assay (**Subheading 3.6.**).

## 3.5. Primary Inhibitor Screening Based on the Basal Eg5 ATPase Activity

1. We have performed our first screens at an inhibitor concentration of 100 μ*M*, but reduced it subsequently to 25 μ*M* for this in vitro assay.
2. Thaw 10 mL screening buffer, warm to room temperature, and pipet 92.5 μL into each well of a half-area 96-well μclear plate.
3. As a negative control (no inhibitor is added), transfer 2.5 μL Me₂SO into the wells of the first column (A1–H1).
4. Transfer 2.5 μL inhibitors (1 m*M* stock solution) from the first plate of the compound library into the equivalent wells of the 96-well plate, to give the desired inhibitor concentration (25 μ*M*).
5. As a positive inhibitor control, transfer 2.5 μL STLC to a final concentration of 25 μ*M* into the wells of the last column (A12–H12).
6. Measure the optical density at 340 nm to identify wells that are saturated (OD greater than 2.0). If necessary dilute the inhibitor concentration of these wells by a factor 2 using the screening buffer left over from **step 2**.
7. Add 5 μL Eg5 (0.7 mg/mL stock solution) at the previously defined optimal concentration to all wells of the 96-well plate, mix gently.
8. Record the decrease in absorbance at 340 nm over a period of about 10 min.
9. Determine the slope (Δ$A_{340}$/min) of the linear region for each well (often automatically performed by the software that is installed on the 96-well spectrophotometer.

10. Represent data graphically (Kaleidagraph, Excel, and so on) as shown in **Fig. 3B**.
11. Select potential Eg5 inhibitors for secondary screen.
12. Repeat **steps 1–12** until all inhibitor plates of the compound library have been measured. Molecules that inhibited the basal Eg5 ATPase activity are validated using the CytoPhos phosphate assay (**Subheading 3.6.**).

### 3.6. Secondary Screen Using the CytoPhos Phosphate Endpoint Assay

The molecules from the compound library that have been identified in the previous steps to inhibit either the MT-stimulated or basal Eg5 ATPase activity may act on Eg5, but also on MTs or the other enzyme components of the enzyme-coupled assay, LDH and PK. These initial drug candidates have to be validated in more detail using a secondary counter screen, which must not contain these enzymes. We use the CytoPhos phosphate assay *(18)*, which is a one-step version based on the multistep molybdate/malachite reagent *(20)*. The inhibition of the basal Eg5 ATPase activity is used (*see* **Note 6**). The calculations next are based on a small molecule library obtained at a concentration of 1 m$M$ and to be screened at a final inhibitor concentration of 25 μ$M$.

1. Add 7.5 μL ATP (*see* **Note 7**) (from a 200 m$M$ stock solution) to a final concentration of 300 μ$M$ to 5 mL buffer A25 supplemented with 300 m$M$ NaCl and mix gently.
2. Distribute 46.25 μL buffer A25 supplemented with 300 m$M$ NaCl and 300 μ$M$ ATP into each well of a U-shaped 96-well plate.
3. Add 1.25 μL Me$_2$SO to a final concentration of 2.5% into the first column of the 96-well plate (well positions A1–H1).
4. Add control inhibitor STLC (1.25 μL from a 1 m$M$ stock solution in Me$_2$SO) to a final concentration of 25 μ$M$ into the last column of the 96-well plate (well positions A12–H12).
5. Dispense 1.25 μL of inhibitors that inhibited the MT-stimulated or basal Eg5 ATPase activity to the remaining 10 columns (from A2–H2 to A11–H11).
6. Start the reaction by adding 2.5 μL (1.5 μ$M$) Eg5 (*see* **Note 8**) (stock solution: 32 μ$M$) to each well of the 96-well plate and mix gently.
7. Allow the reaction to proceed during 10 min at room temperature.
8. In the meantime distribute 70 μL of the CytoPhos phosphate assay (room temperature) to each well of a new half-area 96-well plate.

---

Fig. 3. *(opposite page)* Examples of the quality of inhibition of Eg5 ATPase activity in 96-well format. Filled triangles represent control measurements of Eg5 activity in the absence of any inhibitor; filled squares represent Eg5 activity inhibited by *S*-trityl-L-cysteine. Filled circles represent the Eg5 activity measured in the presence of small molecules from compound libraries. **(A)** Inhibition of MT-stimulated Eg5 ATPase activity using the enzyme-coupled assay. **(B)** Inhibition of basal Eg5 ATPase activity (enzyme-coupled assay). **(C)** Inhibition of basal Eg5 ATPase activity using the CytoPhos phosphate assay. (Please note that the figure presents three independent experiments

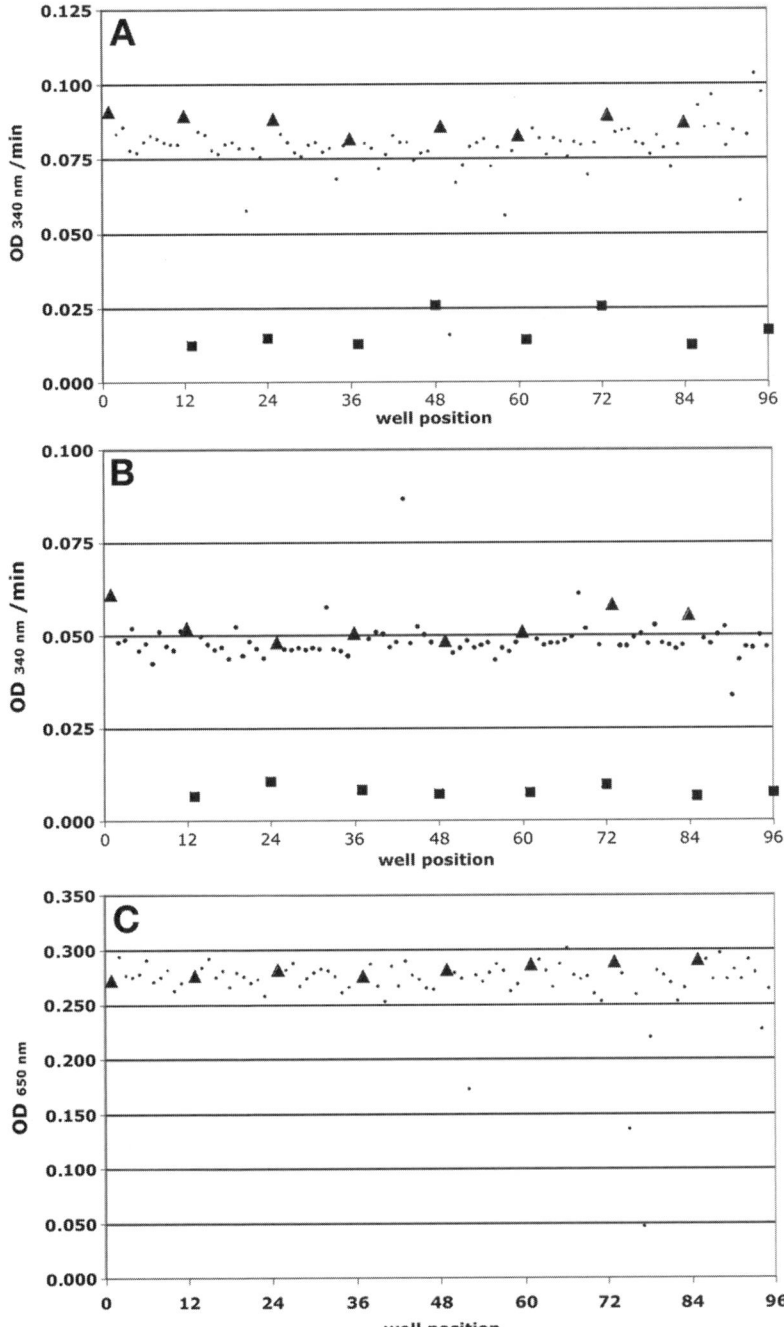

9. After 10 min transfer 30 μL from the 50 μL reaction solution of the U-shaped 96-well plate to the new half-area plate previously filled with CytoPhos phosphate assay solution and mix gently.
10. Wait for another 10 min.
11. Measure the absorbance of the 96-well plate at $OD_{650nm}$ using a 96-well spectrophotometer.
12. Represent data graphically (Kaleidagraph, Excel, and so on) as shown for example in **Fig. 3C**.
13. Select compounds that inhibit Eg5 motor activity. Molecules that inhibit the ATPase activity in both assays are potential Eg5 inhibitors and can subsequently be tested in cell-based assays for their ability to induce mitotic arrest.

### 3.7. Phenotype Cell-Based Assays

The methods described next outline: (1) cell culture of HeLa cells as optimized for (2) induction, (3) detection, and (4) quantitative analysis of the induction of monoastral spindles by immunofluorescence microscopy following treatment with putative motor inhibitors **(Fig. 4A)**.

### 3.7.1. Culture of HeLa Cells for Analysis of Eg5 Inhibition

For immunofluorescence microscopy (*see* **Subheading 3.7.3.**), cells are grown on 24-well plates on cover slips as detailed in **steps 1** and **2**.

1. Treat cover slips with 0.1 mg/mL poly-D-lysine for 30 min to promote the adhesion of cells to cover slips during mitosis.
2. Rinse generously with distilled $H_2O$, sterilize 30 min with 70% ethanol in a laminar flow hood, rinse with absolute ethanol, and allow to air-dry. Keep cover slips in a sterile container.
3. Using fine-tip forceps, place a cover slip in each of the well in the 24-well plate at the desired density. Cells should be replated such that cells are subconfluent after 36–48 h (*see* **Note 10**). At this point, cells can be treated with a range of different concentrations of the inhibitors.

### 3.7.2. Induction of Monoastral Spindles

The degree of induction of monoastral spindle assembly can be modulated by exposure of cells to increasing concentrations of the inhibitors.

Treat cells with the inhibitors for up to 8 h. Treatment for longer than 8 h may result in the loss of mitotic cells from the cover slips because mitotic cells detach relatively easily. Longer incubation times may also lead to cell death.

### 3.7.3. Immunofluorescence Microscopy to Analyze Monoastral Spindle Assembly

This protocol can be utilized to evaluate if various inhibitors of the Eg5 motor identified by in vitro assays are also able to induce monoastral spindles in cell based assays.

1. Grow cells on cover slips as previously described (**Subheading 3.7.1.**) (*see* **Note 10**).

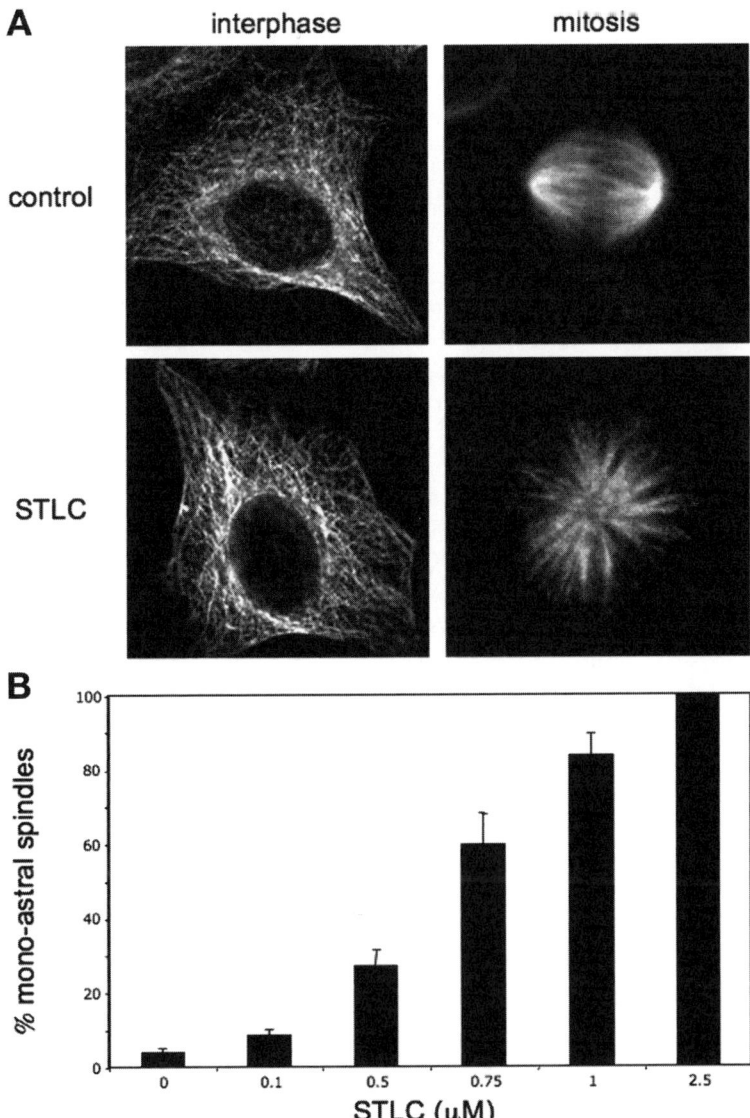

Fig. 4. Phenotype cell-based assay for evaluating the potency of Eg5 inhibitors. (**A**) HeLa cells following 8-h exposure to 5 µ*M* *S*-trityl-ʟ-cysteine (STLC) were fixed and stained with an anti-tubulin antibody and examined by indirect immunofluorescence microscopy. STLC treated cells have a normal interphase microtubule network compared to control, whereas in mitosis, STLC treated cells form a monoastral spindle instead of the normal bipolar spindle present in untreated control cells. (**B**) Concentration dependence of induction of monoastral spindles. Cells following incubation with increasing concentration of the inhibitor were treated as in **A**. Cells with monoastral spindles were scored as a percentage of total mitotic cells.

2. Following drug exposure, fix cells for 3 min at 37°C with 1% paraformaldehyde in PBS (*see* **Note 11**) followed by absolute methanol for 3 min at –20°C. Aspirate medium from the culture dish and add directly the fixative, or transfer cover slips to a separate container for fixation. Following fixation, cells should be placed in PBS. At this point, cells can be stored at 4°C, with the addition of 0.04% $NaN_3$. An alternative fixation protocol can be followed, which includes a 20 min incubation at 37°C with 2% paraformaldehyde in PBS followed by a 3 min permeabilization with 0.2% Triton X-100 in PBS at room temperature.

3. Rinse or wash with PBS.

4. Incubate with primary antibodies (*see* **Subheading 2.7.**) for 1 h at 37°C in antibody buffer (PBS, 3% BSA, 0.05% Tween-20, 0.04% $NaN_3$). Cover slips are placed on Parafilm in a covered container, and 50 mL of diluted antibody solution is applied to each cover slip.

5. Cover slips are washed with PBS three times for 5 min each.

6. Secondary antibodies (FITC-conjugated anti-mouse) are diluted 1/300 in antibody buffer and incubated at 37°C for 30 min. Protect cells can from unnecessary exposure to ambient light by covering the incubation chamber with aluminum foil.

7. Cover slips are washed 5 min with PBS (as described in **step 5**).

8. DNA is stained, with either DAPI (1 mg/mL) or propidium iodide (0.2 mg/mL) in PBS for 5 min. Peak excitations are at 461 nm and 617 nm for DAPI and propidium iodide, respectively. If Vecta sheild with AP1 is used as a mounting medium additional staining with DAPI is not necessary.

9. Wash cover slips with PBS twice for 5 min each (as described in **step 5**).

10. Mount cover slips on microscope slides by inverting over a drop (~10 mL) of Vectashield mounting medium. Excess PBS wash solution should be aspirated from cover slips prior to their inversion.

11. Remove any excess mounting medium from the slide, a folded Kimwipe is placed over the slide, and an extra microscope slide is positioned over the slide that contains the cover slip, then light pressure is applied.

12. Seal the cover slip on the slide with a light coat of nail polish at the edge and allow drying for 5–10 min.

13. The sample can immediately be examined with a fluorescence microscope or can be stored indefinitely at –20°C.

### 3.7.4. Quantification of Mitotic Arrest

The percent of mitotic cells with monoastral spindles at each drug concentration can be determined by visual counting with the aid of an epifluorescence microscope and the data can be plotted as a bar graph (**Fig. 4B**). Low numbers of monoastral spindles are present in untreated cells too, representing those cells that have not yet completed the process of separating their duplicated centrosomes.

## 4. Notes

1. We use monomeric human $Eg5_{2-386}$, a protein construct containing the N-terminal Eg5 motor domain with ATP binding and MT interacting regions, for inhibitor screening *(9,21)*.
2. The yellow reaction solution becomes greenish, if one of the wells is contaminated with phosphate, before Eg5 has been added to start the reaction.
3. To eliminate possible contamination with phosphate, the protein should have been either subjected to gel filtration or diluted in buffer A and subsequently concentrated using Centricons.
4. Compounds from small molecule libraries are commonly solubilized in $Me_2SO$. Because it might modify the ATPase activity, it has to be added to all wells that do not contain library compounds.
5. To access the quality of the assay, the Z-factor *(22)* can be used. It is a statistical parameter to evaluate the quality of high-throughput screening assays and can be used for assay optimization.
6. The CytoPhos phosphate assay has been optimized for a panel of nine kinesin motor proteins measuring the MT-stimulated ATPase activity. Assay conditions can be found elsewhere *(18)*.
7. Used at higher concentrations, phosphate from hydrolyzed ATP increases the background.
8. Eg5 should be either stored without addition of ATP or gel filtrated before used to eliminate phosphate present in the protein purification. Please note that some kinesins will not survive storage without ATP and degrade. These proteins have to be stored in the presence of ATP.
9. An alternative antibody is the rat monoclonal YL1/2 recognizing the tyrosinated a tubulin subunit (ABCAM).
10. Mitotic HeLa cells detach relatively easily from cover slips, particularly after treatment with anti-mitotic agents. Quantification by microscopy is subject to underestimation owing to the fact that detached mitotic cells are not counted. Maximum retention of mitotic cells on cover slips can be achieved by pretreatment of cover slips with poly-D-lysine and by growth of cells on cover slips for 36 h or longer prior to manipulation.
11. Given the cold-lability of microtubules, for optimal preservation of microtubules, one should use fixative solution prewarmed to 37°C and fixation should be carried out at 37°C (in an incubator).

## Acknowledgments

This work was supported by ARC (contract number 5197) and the Rhône-Alpes Region (contract numbers 03013679002 and 0301369001).

## References

1. Hamel, E. (1996) Antimitotic natural products and their interactions with tubulin. *Med. Res. Revs.* **16,** 207–231.

2. Kreis, T. and Vale, R. (1999) *Guidebook to the Cytoskeletal and Motor Proteins.* Oxford University Press, New York.
3. Wood, K. W., Cornell, W. D., and Jackson, J. R. (2001) Past and future of the mitotic spindle as an oncology target. *Curr. Opin. Pharmacol.* **1,** 370–377.
4. Miyamoto, D. T., Perlman, Z. E., Mitchison, T. J., and Shirasu-Hiza, M. (2003) Dynamics of the mitotic spindle: potential therapeutic targets. *Prog. Cell Cycle Res.* **5,** 349–360.
5. Mayer, T. U. (2003) Chemical genetics: tailoring tools for cell biology. *Trends Cell Biol.* **13,** 270–277.
6. Blangy, A., Lane, H. A., d'Herin, P., Harper, M., Kress, M., and Nigg, E. A. (1995) Phosphorylation by p34cdc2 regulates spindle association of human Eg5, a kinesin-related motor essential for bipolar spindle formation in vivo. *Cell* **83,** 1159–1169.
7. Mayer, T. U., Kapoor, T. M., Haggarty, S. J., King, R. W., Schreiber, S. L., and Mitchison, T. J. (1999) Small molecule inhibitor of mitotic spindle bipolarity identified in a phenotype-based screen. *Science* **286,** 971–974.
8. Nakazawa, J., Yajima, J., Usui, T., et al. (2003) A novel action of terpendole e on the motor activity of mitotic Kinesin eg5. *Chem. Biol.* **10,** 131–137.
9. DeBonis, S., Skoufias, D. A., Robin, G., et al. (2004) In vitro screening for inhibitors of the human mitotic kinesin, Eg5, with antimitotic and antitumor activity. *Mol. Cancer Ther.* **3,** 1079–1090.
10. Cox, C. D., Breslin, M. J., Mariano, B. J., et al. (2005) Kinesin spindle protein (KSP) inhibitors. Part 1: The discovery of 3,5-diaryl-4,5-dihydropyrazoles as potent and selective inhibitors of the mitotic kinesin KSP. *Bioorg. Med. Chem. Lett.* **15,** 2041–2045.
11. Hotha, S., Yarrow, J. C., Yang, J. G., et al. (2003) A potent small-molecule probe for the dynamics of cell division. *Angew. Chem. Int. Ed. Engl.* **42,** 2379–2385.
12. Sakowicz, R., Finer, J. T., Beraud, C., et al. (2004) Antitumor activity of a kinesin inhibitor. *Cancer Res.* **64,** 3276–3280.
13. Sakowicz, R., Berdelis, M. S., Ray, K., et al. (1998) A marine natural product inhibitor of kinesin motors. *Science* **280,** 292–295.
14. Hopkins, S. C., Vale, R. D., and Kuntz, I. D. (2000) Inhibitors of kinesin activity from structure-based computer screening. *Biochemistry* **39,** 2805–2814.
15. Yen, T. J., Li, G., Schaar, B. T., Szilak, I., and Cleveland, D.W. (1992) CENP-E is a putative kinetochore motor that accumulates just before mitosis. *Nature* **359,** 536–539.
16. Nislow, C., Lombillo,V. A., Kuriyama, R., and McIntosh, J. R. (1992) A plus-end-directed motor enzyme that moves antiparallel microtubules in vitro localizes to the interzone of mitotic spindles. *Nature* **359,** 543–547.
17. Hackney, D. D. and Jiang, W. (2001) Assays for kinesin microtubule-stimulated ATPase actvity. *Methods Mol. Biol.* **164,** 65–71.
18. Funk, C. J., Davis, A. S., Hopkins, J. A., and Middleton, K. M. (2004) Development of high-throughput screens for discovery of kinesin adenosine triphosphate modulators. *Anal. Biochem.* **329,** 68–76.

19. Andreassen, P. R., Skoufias D. A., and Margolis, R. L. (2004) Analysis of the spindle-assembly checkpoint in HeLa cells. *Methods Mol. Biol.* **281,** 213–225.

20. Webb, M. R. (1992) A continuous spectrophotometric assay for inorganic phosphate and for measuring phosphate release kinetics in biological systems. *Proc. Natl. Acad. Sci. USA* **89,** 4884–4887.

21. DeBonis, S., Simorre, J. P., Crevel, I., et al. (2003) Interaction of the mitotic inhibitor monastrol with human kinesin Eg5. *Biochemistry* **42,** 338–349.

22. Zhang, J. H., Chung, T. D. Y., and Oldenburgh, K. R. (1999). A simple statistical parameter for use in evaluation and validation of high throughput screening assays. *J. Biomol. Screening* **4,** 67–73.

# 15

## Clinical Pharmacology and Use of Microtubule-Targeting Agents in Cancer Therapy

### Amelia B. Zelnak

#### Summary

The microtubule-targeting agents have made significant contributions to cancer therapy over the past 50 years. The vinca alkaloids and taxanes have been used to treat a broad range of malignancies, including leukemias and lymphomas and many types of solid tumors. The taxanes have been frequently used in the treatment of advanced ovarian, breast, lung, head and neck, and prostate cancer, and they are increasingly being used in early stage disease. This chapter reviews the pharmacology, clinical indications, and toxicities associated with the vinca alkaloids and taxanes.

**Key Words:** Vincristine; vinblastine; vinorelbine; paclitaxel; docetaxel; pharmacology; toxicity.

## 1. Introduction: Vinca Alkaloids

The vinca alkaloids have been widely used as chemotherapeutic agents for the treatment of both childhood and adult malignancies. The vinca alkaloids were initially isolated from dried leaves of the periwinkle plant *Catharanthus rosea*, formerly called *Vinca rosea*. In the 17th century, periwinkle extracts were used for multiple medicinal purposes such as wound healing, hemorrhage, and toothaches. In 1957, Noble showed that treating rats with certain fractions of periwinkle extracts resulted in granulocytopenia and bone marrow suppression *(1)*. Further research demonstrated that vincristine had antileukemic activity, and the first successful clinical use of vincristine was published in 1962 *(2,3)*.

Since the 1960s, three vinca alkaloids, vincristine, vinblastine, and vinorelbine, have approval for use in the United States. Vinca alkaloids all share a similar structure consisting of a catharanine moiety linked to a vindoline ring (**Figs. 1–3**). Several compounds such as vindesine and vinorelbine are semisynthetic derivatives of the vinca alkaloids. The vinca alkaloids induce

From: *Methods in Molecular Medicine, Vol. 137: Microtubule Protocols*
Edited by: Jun Zhou © Humana Press Inc., Totowa, NJ

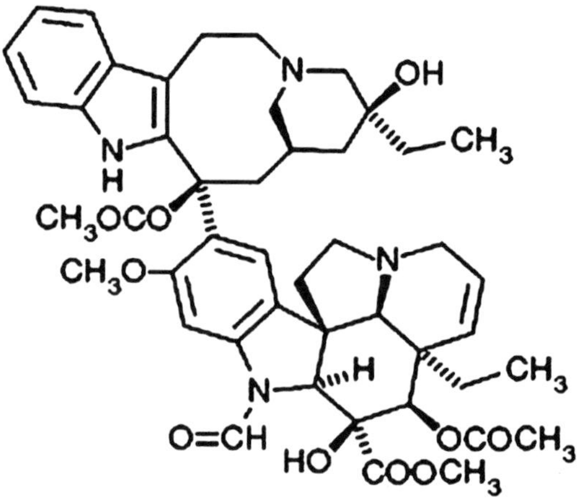

Fig. 1. Structure of vincristine.

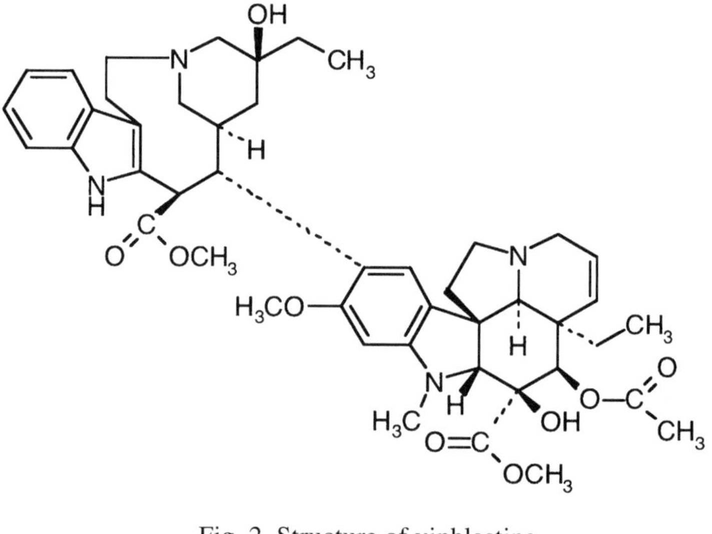

Fig. 2. Structure of vinblastine.

cytotoxicity by binding directly to tubulin and inhibiting microtubule poly-merization. This results in impaired microtubule dynamics and mitotic spindle formation, leading to cell death *(4–8)*. The vinca alkaloids are currently used in a broad spectrum of malignancies, including both solid tumors and hemato-logical malignancies.

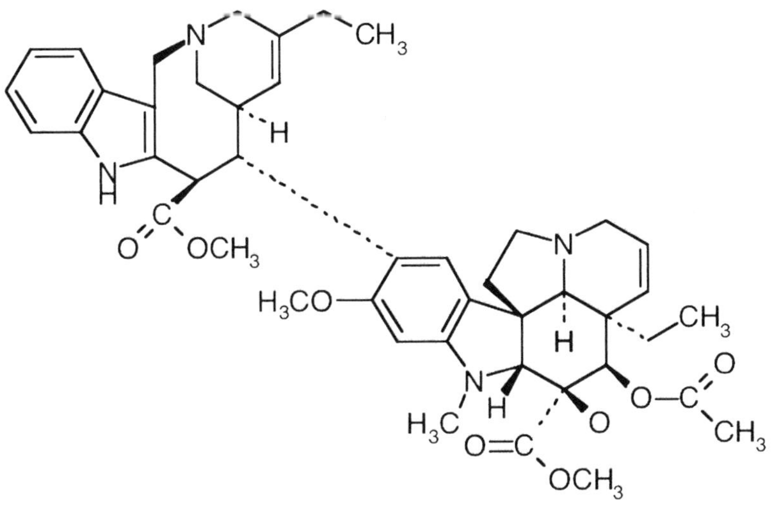

Fig. 3. Structure of vinorelbine.

## 2. Clinical Pharmacology

### 2.1. Vincristine

Pharmakokinetic studies of vincristine are limited by a lack of sensitive assays for measuring plasma concentrations of vincristine. Plasma clearance is rapid owing to extensive tissue binding and large volume of distribution. Vincristine is thought to have tri-exponential pharmacokinetics with rapid distribution following bolus injection, β phase distribution of 50–155 min, and an elimination half-life of approx 85 h *(9–13)*. Children appear to have higher plasma clearance than adults. Vincristine is metabolized in the liver by cytochrome P450 3A, and concomitantly administered drugs may either competitively inhibit or induce cytochrome P450 3A clearance of vincristine *(14,15)*. Vincristine accumulates in many tissues such as the lung, liver, kidney, bone marrow, intestinal mucosa, pancreas, and spleen. It is largely excluded from the brain, eye, and adipose tissue *(16)*. Vincristine is excreted as either unchanged drug or as a metabolite in the bile and feces *(17)*. Vincristine has little renal excretion. The relationship between the plasma pharmacokinetics and antitumor effects of vincristine is not fully defined.

Vincristine is typically administered as a bolus intravenous infusion at 1.4 mg/m$^2$ in adults (maximum dose of 2 mg), and at 1.5–2.0 mg/m$^2$ in children (maximum dose of 2.0–2.5 mg). Vincristine may be given up to once weekly, however, the dosing schedule varies based upon the malignancy, response, and other concomitantly administered drugs. Neurotoxicity is the dose-limiting toxicity, and attempts to use continuous infusion instead of bolus injection to reduce neurotoxicity have been inconclusive *(3,18,19)*. The dose of vincristine is reduced in the setting of severe hepatic dysfunction or severe neurotoxicity. Typically patients will receive 50% of the planned dose for moderate hyperbilirubinemia, and 25% for severe hyperbilirubinemia *(20)*.

## 2.2. Vinblastine

Pharmacokinetic studies of patients treated with a bolus injection of vinblastine are consistent with a triexponental pharmacokinetic model, similar to vincristine. Vinblastine is rapidly distributed from the plasma to tissues, primarily to the lung, liver, spleen, and kidneys. The elimination half-life of vinblastine is approx 29 h, with very little drug remaining in the body at 48 h *(21,22)*. Similar to the other vinca alkaloids, vinblastine is metabolized by the hepatic cytochrome P450 3A enzyme. This pathway may be impaired in patients with hepatic dysfunction and may be affected by other medications, which either induce or inhibit the activity of cytochrome P450 3A. Vinblastine is largely excreted in the bile and feces, with little renal excretion *(22)*. The dose of vinblastine in children and adults is typically 6 mg/m$^2$, with modifications for hepatic dysfunction and hematologic tolerance.

## 2.3. Vinorelbine

The pharmacokinetics of vinorelbine are well-defined, showing rapid plasma clearance, an elimination half-life of 40 h, and a large volume of distribution. Vinorelbine binds avidly to platelets and plasma proteins with less than 2% remaining as free drug within 2 h of administration. Vinorelbine diffuses freely into tissues, and is highly concentrated in the liver, spleen, kidney, lung, muscle, and heart *(23)*. The cytochrome P450 pathway in the liver is the main site of vinorelbine metabolism. This leads to potential interactions with other medications that are also metabolized by cytochrome P450. The majority of vinorelbine is excreted in the feces, with only 11% of the drug eliminated by the kidneys *(24,25)*.

Vinorelbine is generally administered as an intravenous infusion into a running saline solution over 10 min, followed by additional saline flush to minimize the risk of vessel irritation. Initial phase I studies of vinorelbine in France show that the dose-limiting toxicity at 27.5 mg/m$^2$ is neutropenia, with grade 3 neutropenia seen in 14% of cycles given *(26)*. Additional dose-finding studies, beginning with a dose of 30 mg/m$^2$ and escalating in 5 mg/m$^2$ increments, demonstrate that the maximum tolerated dose of vinorelbine is 45 mg/m$^2$ *(27)*. Multiple clinical trials show excellent efficacy using a dose of 25–30 mg/m$^2$ on days 1 and 8 of a 21-d cycle. Although there are no data on patients with severe hepatic dysfunction, the importance of hepatic metabolism of vinorelbine suggests that doses should be modified in these patients based upon the degree of bilirubin elevation *(24)*.

## 3. Clinical Indications

### 3.1. Vincristine

#### 3.1.1. Pediatric Malignancies

Vincristine was first used to treat childhood malignancies in the 1960s in patients who had become refractory to conventional therapy. Response rates to single agent vincristine were high, especially among hematological malignancies such as acute lymphoblastic leukemia (ALL) and Hodgkin's disease *(28)*. However, responses were often partial and brief, leading to the development of combination chemotherapy *(29)*. Vincristine has remained a key drug in combination chemotherapy for ALL *(30)* and non-Hodgkin's lymphoma *(31)*. Nearly all pediatric and adult patients with Hodgkin's disease have been treated with a vinca alkaloid, either vincristine or vinblastine *(32)*. Vincristine has been routinely used as part of combination therapy for Wilm's tumor *(33)*. Vincristine has also been used in the treatment of pediatric sarcomas, including rhabdomyosarcoma, neuroblastoma, and Ewing sarcoma *(34–36)*.

#### 3.1.2. Adult Malignancies

First used as a single agent in adult malignancies in the 1960s, vincristine is now part of virtually all protocols for frontline treatment of adult ALL *(29)*. However, the response rates are typically not as high as in the pediatric population. Combination chemotherapy including vincristine such as CHOP (cyclophosphamide, doxorubicin, vincristine, and prednisone) remains the standard of care for patients with aggressive non-Hodgkin's lymphoma *(37)*. Multimodality treatment of Hodgkin's disease also incorporates a vinca alkaloid, either vincristine or vinblastine *(38)*. Vincristine (Oncovin) in combination with nitrogen mustard, procarbazine, and prednisone (MOPP) induces complete remissions in 80% of patients with advanced Hodgkin's disease *(39)*. More than 50% of these patients are still alive at a 20-yr follow-up report *(40)*. Vincristine is also incorporated into several other combinations such as the Stanford V regimen consisting of doxorubicin, vinblastine, mechlorethamine, vincrisitne, bleomycin, etoposide, and prednisone *(41)*, and HyperCVAD (cyclophosphamide, mesna, vincristine, doxorubicin, and dexamethasone) *(42)*. Vincristine is commonly used in combination therapy for newly diagnosed and refractory multiple myeloma as part of the VAD regimen (vincristine, doxorubicin, and dexamethasone) *(43)*. Vincristine is a key component of therapy for AIDS-associated Kaposi's sarcoma *(44,45)*, but is not commonly used in the treatment of other adult solid tumors.

## 3.2. Vinblastine

Vinblastine was shown to be highly active as a single-agent in treatment of advanced Hodgkin's disease, with complete and overall response rates of 33 and 65% of patients, respectively *(46)*. The responses were often short-lived, leading to the development of combination chemotherapy with vinblastine. Modified MOPP regimens often incorporated vinblastine rather than vincristine *(29)*. Several regimens were developed in the attempt to improve upon outcomes with MOPP chemotherapy. ABVD, consisting of doxorubicin, bleomycin, vinblastine, and dacarbazine, has become one of the most frequently used combinations for treatment of Hodgkin's disease *(47,48)*. Vinblastine has also been incorporated into the Stanford V regimen outlined above for Hodgkin's disease *(41)*. Studies have shown that vinblastine has activity as a single-agent in non-Hodgkin's lymphoma, however, vincristine has more commonly been used in the first-line setting *(49,50)*. Vinblastine has been used as salvage treatment in vincristine-refractory patients where there has been evidence of a lack of cross-resistance *(51)*.

Treatment of transitional cell carcinomas of the genitourinary tract has incorporated vinblastine in combination chemotherapy. The M-VAC regimen combining methotrexate, vinblastine, doxorubicin, and cyclophosphamide induced complete responses in 37% of patients with advanced disease *(52,53)*. A recent clinical trial demonstrated that neoadjuvant M-VAC improved overall survival among patients with locally advanced bladder cancer *(54)*. Like vincristine, vinblastine has also been used in treating AIDS-associated Kaposi's sarcoma (KS). Vinblastine has been administered intravenously for KS as well as intralesionally *(55,56)*. Vinblastine has clinical activity is several other solid tumors such as breast cancer, germ-cell tumors, and nonsmall cell lung cancer, however, has not commonly been used in the treatment of these diseases.

## 3.3. Vinorelbine

Numerous clinical trials have been published using vinorelbine as a single-agent or in combination with other chemotherapeutic drugs for the treatment of nonsmall cell lung cancer (NSCLC). When used as a single-agent, the overall response rate in these trials has been approx 25% *(57)*. Vinorelbine has generally been well tolerated, making it an attractive treatment option for patients who were not candidates for polychemotherapy. A phase III trial of elderly patients with advanced NSCLC demonstrated that vinorelbine, when compared to best supportive care, improved median survival (21 vs 28 wk) *(58)*. Vinorelbine has also been used in combination chemotherapy, usually with cisplatin, for treatment of NSCLC. In a SWOG phase III trial, 415 patients were randomized to either cisplatin alone or cisplatin in combination with vinorelbine. The combination arm resulted in statistically significant improvements in response rate (12 vs 26%) and 1-yr survival (20 vs 36%) *(59)*. Until recently, the role of combination chemotherapy in patients with completely resected NSCLC had been controversial. In the International Adjuvant Lung Cancer Trial (IALT), 1867 patients with completely resected NSCLC were randomized to either observation or cisplatin-based chemotherapy. Patients received cisplatin in combination with etoposide (56%), vinorelbine (27%), vinblastine (11%), or vindesine (6%). Patients in the treatment arm had significantly higher overall survival at 5 yr (44.5 vs 40.4%) *(60)*. Based upon the results of IALT and other trials, adjuvant combination chemotherapy, often incorporating vinorelbine, has now become the standard of care for patients with well-staged and completely resected NSCLC.

Vinorelbine has been frequently used in the treatment of breast cancer. In a large phase II clinical trial on single-agent vinorelbine, 145 women with metastatic breast cancer received vinorelbine 30 mg/m$^2$ weekly as first-line treatment. The overall response rate was 41%, equivalent to other drugs commonly used as front-line therapy *(61)*. A recent phase II trial of trastuzumab and vinorelbine for HER2 overexpressing metastatic breast cancer demonstrated a 68% overall response rate *(62)*. Vinorelbine has also been studied in combination with other chemotherapeutic drugs such as the anthracyclines and taxanes with excellent overall response rates *(57)*. The role of vinorelbine in treatment of metastatic breast cancer has been well established. Additional clinical trials have been designed to further define the role for vinorelbine in the neoadjuvant and adjuvant treatment of invasive breast cancer.

## 4. Side Effects

### 4.1. Vincristine

Neurotoxicity is the dose-limiting side effect of vincristine. Vincristine-induced neuropathy is a cumulative toxicity; however, some symptoms develop within the first few weeks of treatment. Initial neurotoxic signs and symptoms include symmetrical sensory impairment and parasthesias. Patients may later have loss of deep tendon reflexes, develop gross motor abnormalities such as foot or wrist drop, or have a decrease in fine motor skills such as writing. Autonomic polyneuropathy is seen in some patients, manifested by constipation, paralytic ileus, bladder dysfunction, and impotence. Many of the symptoms resolve within weeks to months of discontinuation of therapy, however residual neurotoxicity has been documented *(3,18,19)*. The severity of neurotoxicity may be influenced by the dosage and frequency of administration. Doses of vincristine are typically capped at 2 or 2.5 mg because of concerns that autonomic neurotoxicity is more affected by the size of a single dose rather than cumulative dose *(19,63,64)*. Vincristine-induced neurotoxicity is greatest in infants and the elderly and may be related to dose calculations. A correlation between neurotoxicity and obstructive liver disease has also been shown, likely a result of impaired biliary excretion of vincristine *(64)*. Concomitant administration of radiation therapy or chemotherapeutic agents such as L-aspariginase may also worsen the neurotoxicity associated with vincristine *(65)*. Of note, severe central nervous system toxicity has been reported in patients who were given high doses or who have a disrupted blood brain barrier. Intrathecal administration of vincristine is almost always fatal, and therefore this must be carefully avoided *(66)*.

In addition to neurotoxicity, patients will experience mild neutropenia and anemia following administration of vincristine. This is readily reversible and does not usually result in treatment delay *(67)*. The development of alopecia and rash following vincristine is variable depending upon the dose and duration of treatment. Common gastrointestinal side effects include constipation, abdominal cramping, nausea and vomiting *(29)*. Patients may also complain of urinary symptoms secondary to polyuria, dysuria, or bladder rentention *(68)*.

### 4.2. Vinblastine

Myelosuppresion is the dose-limiting toxicity of vinblastine. Neutropenia is the most common manifestation of myelosuppresion, with anemia and thrombocytopenia being less frequent. Neutropenia occurs approx 4–10 d following administration of the drug, and counts usually recover within 7–21 d of administration *(29)*. Neurotoxicity is less common with vinblastine than with vincristine, and usually occurs after prolonged administration or in combination regimens *(69,70)*. Patients may complain of gastrointestinal side effects such as mucositis and stomatitis; nausea and vomiting may occur but are less common. Mild alopecia is seen and is reversible. There are case reports of acute hypertension and pulmonary edema, but these are infrequent side effects of vinblastine *(29)*.

### 4.3. Vinorelbine

Vinorelbine is generally a well-tolerated chemotherapeutic agent. Neutropenia is the dose-limiting toxicity associated with vinorelbine. Grade 3–4 neutropenia has been observed in 14–52% of patients treated with vinorelbine, with up to 70% of patients requiring dose adjustments *(71)*. Vinorelbine, similar to the other vinca alkaloids, has been associated with neurotoxicity. Up to 30% of patients receiving vinorelbine have developed peripheral neuropathy; however, only 1% of patients developed severe neuropathy grade III or above *(57)*. Gastrointestinal side effects such as nausea, vomiting, and constipation, have been reported with vinorelbine; however, severe effects are relatively infrequent. There have been several case reports of adverse cardiac events, including severe cardiac events and deaths, in the literature. A recent meta-analysis of 19 trials involving nearly 2500 patients treated with vinorelbine did not find a statistically significant difference in event rate between vinorelbine and other chemotherapeutic agents. The authors concluded that vinorelbine did not present a higher cardiac risk than other chemotherapeutic agents for similar indications *(72)*. Vinorelbine has been associated with pain at the injection site or thrombophlebitis, which can be alleviated by flushing the vein with normal saline following injection *(73)*. Alopecia has been reported in approx 10% of patients treated with vinorelbine, and has typically been mild.

## 1. Introduction: Taxanes

The taxanes have been widely used in the treatment of early stage and advanced malignancies, including ovarian, lung, breast, head and neck, bladder, and esophageal cancers. Paclitaxel was discovered by the National Cancer Institute in the late 1960s during a large scale screening program of natural products for anticancer activity *(74)*. Paclitaxel was identified as the active constituent of crude extracts derived from the bark of the pacific yew tree, *Taxus brevifolia* (**Fig. 4**). The development of paclitaxel was hampered by its scarcity and its lack of aqueous solubility. An alternate semisynthetic process using 10-deacetylbacatin III, a precursor derived from a more abundant yew species, has currently met the commercial demands for paclitaxel. Docetaxel has also been semi-synthetically derived from this precursor *(75)* (**Fig. 5**).

The mechanism of action of paclitaxel was first elucidated in 1979 when it was found to bind preferentially to polymerized microtubules rather than tubulin dimers *(76)*. The paclitaxel binding site on microtubules was distinct from those of GTP, colchicine, and vinblastine *(77)*. Additionally, upon binding, paclitaxel stabilized microtubules and shifted the dynamic equilibrium toward polymerization *(78)*. Taxane binding to microtubules inhibited cell proliferation by inducing a metaphase arrest. This disruption of the mitotic spindle led to apoptotic cell death; however, the exact mechanism has not been fully understood *(79)*. Resistance to taxanes has been attributed to the amplification of membrane phosphoglycoproteins that act as drug efflux pumps *(80)*. Taxane resistance has also been attributed to differential expression of tubulin subtypes *(81,82)*. Taxane resistance has been reported *in vitro* after the development of point mutations in the paclitaxel-binding site *(83)*.

## 2. Clinical Pharmacology

### 2.1. Paclitaxel

Pharmacokinetic studies of paclitaxel were initially slowed by its aqueous insolubility. However, highly sensitive and specific HPLC assays were devel-

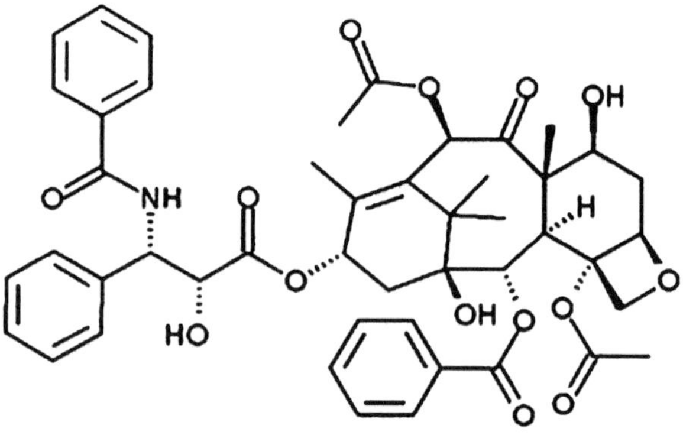

Fig. 4. Structure of paclitaxel.

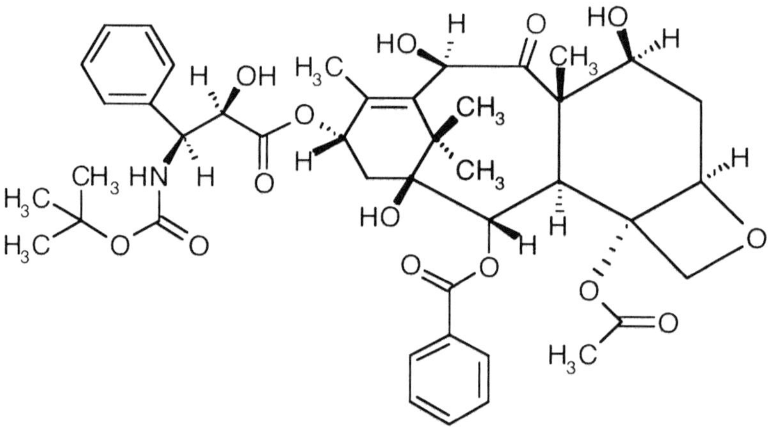

Fig. 5. Structure of docetaxel.

oped during early clinical trials to monitor plasma concentrations of paclitaxel. Paclitaxel has extensive binding to plasma proteins with a large steady-state volume of distribution. Paclitaxel has been shown to have biexponential plasma elimination, with a beta phase distribution between 1.3 and 8.6 h *(84–88)*. Initial pharmacokinetic studies of paclitaxel were performed on prolonged continuous infusion schedules. Additional studies of shorter, more frequently used infusion schedules demonstrated that paclitaxel has nonlinear systemic clearance. Renal clearance of paclitaxel was minimal, based upon insignificant levels of total urinary excretion. Paclitaxel has been shown to be metabolized by the hepatic cytochrome P450 system and excreted in the bile and feces as either unchanged drug or metabolite *(89)*. The hepatic metabolism of paclitaxel has been further supported by studies in rats demonstrating the presence of paclitaxel and its metabolites in rat bile *(90)*.

Initial phase I clinical studies of paclitaxel indicate that a 24-h continuous infusion is associated with a lower frequency of acute hypersensitivity reactions compared to shorter infusions. The prophylactic administration of dexamethasone, diphenhydramine, and an H2-antagonist 30 min prior to treatment significantly lowers the risk of hypersensitivity reactions *(74,91)*. The dose of paclitaxel typically administered is 175 mg/m² over 3 h every 3 wk or 80 mg/m² weekly. The dose of paclitaxel is reduced in patients with hepatic dysfunction. Patients with moderate to severe hyperbilirubinemia or elevation in transaminases typically have their dose reduced by at least 50% *(89)*.

### *2.2. Docetaxel*

Studies of docetaxel demonstrate linear pharmacokinetic behavior. Like paclitaxel, docetaxel is highly bound to plasma proteins and has triexponential plasma elimination. Docetaxel is metabolized by the hepatic cytochrome P450 system and is primarily excreted in the bile and feces *(92)*. Renal excretion accounts for less than 10% of docetaxel clearance. Dose reductions are recommended for patients with hepatic dysfunction or hyperbilirubinemia. Docetaxel is typically administered every 21 d at a dose of 60–100 mg/m² as a 1-h i.v. infusion. Patients typically receive dexamethasone, diphenhydramine, and an H2-antagonist as premedication to prevent hypersensitivity reactions *(93,94)*.

## 3. Clinical Indications

### *3.1. Ovarian Cancer*

Paclitaxel was initially approved for clinical use for the treatment of epilethial ovarian cancer based upon response rates in women with recurrent or refractory disease. Trials using paclitaxel as a single-agent given as a 24-h infusion demonstrated overall response rates (complete response + partial response, greater than 50% reduction in disease) ranging between 20 and 48% *(95–97)*. Many of these patients were refractory to platinum analogs, the standard chemotherapy at the time. The combination of cisplatin and paclitaxel was then studied and shown to be safe *(98)*. A randomized clinical trial was then performed to compare the combination of paclitaxel and cisplatin to the standard regimen of cisplatin and cyclophosphamide in previously untreated patients with Stage III or IV ovarian cancer *(99)*. Clinical response rates were 73 and 60%, respectively ($p = 0.01$), favoring the paclitaxel-containing arm. Progression-free survival also favored the group treated with paclitaxel, 18 vs 13 mo, and most importantly, overall survival was significantly improved with paclitaxel, 38 vs 24 mo ($p < 0.001$). Based upon results from this trial, the combination of cisplatin and paclitaxel became the new standard of care for treatment of advanced ovarian cancer.

Additional clinical trials have been performed to determine the optimal dose and schedule of administration of paclitaxel. A phase III study randomized patients to treatment with two different doses of paclitaxel (135 vs 175 mg/m$^2$) and two different infusion schedules (24 vs 3 h) *(100)*. The high-dose group had significantly longer progression-free survivals, 19 vs 14 wk ($p = 0.02$), however, there was no significant difference in overall survival between the groups. Paclitaxel was nitially approved at the dose of 135 mg/m$^2$ given as a 24-h infusion, however, this trial led to the subsequent approval of the 175 mg/m$^2$ dose given over 3 h. Three large phase III clinical trials have compared the combination of carboplatin and paclitaxel with cisplatin and paclitaxel. These trials demonstrated that the carboplatin-containing regimen was not inferior to the cisplatin arm, and it was associated with significantly less toxicity *(101–103)*.

Docetaxel has also been studied in the treatment of advanced ovarian cancer. Similar to paclitaxel, docetaxel has been active in platinum-refractory disease. Several phase II studies demonstrated overall response rates between 25 and 41% *(104–107)*. A large randomized phase III clinical trial comparing the combination of carboplatin with either paclitaxel or docetaxel, demonstrated similar progression-free survival and overall survival rates for the two treatment arms *(108)*. Docetaxel has also been studied as single-agent therapy for patients with recurrent disease, demonstrating response rates of 22% and median overall survival of 13.7 mo *(109)*.

### 3.2. Breast Cancer

Paclitaxel and docetaxel have both been widely used in the treatment of early stage and advanced breast cancer. Paclitaxel was initially studied in patients with metastatic breast cancer who had received one prior chemotherapy regimen. Patients were given paclitaxel at 250 mg/m$^2$ as a 24-h infusion, demonstrating an overall response rate of 56% and median progression-free survival of 9 mo *(110)*. Excellent response rates have also been seen with lower dose paclitaxel (175 mg/m$^2$) given over 3 h to metastatic breast cancer patients who were previously treated with an anthracycline *(111)*. Paclitaxel has also been studied in the first-line treatment of patients with metastatic breast cancer, demonstrating overall response rate or 29%, median time to progression of 5.3 mo and overall survival of 17.3 mo *(112)*. Docetaxel has been studied in the treatment of metastatic breast cancer, with response rates ranging from 52 to 68%, median time to progression of 4.9 mo, and overall survival of 16.4 mo *(113–116)*. Although the response rate is higher for first-line docetaxel compared to paclitaxel, the overall survival is similar between the two drugs.

Paclitaxel given weekly was recently shown to be superior to paclitaxel given every 3 wk to patients with metastatic breast cancer. Median time to progression was 9 vs 5 mo, and overall survival was 24 vs 16 mo, favoring weekly administration of paclitaxel *(117)*.

The taxanes have also been studied in the treatment of early stage breast cancer based upon their excellent response rates in the metastatic setting. In a large phase III trial, patients with lymph-node positive, early stage breast cancer were randomized to receive either four cycles of adriamycin plus cytoxan (AC) alone or four cycles of AC followed by four cycles of paclitaxel. Paclitaxel was given at a dose of 175 mg/m$^2$ over 3 h every 3 wk. At 5 yr, the disease free survival was 65 and 70%, and overall survival was 77 and 80% after AC alone or AC followed by paclitaxel *(118)*. Docetaxel has also been studied in the adjuvant treatment of node-positive breast cancer. In a large randomized phase III trial, patient either received docetaxel in combination with adriamycin and cyclophosphamide (TAC) or fluorouracil plus adriamycin and cyclophosphamide (FAC). At a median follow-up of 55 mo, estimated 5 yr disease-free survival rates were 75 and 68%, and estimated overall survival rates were 87 and 81%, respectively *(119)*. Based upon these and other clinical trials, the addition of a taxane to anthracyline-containing regimens has become standard of care of patients with node-positive, early stage breast cancer.

### 3.3. Lung Cancer

Paclitaxel and docetaxel have both been used in the treatment of lung cancer. In an early phase II study in previously untreated Stage IIIB and IV nonsmall cell lung cancer (NSCLC), paclitaxel (200 mg/m$^2$) was shown to have a response rate of 24% and a median survival of 40 wk *(120)*. Phase II studies of docetaxel (100 mg/m$^2$) demonstrated response rates of 31% in previously untreated patients and 19% in pretreated patients *(121–124)*. Based upon impressive single-agent activity in NSCLC, combinations of paclitaxel and docetaxel with platinum compounds have been investigated. A large phase III trial comparing carboplatin plus paclitaxel to cisplatin plus vinorelbine demonstrated that response rates and median survival were similar, however the carboplatin plus paclitaxel arm was less toxic and better tolerated. Objective response rates were 25 and 28%, respectively, and median survival was 8 mo in both arms with 1-yr survival rates of 36 and 38%, respectively *(125)*. Docetaxel has been studied as second-line therapy for advanced NSCLC. In a randomized trial comparing docetaxel (75 or 100 mg/m$^2$) to best supportive care, docetaxel improved both time to progression (10.6 vs 6.7 wk, $p < 0.001$) and median survival (7.0 vs 4.6 mo, $p = 0.047$) *(126)*.

Since paclitaxel has been shown to enhance the effects of ionizing radiation, paclitaxel has been used with carboplatin in the treatment of locally advanced

NSCLC in combination with radiation therapy. In a phase II study of unresectable NSCLC patients, weekly carboplatin and paclitaxel in combination with radiation therapy followed by two cycles of carboplatin and paclitaxel given every 3 wk resulted in a median survival of 16.3 mo *(127)*. The combination of paclitaxel with platinum compounds in the adjuvant setting has also resulted in improved disease-free and overall survival *(60)*. The CALGB 9633 trial randomized patients with completely resected Stage IB NSCLC to either four cycles of paclitaxel (200 mg/m$^2$) and carboplatin (AUC 6) every 3 wk or to observation. A preliminary report of a planned interim analysis demonstrated significantly higher 4-yr overall survival rates for the treatment arm compared to observation (71 vs 59%) *(128)*. The taxanes have also shown activity small cell lung cancer and have been commonly used as second-line therapy with response rates ranging from 23 to 29% and median overall survival ranging between 3.3 and 9 mo *(129,130)*.

### 3.4. Prostate Cancer

Two large phase III clinical trials investigating the role of docetaxel in the treatment of hormone-refractory prostate cancer (HRPC) were recently reported. Mitoxantrone plus prednisone had been the previous standard of care for HRPC, based upon decreased pain and improved quality of life. However, mitoxantrone had not been shown to have a benefit on overall survival. In the trial comparing docetaxel plus prednisone to mitoxantrone plus prednisone, overall survival was 18.9 vs 16.5 mo, respectively *(131)*. Patients also reported improved quality of life in the docetaxel-containing arm. The other trial compared docetaxel plus estramustine to mitoxantrone plus prednisone, again demonstrating an overall survival benefit of 2 mo *(132)*. Docetaxel has become the standard of care for first-line treatment of hormone-refractory prostate cancer based upon the positive results from these trials.

### 3.5. Other Cancers

The taxanes are widely used in the treatment of many additional types of cancer. Paclitaxel shows a 26% response rate in a phase II clinical trial of patients with cisplatin-resistance germ cell tumors *(133)*. Combination regimens containing paclitaxel, such as ifosfamide, cisplatin, and paclitaxel are commonly used as second-line chemotherapy for testicular cancer with a complete response rate of 70% and a 2-yr progression-free survival rate of 65% *(134)*. Paclitaxel is also beneficial in the treatment of transitional cell carcinoma of the bladder. A phase II clinical trial of paclitaxel for previously untreated metastatic TCC bladder patients demonstrates a 42% response rate and median survival of 8.4 mo *(135)*. Combination regimens including paclitaxel have higher response rates and median survival; however, the regimens are not as well tolerated owing to increased toxicities. Paclitaxel is often used in the treatment of stage III or IV endometrial cancer. Patients with advanced or recurrent endometrial cancer have a 56–67% response rate to paclitaxel in combination with platinum analogs *(136,137)*. Paclitaxel is also used in the treatment of Kaposi's sarcoma, with response rates of 59% and a median duration of response of 10.4 mo *(138)*. The combination of gemcitabine and docetaxel is also used as second-line treatment for uterine leiomyosarcomas with an overall response rate of 53% *(139)*. Paclitaxel in combination with platinum analogs is frequently used in the treatment of esophageal cancer in the preoperative *(140)* and metastatic setting. A phase II trial of carboplatin plus paclitaxel in patients with advanced esophageal cancer, demonstrates a response rate 43% with a median survival of 9 mo *(141)*. Docetaxel and paclitaxel in combination with platinum analogs are commonly used in the treatment of metastatic or recurrent head and neck cancer with response rates ranging from 38 to 57% *(142,143)*. The taxanes have one of the broadest antitumor activities of all the cytotoxic chemotherapeutic agents. Further investigation into combinations with other cytotoxic drugs or targeted therapies will continue to improve overall survival in advanced cancers and impact long-term survivorship in early stage disease.

## 4. Side Effects

### 4.1. Paclitaxel

Myelosuppression, specifically neutropenia, is the principal dose-limiting toxicity of paclitaxel *(144)*. The nadir neutrophil count usually occurs between days 8 and 11, followed by a rapid recovery. Because the duration of neutropenia is short, neutropenic fever and sepsis remain infrequent. Hypersensitivity reactions manifested as hypotension, bronchospasm, and urticaria, were documented in the initial clinical studies of paclitaxel. The reactions typically occur within the first 2–3 min of the first treatment and resolve after the infusion was discontinued *(145)*. Paclitaxel's Cremophor vehicle is thought to induce histamine release from basophils and mast cells, causing the hypersensitivity reactions. Patients currently receive corticosteroids and histamine H1- and H2-antagonists as prophylaxis against hypersensitivity reactions. Premedication with these agents reduces the rate of reactions to approx 1% *(100,144)*. Paclitaxel may also induce a peripheral neuropathy, typically manifested as numbness and paresthesia in a stocking-glove distribution. Symptoms may be seen in the days immediately following administration; however, they usually occur after a patient receives multiple cycles of treatment with paclitaxel. Patients with diabetes mellitus or preexisting peripheral neuropathies are at higher risk of developing neurotoxicity. The neuropathy usually improves after paclitaxel is discontinued *(144,146)*. Paclitaxel is associated with bradyarrhythmias in clinical trials. The significance of these effects is unknown, and routine cardiac monitoring during paclitaxel treatment is not currently recommended *(144,147)*. Patients may also describe transient flu-like symptoms such as myalgias and arthralgias 2–5 d following administration. Paclitaxel is associated with reversible alopecia, however, does not frequently induce nausea and vomiting *(130)*.

## 4.2. Docetaxel

Neutropenia is the dose-limiting toxicity of docetaxel. The initial phase II clinical trials used a dose of 100 mg/m$^2$, however recent trials administer a dose of 75 mg/m$^2$ to reduce the incidence of neutropenia and infectious complications *(93,94)*. Hypersensitivity reactions similar to those caused by paclitaxel, can also occur following docetaxel administration. They typically occur during the first two treatments and resolve with discontinuation of the infusion. Premedication with corticosteroids and H1- and H2-antagonists significantly reduces the incidence of hypersensitivity reactions *(93,94)*. Fluid retention is a side effect unique to docetaxel, likely secondary to increased capillary permeability. This is commonly manifested as pleural effusions, ascites, and peripheral edema. Premedication with corticosteroids reduces the severity of fluid retention. Docetaxel may induce a peripheral neuropathy similar to paclitaxel, with symptoms occurring after receiving multiple cycles of therapy. Additional toxicities include alopecia, rash, and nail changes. Asthenia is often reported by patients receiving docetaxel, and appears to be more common in women than in men. Patients may experience nausea, vomiting, and diarrhea, however, severe gastrointestinal toxicity is uncommon *(93,94,130)*.

## 5. Future Directions

The vinca alkaloids and taxanes are used in the treatment of a broad range of malignancies, and have contributed significantly to prolongation of survival in advanced disease and overall survival in early stage disease. The success of these agents is leading to further research into the potential role of microtubule-targeting agents in the treatment of cancer. Abraxane, a new nanoparticle form of paclitaxel, has significantly higher response rates and longer time to progression than conventional paclitaxel in the treatment of metastatic breast cancer patients, leading to the recent FDA-approval of this drug. Clinical trials are underway to further define the role of abraxane in early stage breast cancer. New microtubule-stabilizing agents such as the epotheliones, laulimalide analogues, and discodermolide are also currently in clinical development. As new drugs are discovered and additional clinical trials are completed, the impact of antimicrotubule agents on the treatment of cancer will continue to expand.

# References

1. Noble, R., Beer, C. T., and Cutts, J. H. (1958) Role of change observations in chemotherapy: *Vinca rosea. Ann. N.Y. Acad. Sci.* **76,** 882–894.
2. Karon, M., Freireich, E. J., and Frei, E. I. (1962) A preliminary report on vincristine sulfate; a new active agent for the treatment of acute leukemia. *Pediatrics* **30,** 791–802.
3. Costa, G., Hreshchyshyn, M. M., and Holland, J. F. (1962) Initial clinical studies with vincristine. *Cancer Chemother. Rep.* **24,** 39–44.
4. Donoso, J. A., Haskins, K. M., and Himes, R. H. (1979) Effect of microtubule-associated proteins on the interaction of vincristine with microtubules and tubulin. *Cancer Res.* **39,** 1604–1610.
5. George, P., Journey, L. J., and Goldstein, M. N. (1965) Effect of vincristine on the fine structure of HeLa cells during mitosis. *J. Natl. Cancer Inst.* **35,** 355–375.
6. Himes, R. H., Kersey, R. N., Heller-Bettinger, I., and Samson, F. E. (1976) Action of the vinca alkaloids vincristine, vinblastine, and desacetyl vinblastine amide on microtubules in vitro. *Cancer Res.* **36,** 3798–3802.
7. Jordan, M. A., Thrower, D., and Wilson, L. (1991) Mechanism of inhibition of cell proliferation by Vinca alkaloids. *Cancer Res.* **51,** 2212–2222.
8. Owellen, R. J., Hartke, C. A., Dickerson, R. M., and Hains, F. O. (1976) Inhibition of tubulin-microtubule polymerization by drugs of the Vinca alkaloid class. *Cancer Res.* **36,** 1499–1502.
9. Rahmani, R. and Zhou, X. J. (1993) Pharmacokinetics and metabolism of vinca alkaloids. *Cancer Surv.* **17,** 269–281.
10. Owellen, R. J., Root, M. A., and Hains, F. O. (1977) Pharmacokinetics of vindesine and vincristine in humans. *Cancer Res.* **37,** 2603–2607.
11. Sethi, V. S., Jackson, D. V., Jr., White, D. R., et al. (1981) Pharmacokinetics of vincristine sulfate in adult cancer patients. *Cancer Res.* **41,** 3551–3555.
12. Sethi, V. S. and Kimball, J. C. (1981) Pharmacokinetics of vincristine sulfate in children. *Cancer Chemother. Pharmacol.* **6,** 111–115.
13. Bender, R. A., Castle, M. C., Margileth, D. A., and Oliverio, V. T. (1977) The pharmacokinetics of [3H]-vincristine in man. *Clin. Pharmacol. Ther.* **22,** 430–435.
14. Zhou-Pan, X. R., Seree, E., Zhou, X. J., et al. (1993) Involvement of human liver cytochrome P450 3A in vinblastine metabolism: drug interactions. *Cancer Res.* **53,** 5121–5126.
15. Zhou, X. J., Zhou-Pan, X. R., Gauthier, T., Placidi, M., Maurel, P., and Rahmani, R. (1993) Human liver microsomal cytochrome P450 3A isozymes mediated vindesine biotransformation. Metabolic drug interactions. *Biochem. Pharmacol.* **45,** 853–861.
16. El Dareer, S. M., White, V. M., Chen, F. P., Mellet, L. B., and Hill, D. L. (1977) Distribution and metabolism of vincristine in mice, rats, dogs, and monkeys. *Cancer Treat. Rep.* **61,** 1269–1277.
17. Jackson, D. V., Jr., Castle, M. C., and Bender, R. A. (1978) Biliary excretion of vincristine. *Clin. Pharmacol. Ther.* **24,** 101–107.

18. Rosenthal, S. and Kaufman, S. (1974) Vincristine neurotoxicity. *Ann. Intern. Med.* **80,** 733–737.

19. Sandler, S. G., Tobin, W., and Henderson, E. S. (1969) Vincristine-induced neuropathy. A clinical study of fifty leukemic patients. *Neurology* **19,** 367–374.

20. Van den Berg, H. W., Desai, Z. R., Wilson, R., Kennedy, G., Bridges, J. M., and Shanks, R. G. (1982) The pharmacokinetics of vincristine in man: reduced drug clearance associated with raised serum alkaline phosphatase and dose-limited elimination. *Cancer Chemother. Pharmacol.* **8,** 215–219.

21. Owellen, R. J. and Hartke, C. A. (1975) The pharmacokinetics of 4-acetyl tritium vinblastine in two patients. *Cancer Res.* **35,** 975–980.

22. Owellen, R. J., Hartke, C. A., and Hains, F. O. (1977) Pharmacokinetics and metabolism of vinblastine in humans. *Cancer Res.* **37,** 2597–2602.

23. Marquet, P., Lachatre, G., Debord, J., Eichler, B., Bonnaud, F., and Nicot, G. (1992) Pharmacokinetics of vinorelbine in man. *Eur. J. Clin. Pharmacol.* **42,** 545–547.

24. Leveque, D., Jehl, F., Quoix, E., and Breillout, F. (1992) Clinical pharmacokinetics of vinorelbine alone and combined with cisplatin. *J. Clin. Pharmacol.* **32,** 1096–1098.

25. Bore, P., Rahmani, R., van Cantfort, J., Focan, C., and Cano, J. P. (1989) Pharmacokinetics of a new anticancer drug, navelbine, in patients. Comparative study of radioimmunologic and radioactive determination methods. *Cancer Chemother. Pharmacol.* **23,** 247–251.

26. Mathe, G. and Reizenstein, P. (1985) Phase I pharmacologic study of a new Vinca alkaloid: navelbine. *Cancer Lett.* **27,** 285–293.

27. Khayat, D., Rixe, O., Brunet, R., et al. (2004) Pharmacokinetic linearity of i.v. vinorelbine from an intra-patient dose escalation study design. *Cancer Chemother. Pharmacol.* **54,** 193–205.

28. Gidding, C. E., Kellie, S. J., Kamps, W. A., and de Graaf, S. S. (1999) Vincristine revisited. *Crit. Rev. Oncol. Hematol.* **29,** 267–287.

29. Rowinsky, E. K. and Donehower, R. C. (1991) The clinical pharmacology and use of antimicrotubule agents in cancer chemotherapeutics. *Pharmacol. Ther.* **52,** 35–84.

30. Lampert, F. and Henze, G. (1997) Acute lymphoblastic leukaemia, in *Pediatric Oncology: Clinical Practice and Controversies,* (Pinkerton, C. and Plowman, P. N., ed.), Chapman & Hall Medical, London, UK, pp. 258–278.

31. Patte, C. (1997) Childhood non-Hodgkin's lymphoma, in *Pediatric Oncology: Clinical Practice and Controversies,* (Pinkerton, C. and Plowman, P. N., ed.), Chapman & Hall Medical, London, UK, pp. 278–291.

32. Oberlin, O. and McDowell, H. P. (1997) Hodgkin's disease, in *Pediatric Oncology: Clinical Practice and Controversies,* (Pinkerton, C. and Plowman, P. N., ed.), Chapman & Hall Medical, London, UK, pp. 296–316.

33. D'Angio, G. J., Evans, A. E., Breslow, N., et al. (1976) The treatment of Wilms' tumor: Results of the national Wilms' tumor study. *Cancer* **38,** 633–646.

34. Ninane, J. and Pearson, A. D. J. (1997) Neuroblastoma, in *Pediatric Oncology: Clinical Practice and Controversies,* (Pinkerton, C. and Plowman, P. N., ed.), Chapman & Hall Medical, London, UK, pp. 443–446.

35. Wexler, L. and Helman, L. J. (1997) Rhabdomyosarcoma and the undifferentiated sarcomas, in *Principles and Practice of Pediatric Oncology,* (Pizzo, P. and Poplack, D. G., ed.), Lippincott-Raven, Philadelphia, PA, pp. 799–830.

36. Horowitz, M., Malawer, M. M., and Woo, S. Y. (1997) Hicks JM Ewing's sarcoma family of tumors: Ewing's neuroectodermal tumors, in *Principles and Practice of Pediatric Oncology,* (Pizzo, P. and Poplack, D. G., ed.), Lippincott-Raven, Philadelphia, PA, pp. 831–863.

37. Armitage, J., Mauch, P. M., Harris, N. L., and Bierman, P. (2001) Non-Hodgkin's lymphomas, in *Cancer: Principles and Practice of Oncology,* (DeVita, Jr, V. T., Hellman, S., and Rosenberg, S. A., ed.), Lippincott Williams & Wilkins, Philadelphia, PA, pp. 2256–2315.

38. Diehl, V., Mauch, P. M., and Harris, N. L. (2001) Hodgkin's Disease, in *Cancer: Principles and Practice of Oncology,* (DeVita, Jr, V. T., Hellman, S., and Rosenberg, S. A., ed.), Lippincott Williams & Wilkins, Philadelphia, PA, pp. 2339–2388.

39. DeVita, V., Serpick, A. A., and Carbone, P. P. (1970) Combination chemotherapy in the treatment of advanced Hodgkin's disease. *Ann. Intern. Med.* **73,** 891–895.

40. Longo, D. L., Young, R. C., Wesley, M., et al. (1986) Twenty years of MOPP therapy for Hodgkin's disease. *J. Clin. Oncol.* **4,** 1295–1306.

41. Horning, S. J., Hoppe, R. T., Breslin, S., Bartlett, N. L., Brown, B. W., and Rosenberg, S. A. (2002) Stanford V and radiotherapy for locally extensive and advanced Hodgkin's disease: mature results of a prospective clinical trial. *J. Clin. Oncol.* **20,** 630–637.

42. Kantarjian, H. M., O'Brien, S., Smith, T. L., et al. (2000) Results of treatment with hyper-CVAD, a dose-intensive regimen, in adult acute lymphocytic leukemia. *J. Clin. Oncol.* **18,** 547–561.

43. Samson, D., Gaminara, E., Newland, A., et al. (1989) Infusion of vincristine and doxorubicin with oral dexamethasone as first-line therapy for multiple myeloma. *Lancet* **2,** 882–885.

44. Kaplan, L., Abrams, D., and Volberding, P. (1986) Treatment of Kaposi's sarcoma in acquired immunodeficiency syndrome with an alternating vincristine-vinblastine regimen. *Cancer Treat. Rep.* **70,** 1121–1122.

45. Gill, P., Rarick, M., Bernstein-Singer, M., et al. (1990) Treatment of advanced Kaposi's sarcoma using a combination of bleomycin and vincristine. *Am. J. Clin. Oncol.* **13,** 315–319.

46. Livingston, R. and Carter, S. K. (1970) *Single Agents in Cancer Chemotherapy,* IFI/Plenum, New York.

47. Santoro, A. and Bonadonna, G. (1979) Prolonged disease-free survival in MOPP-resistant Hodgkin's disease after treatment with adriamycin, bleomycin, vinblastine and dacarbazine (ABVD). *Cancer Chemother. Pharmacol.* **2,** 101–105.

48. Papa, G., Mandelli, F., Anselmo, A. P., et al. (1982) Treatment of MOPP-resistant Hodgkin's disease with adriamycin, bleomycin, vinblastine and dacarbazine (ABVD). *Eur. J. Cancer Clin. Oncol.* **18,** 803–806.

49. Carbone, P. P., Spurr, C., Schneiderman, M., Scotto, J., Holland, J. F., and Shnider, B. (1968) Management of patients with malignant lymphoma: a comparative study with cyclophosphamide and vinca alkaloids. *Cancer Res.* **28**, 811–822.

50. Stutzman, L., Ezdinli, E. Z., and Stutzman, M. A. (1966) Vinblastine sulfate vs cyclophosphamide in the therapy for lymphoma. *JAMA* **195**, 173–178.

51. Jackson, D. V., Jr., Spurr, C. L., Caponera, M. E., et al. (1987) Vinblastine infusion in non-Hodgkin's lymphomas: lack of total cross-resistance with vincristine. *Cancer Invest.* **5**, 535–539.

52. Sternberg, C. N., Yagoda, A., Scher, H. I., et al. (1989) Methotrexate, vinblastine, doxorubicin, and cisplatin for advanced transitional cell carcinoma of the urothelium. Efficacy and patterns of response and relapse. *Cancer* **64**, 2448–2458.

53. Sternberg, C. N., Yagoda, A., Scher, H. I., et al. (1988) M-VAC (methotrexate, vinblastine, doxorubicin and cisplatin) for advanced transitional cell carcinoma of the urothelium. *J. Urol.* **139**, 461–469.

54. Grossman, H. B., Natale, R. B., Tangen, C. M., et al. (2003) Neoadjuvant chemotherapy plus cystectomy compared with cystectomy alone for locally advanced bladder cancer. *N. Engl. J. Med.* **349**, 859–866.

55. Epstein, J. B. and Scully, C. (1989) Intralesional vinblastine for oral Kaposi sarcoma in HIV infection. *Lancet* **2**, 1100–1101.

56. Gill, P. S., Akil, B., Colletti, P., et al. (1989) Pulmonary Kaposi's sarcoma: clinical findings and results of therapy. *Am. J. Med.* **87**, 57–61.

57. Gregory, R. K. and Smith, I. E. (2000) Vinorelbine: a clinical review. *Br. J. Cancer* **82**, 1907–1913.

58. The Elderly Lung Cancer Vinorelbine Italian Study Group. (1999) Effects of vinorelbine on quality of life and survival of elderly patients with advanced non-small-cell lung cancer. *J. Natl. Cancer Inst.* **91**, 66–72.

59. Wozniak, A. J., Crowley, J. J., Balcerzak, S. P., et al. (1998) Randomized trial comparing cisplatin with cisplatin plus vinorelbine in the treatment of advanced non-small-cell lung cancer: a Southwest Oncology Group study. *J. Clin. Oncol.* **16**, 2459–2465.

60. Arriagada, R., Bergman, B., Dunant, A., Le Chevalier, T., Pignon, J. P., and Vansteenkiste, J. (2004) Cisplatin-based adjuvant chemotherapy in patients with completely resected non-small-cell lung cancer. *N. Engl. J. Med.* **350**, 351–360.

61. Fumoleau, P., Delgado, F. M., Delozier, T., et al. (1993) Phase II trial of weekly intravenous vinorelbine in first-line advanced breast cancer chemotherapy. *J. Clin. Oncol.* **11**, 1245–1252.

62. Burstein, H. J., Harris, L. N., Marcom, P. K., et al. (2003) Trastuzumab and vinorelbine as first-line therapy for HER2-overexpressing metastatic breast cancer: multicenter phase II trial with clinical outcomes, analysis of serum tumor markers as predictive factors, and cardiac surveillance algorithm. *J. Clin. Oncol.* **21**, 2889–2895.

63. Casey, E. B., Jellife, A. M., Le Quesne, P. M., and Millett, Y. L. (1973) Vincristine neuropathy. Clinical and electrophysiological observations. *Brain* **96**, 69–86.

64. Bradley, W. G., Lassman, L. P., Pearce, G. W., and Walton, J. N. (1970) The neuromyopathy of vincristine in man. Clinical, electrophysiological and pathological studies. *J. Neurol. Sci.* **10**, 107–131.

65. Weiss, H. D., Walker, M. D., and Wiernik, P. H. (1974) Neurotoxicity of commonly used antineoplastic agents (first of two parts). *N. Engl. J. Med.* **291**, 75–81.

66. Slyter, H., Liwnicz, B., Herrick, M. K., and Mason, R. (1980) Fatal myeloencephalopathy caused by intrathecal vincristine. *Neurology* **30**, 867–871.

67. Steurer, G., Kuzmits, R., Pavelka, M., Sinzinger, H., Fritz, E., and Ludwig, H. (1989) Early-onset thrombocytopenia during combination chemotherapy in testicular cancer is induced by vinblastine. *Cancer* **63**, 51–58.

68. Gottlieb, R. J. and Cuttner, J. (1971) Vincristine-induced bladder atony. *Cancer* **28**, 674–675.

69. Hansen, S. W. (1990) Autonomic neuropathy after treatment with cisplatin, vinblastine, and bleomycin for germ cell cancer. *BMJ* **300**, 511–512.

70. Bostrom, B. (1988) Severe ileus from cisplatin and vinblastine infusion in neuroblastoma. *J. Clin. Oncol.* **6**, 1356.

71. Cvitkovic, E. and Izzo, J. (1992) The current and future place of vinorelbine in cancer therapy. *Drugs* **44**, 36–45.

72. Lapeyre-Mestre, M., Gregoire, N., Bugat, R., and Montastruc, J. L. (2004) Vinorelbine-related cardiac events: a meta-analysis of randomized clinical trials. *Fundam Clin. Pharmacol.* **18**, 97–105.

73. Besenval, M., Delgado, M., Demarez, J. P., and Krikorian, A. (1989) Safety and tolerance of Navelbine in phase I-II clinical studies. *Semin. Oncol.* **16**, 37–40.

74. Rowinsky, E. K., Onetto, N., Canetta, R. M., and Arbuck, S. G. (1992) Taxol: the first of the taxanes, an important new class of antitumor agents. *Semin. Oncol.* **19**, 646–662.

75. Bissery, M. C., Nohynek, G., Sanderink, G. J., and Lavelle, F. (1995) Docetaxel (Taxotere): a review of preclinical and clinical experience. Part I: Preclinical experience. *Anticancer Drugs* **6**, 339–368.

76. Schiff, P. B., Fant, J., and Horwitz, S. B. (1979) Promotion of microtubule assembly in vitro by taxol. *Nature* **277**, 665–667.

77. Kumar, N. (1981) Taxol-induced polymerization of purified tubulin. Mechanism of action. *J. Biol. Chem.* **256**, 10,435–10,441.

78. Manfredi, J. J. and Horwitz, S. B. (1984) Taxol: an antimitotic agent with a new mechanism of action. *Pharmacol. Ther.* **25**, 83–125.

79. Jordan, M. A., Wendell, K., Gardiner, S., Derry, W. B., Copp, H., and Wilson, L. (1996) Mitotic block induced in HeLa cells by low concentrations of paclitaxel (Taxol) results in abnormal mitotic exit and apoptotic cell death. *Cancer Res.* **56**, 816–825.

80. Horwitz, S. B., Cohen, D., Rao, S., Ringel, I., Shen, H. J., and Yang, C. P. (1993) Taxol: mechanisms of action and resistance. *J. Natl. Cancer Inst. Monogr.* 55–61.

81. Dumontet, C., Duran, G. E., Steger, K. A., Beketic-Oreskovic, L., and Sikic, B. I. (1996) Resistance mechanisms in human sarcoma mutants derived by single-step exposure to paclitaxel (Taxol). *Cancer Res.* **56**, 1091–1097.

82. Rowinsky, E. K., Donehower, R. C., Jones, R. J., and Tucker, R. W. (1988) Microtubule changes and cytotoxicity in leukemic cell lines treated with taxol. *Cancer Res.* **48,** 4093–4100.

83. Giannakakou, P., Sackett, D. L., Kang, Y. K., et al. (1997) Paclitaxel-resistant human ovarian cancer cells have mutant beta-tubulins that exhibit impaired paclitaxel-driven polymerization. *J. Biol. Chem.* **272,** 17,118–17,125.

84. Rowinsky, E. K., Burke, P. J., Karp, J. E., Tucker, R. W., Ettinger, D. S., and Donehower, R. C. (1989) Phase I and pharmacodynamic study of taxol in refractory acute leukemias. *Cancer Res.* **49,** 4640–4647.

85. Wiernik, P. H., Schwartz, E. L., Strauman, J. J., Dutcher, J. P., Lipton, R. B., and Paietta, E. (1987) Phase I clinical and pharmacokinetic study of taxol. *Cancer Res.* **47,** 2486–2493.

86. Wiernik, P. H., Schwartz, E. L., Einzig, A., Strauman, J. J., Lipton, R. B., and Dutcher, J. P. (1987) Phase I trial of taxol given as a 24-hour infusion every 21 days: responses observed in metastatic melanoma. *J. Clin. Oncol.* **5,** 1232–1239.

87. Grem, J. L., Tutsch, K. D., Simon, K. J., et al. (1987) Phase I study of taxol administered as a short i.v. infusion daily for 5 days. *Cancer Treat. Rep.* **71,** 1179–1184.

88. Donehower, R. C., Rowinsky, E. K., Grochow, L. B., Longnecker, S. M., and Ettinger, D. S. (1987) Phase I trial of taxol in patients with advanced cancer. *Cancer Treat. Rep.* **71,** 1171–1177.

89. Rowinsky, E. K. (1995) Phamacology and metabolism, in *Paclitaxel in Cancer Treatment,* (McGuire, W. and Rowinsky, E. K., ed.), Dekker, New York, pp. 91–120.

90. Monsarrat, B., Mariel, E., Cros, S., et al. (1990) Taxol metabolism. Isolation and identification of three major metabolites of taxol in rat bile. *Drug Metab. Dispos.* **18,** 895–901.

91. Rowinsky, E. K. and Donehower, R. C. (1995) Paclitaxel (taxol). *N. Engl. J. Med.* **332,** 1004–1014.

92. Marre, F., Sanderink, G. J., de Sousa, G., Gaillard, C., Martinet, M., and Rahmani, R. (1996) Hepatic biotransformation of docetaxel (Taxotere) in vitro: involvement of the CYP3A subfamily in humans. *Cancer Res.* **56,** 1296–1302.

93. Cortes, J. E. and Pazdur, R. (1995) Docetaxel. *J. Clin. Oncol.* **13,** 2643–2655.

94. van Oosterom, A. T., Schrijvers, D., and Schriivers, D. (1995) Docetaxel (Taxotere), a review of preclinical and clinical experience. Part II: Clinical experience. *Anticancer Drugs* **6,** 356–368.

95. Kohn, E. C., Sarosy, G., Bicher, A., et al. (1994) Dose-intense taxol: high response rate in patients with platinum-resistant recurrent ovarian cancer. *J. Natl. Cancer Inst.* **86,** 18–24.

96. Einzig, A. I., Wiernik, P. H., Sasloff, J., Runowicz, C. D., and Goldberg, G. L. (1992) Phase II study and long-term follow-up of patients treated with taxol for advanced ovarian adenocarcinoma. *J. Clin. Oncol.* **10,** 1748–1753.

97. Thigpen, J. T., Blessing, J. A., Ball, H., Hummel, S. J., and Barrett, R. J. (1994) Phase II trial of paclitaxel in patients with progressive ovarian carcinoma after platinum-based chemotherapy: a Gynecologic Oncology Group study. *J. Clin. Oncol.* **12,** 1748–1753.

98. Rowinsky, E. K., Gilbert, M. R., McGuire, W. P., et al. (1991) Sequences of taxol and cisplatin: a phase I and pharmacologic study. *J. Clin. Oncol.* **9,** 1692–1703.
99. McGuire, W. P., Hoskins, W. J., Brady, M. F., et al. (1996) Cyclophosphamide and cisplatin compared with paclitaxel and cisplatin in patients with stage III and stage IV ovarian cancer. *N. Engl. J. Med.* **334,** 1–6.
100. Eisenhauer, E. A., ten Bokkel Huinink, W. W., Swenerton, K. D., et al. (1994) European-Canadian randomized trial of paclitaxel in relapsed ovarian cancer: high-dose versus low-dose and long versus short infusion. *J. Clin. Oncol.* **12,** 2654–2666.
101. Ozols, R. F., Bundy, B. N., Greer, B. E., et al. (2003) Phase III trial of carboplatin and paclitaxel compared with cisplatin and paclitaxel in patients with optimally resected stage III ovarian cancer: a Gynecologic Oncology Group study. *J. Clin. Oncol.* **21,** 3194–3200.
102. Neijt, J. P., Engelholm, S. A., Tuxen, M. K., et al. (2000) Exploratory phase III study of paclitaxel and cisplatin versus paclitaxel and carboplatin in advanced ovarian cancer. *J. Clin. Oncol.* **18,** 3084–3092.
103. du Bois, A., Luck, H. J., Meier, W., et al. (2003) A randomized clinical trial of cisplatin/paclitaxel versus carboplatin/paclitaxel as first-line treatment of ovarian cancer. *J. Natl. Cancer Inst.* **95,** 1320–1329.
104. Francis, P., Schneider, J., Hann, L., et al. (1994) Phase II trial of docetaxel in patients with platinum-refractory advanced ovarian cancer. *J. Clin. Oncol.* **12,** 2301–2308.
105. Kaye, S. B., Piccart, M., Aapro, M., and Kavanagh, J. (1995) Docetaxel in advanced ovarian cancer: preliminary results from three phase II trials. EORTC Early Clinical Trials Group and Clinical Screening Group, and the MD Anderson Cancer Center. *Eur. J. Cancer* **31A,** S14–S17.
106. Kavanagh, J. J., Kudelka, A. P., de Leon, C. G., et al. (1996) Phase II study of docetaxel in patients with epithelial ovarian carcinoma refractory to platinum. *Clin. Cancer Res.* **2,** 837–842.
107. Piccart, M. J., Gore, M., Ten Bokkel Huinink, W., et al. (1995) Docetaxel: an active new drug for treatment of advanced epithelial ovarian cancer. *J. Natl. Cancer Inst.* **87,** 676–681.
108. Vasey, P. A. (2003) Role of docetaxel in the treatment of newly diagnosed advanced ovarian cancer. *J. Clin. Oncol.* **21,** 136–144.
109. Niwa, Y., Nakanishi, T., Kuzuya, K., Nawa, A., and Mizutani, S. (2003) Salvage treatment with docetaxel for recurrent epithelial ovarian cancer. *Int. J. Clin. Oncol.* **8,** 343–347.
110. Holmes, F. A., Walters, R. S., Theriault, R. L., et al. (1991) Phase II trial of taxol, an active drug in the treatment of metastatic breast cancer. *J. Natl. Cancer Inst.* **83,** 1797–1805.
111. Gianni, L., Munzone, E., Capri, G., et al. (1995) Paclitaxel in metastatic breast cancer: a trial of two doses by a 3-hour infusion in patients with disease recurrence after prior therapy with anthracyclines. *J. Natl. Cancer Inst.* **87,** 1169–1175.

112. Bishop, J. F., Dewar, J., Toner, G. C., et al. (1999) Initial paclitaxel improves outcome compared with CMFP combination chemotherapy as front-line therapy in untreated metastatic breast cancer. *J. Clin. Oncol.* **17,** 2355–2364.

113. Chevallier, B., Fumoleau, P., Kerbrat, P., et al. (1995) Docetaxel is a major cytotoxic drug for the treatment of advanced breast cancer: a phase II trial of the Clinical Screening Cooperative Group of the European Organization for Research and Treatment of Cancer. *J. Clin. Oncol.* **13,** 314–322.

114. Dieras, V., Chevallier, B., Kerbrat, P., et al. (1996) A multicentre phase II study of docetaxel 75 mg m-2 as first-line chemotherapy for patients with advanced breast cancer: report of the Clinical Screening Group of the EORTC. European Organization for Research and Treatment of Cancer. *Br. J. Cancer* **74,** 650–656.

115. Hudis, C. A., Seidman, A. D., Crown, J. P., et al. (1996) Phase II and pharmacologic study of docetaxel as initial chemotherapy for metastatic breast cancer. *J. Clin. Oncol.* **14,** 58–65.

116. Trudeau, M. E., Eisenhauer, E. A., Higgins, B. P., et al. (1996) Docetaxel in patients with metastatic breast cancer: a phase II study of the National Cancer Institute of Canada-Clinical Trials Group. *J. Clin. Oncol.* **14,** 422–428.

117. Seidman, A., Berry, D., Cirrincione, C., et al. (2004) *American Society of Clinical Oncology.* New Orleans, LA.

118. Henderson, I. C., Berry, D. A., Demetri, G. D., et al. (2003) Improved outcomes from adding sequential Paclitaxel but not from escalating Doxorubicin dose in an adjuvant chemotherapy regimen for patients with node-positive primary breast cancer. *J. Clin. Oncol.* **21,** 976–983.

119. Martin, M., Pienkowski, T., Mackey, J., et al. (2005) Adjuvant docetaxel for node-positive breast cancer. *N. Engl. J. Med.* **352,** 2302–2313.

120. Murphy, W. K., Fossella, F. V., Winn, R. J., et al. (1993) Phase II study of taxol in patients with untreated advanced non-small-cell lung cancer. *J. Natl. Cancer Inst.* **85,** 384–388.

121. Fossella, F. V., Lee, J. S., Shin, D. M., et al. (1995) Phase II study of docetaxel for advanced or metastatic platinum-refractory non-small-cell lung cancer. *J. Clin. Oncol.* **13,** 645–651.

122. Fossella, F. V., Lee, J. S., Murphy, W. K., et al. (1994) Phase II study of docetaxel for recurrent or metastatic non-small-cell lung cancer. *J. Clin. Oncol.* **12,** 1238–1244.

123. Francis, P. A., Rigas, J. R., Kris, M. G., et al. (1994) Phase II trial of docetaxel in patients with stage III and IV non-small-cell lung cancer. *J. Clin. Oncol.* **12,** 1232–1237.

124. Cerny, T., Kaplan, S., Pavlidis, N., et al. (1994) Docetaxel (Taxotere) is active in non-small-cell lung cancer: a phase II trial of the EORTC Early Clinical Trials Group (ECTG). *Br. J. Cancer* **70,** 384–387.

125. Kelly, K., Crowley, J., Bunn, P. A., Jr., et al. (2001) Randomized phase III trial of paclitaxel plus carboplatin versus vinorelbine plus cisplatin in the treatment of patients with advanced non-small-cell lung cancer: a Southwest Oncology Group trial. *J. Clin. Oncol.* **19,** 3210–3218.

126. Shepherd, F. A., Dancey, J., Ramlau, R., et al. (2000) Prospective randomized trial of docetaxel versus best supportive care in patients with non-small-cell lung cancer previously treated with platinum-based chemotherapy. *J. Clin. Oncol.* **18,** 2095–2103.

127. Belani, C. P., Choy, H., Bonomi, P., et al. (2005) Combined chemoradiotherapy regimens of paclitaxel and carboplatin for locally advanced non-small-cell lung cancer: a randomized phase II locally advanced multi-modality protocol. *J. Clin. Oncol.* **23,** 5883–5891.

128. Strauss, G., Herndon, J., Maddaus, M. A., et al. (2004) Randomized clinical trial of adjuvant chemotherapy with paclitaxel and carboplatin following resection in stage IB non-small cell lung cancer (NSCLC): report of Cancer and Leukemia Group B (CALGB) protocol 9633 (abstract). *Proc. Am. Soc. Clin. Oncol.* **23,** 621s.

129. Smyth, J. F., Smith, I. E., Sessa, C., et al. (1994) Activity of docetaxel (Taxotere) in small cell lung cancer. The Early Clinical Trials Group of the EORTC. *Eur. J. Cancer* **30A,** 1058–1060.

130. Rowinsky, E. K. (1997) The development and clinical utility of the taxane class of antimicrotubule chemotherapy agents. *Annu. Rev. Med.* **48,** 353–374.

131. Tannock, I. F., de Wit, R., Berry, W. R., et al. (2004) Docetaxel plus prednisone or mitoxantrone plus prednisone for advanced prostate cancer. *N. Engl. J. Med.* **351,** 1502–1512.

132. Petrylak, D. P., Tangen, C. M., Hussain, M. H., et al. (2004) Docetaxel and estramustine compared with mitoxantrone and prednisone for advanced refractory prostate cancer. *N. Engl. J. Med.* **351,** 1513–1520.

133. Motzer, R. J., Bajorin, D. F., Schwartz, L. H., et al. (1994) Phase II trial of paclitaxel shows antitumor activity in patients with previously treated germ cell tumors. *J. Clin. Oncol.* **12,** 2277–2283.

134. Kondagunta, G. V., Bacik, J., Donadio, A., et al. (2005) Combination of paclitaxel, ifosfamide, and cisplatin is an effective second-line therapy for patients with relapsed testicular germ cell tumors. *J. Clin. Oncol.* **23,** 6549–6555.

135. Roth, B. J., Dreicer, R., Einhorn, L. H., et al. (1994) Significant activity of paclitaxel in advanced transitional-cell carcinoma of the urothelium: a phase II trial of the Eastern Cooperative Oncology Group. *J. Clin. Oncol.* **12,** 2264–2270.

136. Dimopoulos, M. A., Papadimitriou, C. A., Georgoulias, V., et al. (2000) Paclitaxel and cisplatin in advanced or recurrent carcinoma of the endometrium: long-term results of a phase II multicenter study. *Gynecol. Oncol.* **78,** 52–57.

137. Hoskins, P. J., Swenerton, K. D., Pike, J. A., et al. (2001) Paclitaxel and carboplatin, alone or with irradiation, in advanced or recurrent endometrial cancer: a phase II study. *J. Clin. Oncol.* **19,** 4048–4053.

138. Gill, P. S., Tulpule, A., Espina, B. M., et al. (1999) Paclitaxel is safe and effective in the treatment of advanced AIDS-related Kaposi's sarcoma. *J. Clin. Oncol.* **17,** 1876–1883.

139. Hensley, M. L., Maki, R., Venkatraman, E., et al. (2002) Gemcitabine and docetaxel in patients with unresectable leiomyosarcoma: results of a phase II trial. *J. Clin. Oncol.* **20,** 2824–2831.

140. Urba, S. G., Orringer, M. B., Ianettonni, M., Hayman, J. A., and Satoru, H. (2003) Concurrent cisplatin, paclitaxel, and radiotherapy as preoperative treatment for patients with locoregional esophageal carcinoma. *Cancer* **98,** 2177–2183.
141. El-Rayes, B. F., Shields, A., Zalupski, M., et al. (2004) A phase II study of carboplatin and paclitaxel in esophageal cancer. *Ann. Oncol.* **15,** 960–965.
142. Clark, J. I., Hofmeister, C., Choudhury, A., et al. (2001) Phase II evaluation of paclitaxel in combination with carboplatin in advanced head and neck carcinoma. *Cancer* **92,** 2334–2340.
143. Fountzilas, G., Skarlos, D., Athanassiades, A., et al. (1997) Paclitaxel by three-hour infusion and carboplatin in advanced carcinoma of nasopharynx and other sites of the head and neck. A phase II study conducted by the Hellenic Cooperative Oncology Group. *Ann. Oncol.* **8,** 451–455.
144. Rowinsky, E. K., Eisenhauer, E. A., Chaudhry, V., Arbuck, S. G., and Donehower, R. C. (1993) Clinical toxicities encountered with paclitaxel (Taxol). *Semin. Oncol.* **20,** 1–15.
145. Weiss, R. B., Donehower, R. C., Wiernik, P. H., et al. (1990) Hypersensitivity reactions from taxol. *J. Clin. Oncol.* **8,** 1263–1268.
146. Rowinsky, E. K., Chaudhry, V., Cornblath, D. R., and Donehower, R. C. (1993) Neurotoxicity of Taxol. *J. Natl. Cancer Inst. Monogr.* **15,** 107–115.
147. Rowinsky, E. K., McGuire, W. P., Guarnieri, T., Fisherman, J. S., Christian, M. C., and Donehower, R. C. (1991) Cardiac disturbances during the administration of taxol. *J. Clin. Oncol.* **9,** 1704–1712.

# 16

## Studying Drug–Tubulin Interactions by X-Ray Crystallography

**Audrey Dorleans, Marcel Knossow, and Benoît Gigant**

### Summary

Tubulin, the microtubule building-block, is the target of numerous small molecule compounds that interfere with microtubule dynamics. Several of these ligands are in clinical use as antitumor drugs. There have been numerous studies on these molecules, with two main objectives: to determine their mechanism of action and to find new compounds that would expand the arsenal available for cancer chemotherapy. Although these studies would undoubtedly benefit from structural data on tubulin, this protein has long resisted crystallization attempts. We have used stathmin-like domains (SLDs) of stathmin family proteins as a tool to crystallize tubulin and have obtained three-dimensional crystals of the tubulin:SLD complexes. As many tubulin ligands bind to these complexes, the crystals are valuable tools to study tubulin–drug interactions by X-ray crystallography. They open the way to a structure-based drug design approach.

**Key Words:** Crystallization; colchicine; drug design; microtubule; stathmin; stathmin-like domain; structure; vinca domain.

## 1. Introduction

Structural approaches, and X-ray crystallography in particular, are methods of choice to study the interaction of small molecule ligands with their target proteins. However, in the case of tubulin, these approaches have long been hampered by the failure to obtain well diffracting crystals. This probably arises in part because of the conformational flexibility of this protein and of its instability in solution when it is not embedded in microtubules (*1*). The first structure of tubulin that has been determined was by electron microscopy of two-dimensional (2D) crystals of zinc-induced tubulin sheets (*2*). In this assembly, tubulin's structure is expected to be close to the one it adopts in microtubules, an assumption that has been supported by the observation of molecules that bind to microtubules and favor their polymerization also bind to

From: *Methods in Molecular Medicine, Vol. 137: Microtubule Protocols*
Edited by: Jun Zhou © Humana Press Inc., Totowa, NJ

tubulin sheets. Therefore the interaction with tubulin of ligands that bind to microtubules has been structurally characterized by electron microscopy *(2,3)*, but this method would not be suitable to study compounds that target nonmicrotubular tubulin. Two of the three main classes of tubulin-targeting compounds fall in this category, namely those of molecules that bind to the colchicine- and vinca-sites of tubulin.

In the course of our efforts to study the structure of tubulin by X-ray crystallography, we have developed a crystallization protocol based on the use of the stathmin-like domain (SLD) of the RB3 protein. SLDs form with tubulin a stable complex consisting of 2 tubulin molecules per one SLD (T2:SLD complex) *(4,5)*. The structure of tubulin in this complex reflects that of tubulin in curved assemblies *(6)* and also probably that of unpolymerized tubulin, as numerous small molecule compounds that inhibit microtubule assembly also bind to the T2:SLD complex *(7,8)*. These include ligands of the colchicine- and vinca-sites of tubulin. Here, we present the methods we developed to obtain three-dimensional (3D) crystals of T2:RB3-SLD (T2R) in complex with colchicine and with other colchicine- and vinca-site ligands.

## 2. Materials

All the solutions and buffers are water-based unless otherwise stated.

### 2.1. Recombinant RB3-SLD Expression and Purification

1. BL21(DE3) bacteria transformed with the expression vector pET-3d (formerly pET-8c) containing the RB3-SLD cDNA *(5)*.
2. Luria-Bertani (LB) medium.
3. Isopropyl-β-D-thiogalactopyranoside (IPTG) (Sigma): 1 $M$ solution, sterile-filtered, aliquoted, and stored at −20°C until use.
4. Ampicillin sodium salt (Sigma): 100 mg/mL solution, sterile-filtered, aliquoted, and stored at −20°C.
5. Extraction buffer: 20 m$M$ Tris-HCl pH 8.0, 1 m$M$ ethylene glycol bis(2-aminoethyl ether)-N,N,N'N'-tetraacetic acid (EGTA), 2 m$M$ Tris(2-carboxyethyl)phosphine (TCEP), containing in addition the complete antiprotease cocktail (Roche). All along the purification, TCEP may be replaced by dithiothreitol (DTT).
6. Q sepharose FF anion exchange column (Amersham Biosciences).
7. Anion exchange chromatography buffers. Loading buffer: 20 m$M$ Tris-HCl, pH 8.2, 1 m$M$ EGTA, 2 m$M$ TCEP. Elution buffer: same as loading buffer, supplemented with 2 $M$ NaCl.
8. Superdex 75 26/60 gel filtration column (Amersham Biosciences).
9. Gel filtration buffer: 10 m$M$ HEPES adjusted to pH 7.2 with NaOH, 150 m$M$ NaCl, 2 m$M$ TCEP.

## 2.2. Tubulin Preparation, Complexation With Colchicine-Site Ligands, and Buffer Exchange

1. Guanosine 5'-triphosphate, sodium salt (GTP): 100 m$M$ mother solution, prepared fresh.
2. MgCl$_2$: concentrated (e.g., 1 $M$) stock solution. Store at room temperature.
3. Purified bovine brain tubulin free of microtubule-associated proteins (MAPs) (*see* **Note 1**), stored at –70°C or in liquid nitrogen (*see* **Note 2**), in a buffer consisting of 50 m$M$ Mes adjusted to pH 6.8 with KOH (Mes-K), 0.5 m$M$ EGTA, 0.25 m$M$ MgCl$_2$, 0.1 m$M$ GTP, 33% (v/v) glycerol. Among the protocols available for purification of mammalian brain tubulin, we indifferently use the "classic one" based on cycles of assembly and disassembly followed by chromatography on phosphocellulose (*see* for example **ref. 9**) or a more recently described method where MAPs are removed from tubulin through cycles of polymerization in a high molarity buffer *(10)*.
4. Depolymerization buffer: 80 m$M$ PIPES-K pH 6.8, 0.5 m$M$ EGTA, 2 m$M$ guanosine 5'-diphosphate (GDP), 1 m$M$ MgCl$_2$, 2 m$M$ CaCl$_2$.
5. Colchicine (Fluka) or other colchicine-site ligands: concentrated stock solution (e.g., 50 m$M$) (*see* **Note 3**).
6. Sephadex G25 chromatography column (desalting PD-10 column, Amersham Biosciences).
7. PD-10 buffer: 15 m$M$ PIPES-K pH 6.8, 0.3 m$M$ MgCl$_2$, 0.2 m$M$ EGTA, 0.1 m$M$ GDP. Equilibrate to 4°C before use.

## 2.3. Formation of the Complex With the Stathmin-Like Domain of RB3

1. Purified RB3-SLD protein.
2. Centrifugal Filtering Device Microcon YM-30 (Millipore).

## 2.4. Crystallization

1. Polyethylene glycol (PEG) 20,000: 25% (w/v) stock solution, stored at 4°C in the dark.
2. Crystallization buffer: 50 m$M$ PIPES-K pH 7.5, 5% (v/v) PEG 400, 10% (v/v) ethylene glycol, with added PEG 20,000, as detailed in **Subheading 3.4.**
3. Other crystallization materials: silicone grease, cover slides (either siliconized glass or plastic) (Hampton Research), crystallization plates (e.g., XRL Plate, Molecular Dimensions, Ltd.).

## 2.5. Crystal Recovery and Flash-Cooling

1. Recovery solution: 13% PEG 20,000, 50 m$M$ PIPES-K pH 7.5, 5% PEG 400, 0.1 m$M$ GDP, 0.2 m$M$ MgCl$_2$, 10% ethylene glycol, with added colchicine-site ligands (*see* **Note 11**).
2. Cryo-protecting buffer: same as the recovery solution except ethylene glycol concentration that is 35%.

3. Mounted cryo-loops (Hampton Research).
4. Dewars to handle liquid nitrogen and store crystals. For safety reasons, always wear protective glasses and gloves and work in a ventilated place when handling liquid nitrogen.

## 2.6. Crystal Crosslinking

1. Glutaraldehyde 25% (Sigma). Glutaraldehyde is toxic and, in particular, harmful to the eyes. Manipulate under a fume hood. It should be inactivated by concentrated sodium hydroxide before disposal *(11)*.
2. If dilution of glutaraldehyde is required, it should be done in HCl pH 3.0 *(11)*.

## 3. Methods

The diffracting power of T2R crystals depends on the ligand filling the colchicine-site of tubulin, colchicine giving the best results among the molecules tested so far. The main methods presented below aim at obtaining T2R crystals in complex with colchicine. For complexes with other colchicine-site ligands, variations of the methods are presented in the **Subheading 4.** Structural data on the interaction of tubulin with molecules that bind to the vinca-site are best obtained through crystal soaking experiments as detailed in **Subheading 3.6.**

## 3.1. Recombinant RB3-SLD Expression and Purification

This protocol is adapted from **ref. 5**.

1. All the cultures are done at 37°C. A 10-mL overnight preculture of transformed BL21(DE3) bacteria is used to inoculate 1 L of LB-medium containing 100 µg/mL ampicillin.
2. When the absorbance at 600 nm is in the 0.4 to 0.6 range, RB3-SLD expression is induced by the addition of 0.4 m$M$ IPTG.
3. Three hours later, the culture is cooled and centrifuged for 15 min at 5000$g$, 4°C. At this stage, the bacterial pellets may be stored at −20°C and it is often convenient to do so as freezing bacteria helps to break them.
4. Bacterial pellets are resuspended in the extraction buffer (10 mL per 1 g of bacteria) and sonicated on ice.
5. The lysate is centrifuged for 15 min at 10,000$g$, 4°C. Taking advantage of RB3-SLD stability, the supernatant is heated at 80°C for 5 min, centrifuged for 15 min at 10,000$g$ and then ultracentrifuged for 1 h at 100,000$g$, 4°C.
6. The supernatant is loaded on a Q Sepharose FF anion exchange column, which is then washed with the anion exchange chromatography-loading buffer. After elution with a linear salt gradient, the RB3-SLD-containing fractions, identified by sodium dodecyl sulfate-polyacrylamide gel electrophoresis (SDS-PAGE), are pooled and loaded on a Superdex 75 26/60 gel filtration column which is eluted with the gel filtration buffer. RB3-SLD-containing fractions are pooled, concentrated to 10 mg/mL (*see* **Note 4**) and stored at −70°C until use.

## 3.2. Tubulin Preparation, Complexation With Colchicine-Site Ligands, and Buffer Exchange

1. A cycle of microtubule assembly—disassembly is performed in order to recover fully functional tubulin. To a thawed aliquot of tubulin, add GTP (0.5 m$M$ final) and MgCl$_2$ (6 m$M$ final) (*see* **Note 5**).
2. Incubate at 37°C for 20 min in a water bath. Polymerized tubulin is pelleted for 15 min at 300,000$g$, 30°C.
3. The next steps are performed at 4°C or on ice. Microtubule pellets are resuspended with a potter in the depolymerization buffer containing 1 m$M$ colchicine (*see* **Note 6**). Be careful not to dilute tubulin too much. To this end, as a rule of thumb, the volume of the buffer to be added is a third of the initial volume of the tubulin solution.
4. The solution is left to depolymerize 20 min on ice, then centrifuged 10 min at 300,000$g$, 4°C.
5. The supernatant is loaded on a desalting PD-10 column, previously equilibrated with the PD-10 buffer (*see* **Note 6**). Tubulin-containing fractions are pooled and the tubulin concentration is determined by measuring the absorbance at 278 nm. One milligram per milliliter is taken to an A$_{278}$ equal to 1.23. For preparation of tubulin complexes with other colchicine-site ligands (*see* **Notes 3** and **7**).

## 3.3. Formation of the Tubulin: RB3-SLD Complex

1. The recycled tubulin is mixed with RB3-SLD at a final 1.3:2 RB3-SLD:tubulin ratio (*see* **Note 8**).
2. The complex is concentrated to 20 mg of tubulin per milliliter by ultrafiltration on a microcon 30 centrifugal filtering device and is either used immediately or stored at –70°C (*see* **Note 9**).

## 3.4. Crystallization

Crystals are obtained at 19°C by vapor diffusion. In this method, a droplet consisting of the macromolecule to crystallize and a crystallization buffer is equilibrated against a reservoir containing the crystallization buffer at a concentration higher than in the drop.

1. The rim of the crystallization plate wells is greased with silicone grease. The reservoir is filled with 1 mL of crystallization buffer. The PEG 20,000 concentration is adjusted to each batch of T2R complex and varies between 5 and 7%.
2. Equal volumes (typically 1 µL) of the protein solution and of the crystallization buffer are deposited on the cover slide so that a nearly hemispherical drop is formed. Do not mix the drop but, instead, let it equilibrate by diffusion. The cover slide is overturned with tweezers and placed on the greased rim. Gently press the cover slide to seal the well. The drop is then left to equilibrate with the reservoir buffer by vapor diffusion.
3. Crystals usually appear within 48 h and the maximum size is reached in 1 wk. It is best to leave the plates in a quiet place for 3 d before the first inspection under the microscope. For an alternative crystallization protocol (*see* **Note 10**).

4. The diffracting power of the crystals decreases with time. Therefore, they should be flash-cooled in liquid $N_2$ within 1 mo of their appearance and stored until data collection.

### 3.5. Crystals Recovery and Flash-Cooling

The following steps are taken to avoid damaging the crystals while flash-cooling them *(12)*.

1. Crystals are harvested in the recovery solution with added colchicine-site ligands (*see* **Note 11**). Drops commonly contain numerous entangled crystals. These can be individualized using a soft tool, e.g., a rabbit whisker.
2. The ethylene glycol concentration is gradually increased in 5% steps with the cryo-protecting solution to a final concentration of 25%, whereas the concentration of the other components is maintained constant (*see* **Note 12**). As an example, to crystals harvested in 100 μL of recovery solution the 35% ethylene glycol cryo-protecting buffer is added in three steps (25, 42, and 83 μL).
3. A selected crystal is fished with a nylon cryo-loop, snap-cooled in liquid nitrogen, and stored until data collection. Crystals obtained using this protocol diffract up to 3.5 Å resolution.

### 3.6. Crystal Soaking

This is the method we use to study the interaction with tubulin of compounds that bind to the vinca-site, as opposed to cocrystallization of T2R with colchicine-site ligands as previously described.

1. Crystals of T2R are prepared. For better diffracting crystals, tubulin is complexed with colchicine.
2. Crystals are transferred in a recovery solution to which the ligand to be soaked has been added. Typical soaking time is 24 h and ligand concentration is in the millimolar range (*see* **Note 13**). Crystals are then prepared for flash-cooling, as described in **Subheading 3.5.**

### 3.7. Crystal Crosslinking

Crystals of the T2R complex are fragile. For instance, they frequently crack when they are harvested, but fortunately often reanneal within seconds. If this fragility becomes a problem, for example for the soaking of some ligands, crystals may be strengthened by chemical crosslinking.

Overturn the cover slide on which the crystal-containing drop has been deposited. Put a droplet of glutaraldehyde solution beside this drop and turn the cover slide back over the well. Mild (e.g., crosslinking time of 30 min with a droplet of 1 μL of 25% glutaraldehyde) or stronger crosslinking conditions (1 h with 2 μL of 25% glutaraldehyde) work fine. In the last case, the crystals become hardly soluble in water.

## 4. Notes

1. Alternatively to bovine, other mammalian (e.g., sheep) brain tubulins may be used.
2. Storage for several months is possible but as time goes the quality of tubulin wanes. It is best used within 4–6 mo of its purification.
3. Some ligands are poorly water soluble. The concentration of organic solvents (e.g., dimethyl sulfoxide [DMSO]) should however be kept as low as possible, as they weaken the tubulin:RB3-SLD interaction and interfere with crystallization. It is best to keep DMSO concentration in the tubulin solution lower than 2%.
4. As RB3-SLD has no tryptophan and only one tyrosine, it does not significantly absorb light at 280 nm. The best way to accurately determine an RB3-SLD solution concentration is by amino acid analysis. Alternatively, using a solution of known concentration as a reference, an estimate of the concentration may be obtained by SDS-PAGE.
5. If the tubulin is not highly pure, in particular if it is contaminated by microtubule-associated proteins (MAPs), it may help to add Na-glutamate up to 1 $M$ to the tubulin solution in the polymerization step *(13)*. High molarity buffer is known to favor the dissociation of MAPs from microtubules *(10)*; furthermore, glutamate has a stabilizing effect on tubulin.
6. The binding of colchicine to tubulin is a slow process, which often requires an incubation step at high temperature (for an example, *see* **ref. 14**). We have found that incorporating colchicine in the depolymerization buffer leads to efficient binding even in the cold. Moreover, as colchicine dissociates very slowly tubulin *(15)*, the PD-10 buffer may be devoid of this ligand. To check that formation of the complex is complete, the colchicine concentration may be estimated by measuring absorbance at 351 nm ($E_{351} = 16218$). It is a good practice to do so.
7. For the preparation of tubulin complexes with colchicine-site molecules that do not share the colchicine features mentioned in **Note 6**, the composition of some buffers has to be modified. If the complex is formed right after microtubule disassembly, i.e., if the depolymerization buffer contains the ligand of interest, the latter should also be present in the PD-10 desalting buffer. Alternatively the colchicine-site molecule may be added at the desired concentration to the tubulin solution after the buffer exchange step.
8. It is best to avoid the presence of free tubulin in solution, as it will denature/precipitate quickly in the crystallization buffer. To this end, an excess of RB3-SLD is added. This excess, which is partly removed during the concentration step, does not hinder the crystallization of the complex.
9. Although it is always recommended to freeze proteins in aliquots which are used one at a time, the T2R complex will stand several freeze/thaw cycles without significant loss of crystallizability.
10. We also have had good success with crystallization in the cold room and using a streak seeding technique *(16)*. We use the same buffer as with 19°C crystallizations but adjust the PEG 20,000 concentration, usually between 4 and 5% (w/v).

Just after the drops have been deposited on the cover slides, they are streaked with a fibre (e.g., a rabbit whisker), which has been in contact with previously obtained crystals. We streak several drops sequentially in order to vary the number of crystal seeds deposited in each one.

11. In the particular case of colchicine which binds tightly to tubulin, as mentioned in **Note 6**, we recover the crystals in a solution which contains 10 to 20 μ*M* of colchicine; the PEG 20,000 concentration of the recovery solution may also be lowered down to 7% without affecting crystals quality. Ligands in fast equilibrium with tubulin should be added to the recovery solution at higher concentrations, typically 100–300 μ*M*.

12. In some cases, in particular with soaked crystals, the diffracting power may decline to an unacceptable level during data collection. Although not fully characterized, it appears that substitution of ethylene glycol by glucose as a cryo-protectant limits this problem.

13. As the T2R complex is further stabilized in the crystal, the organic solvent concentration is less of a problem than for cocrystallization (*see* **Note 3**). Some poorly water soluble compounds have been successfully soaked in T2R crystals in a solution containing up to 15% DMSO. In this case, the flash-cooling step has to be modified. The cryo-protecting solution may be adapted for such weakly soluble molecules (a solution with, e.g., 10% DMSO and 20% ethylene glycol works fine). Alternatively, soaked crystals are transferred directly to a cryo-buffer containing 25% ethylene glycol, left there for a short time (e.g., 1 min) to minimise the diffusion of the ligands out of tubulin, and snap-cooled.

## Acknowledgments

This work initiated as a collaboration with Dr. A. Sobel on the interaction of stathmin-like proteins with tubulin. We would like to thank Dr. A. Sobel as well as present and former members of his group (Dr. P. Curmi, Dr. E. Charbaut, Dr. I. Jourdain, Dr. S. Siavoshian, and Mrs. S. Lachkar) for introducing us to the use of stathmin-like domains, Dr. M. -F. Carlier for useful discussions and Ms. F. Jacquinot for excellent technical assistance. We are also grateful to Dr. R. Ravelli, in particular for sharing with us his expertise in crystallography.

## References

1. Schönbrunn, E., Phlippen, W., Trinczek, B., et al. (1999) Crystallization of a macromolecular ring assembly of tubulin ligeded with the anti-mitotic drug podophyllotoxin. *J. Struct. Biol.* **128,** 211–215.
2. Nogales, E., Wolf, S. G., and Downing, K. H. (1998) Structure of the β-tubulin dimer by electron crystallography. *Nature* **391,** 199–203.
3. Nettles, J. H., Li, H., Cornett, B., Krahn, J. M., Snyder, J. P., and Downing, K. H. (2004) The binding mode of epothilone A on β-tubulin by electron crystallography. *Science* **305,** 866–869.

4. Jourdain, L., Curmi, P., Sobel, A., Pantaloni, D., and Carlier, M. -F. (1997) Stathmin: a tubulin sequestering protein which forms a ternary T2S complex with two tubulin molecules. *Biochemistry* **36,** 10,817–10,821.

5. Charbaut, E., Curmi, P. A., Ozon, S., Lachkar, S., Redeker, V., and Sobel, A. (2001) Stathmin family proteins display specific molecular and tubulin binding properties. *J. Biol. Chem.* **276,** 16,146–16,154.

6. Gigant, B., Curmi, P. A., Martin-Barbey, C., et al. (2000) The 4 Å X-ray structure of a tubulin:stathmin-like domain complex. *Cell* **102,** 809–816.

7. Ravelli, R. B., Gigant, B., Curmi, P. A., et al. (2004) Insight into tubulin regulation from a complex with colchicine and a stathmin-like domain. *Nature* **428,** 198–202.

8. Gigant, B., Wang, C., Ravelli, R. B., et al. (2005) Structural basis for the regulation of tubulin by vinblastine. *Nature* **435,** 519–522.

9. Williams, R. C., Jr. and Lee, J. C. (1982) Preparation of tubulin from brain, in *Structural and Contractile Proteins Part B,* (Frederiksen, D. and Cunningham, L., eds.), Academic Press, New York, pp. 376–385.

10. Castoldi, M. and Popov, A. V. (2003) Purification of brain tubulin through two cycles of polymerization-depolymerization in a high-molarity buffer. *Protein Expr. Purif.* **32,** 83–88.

11. Lusty, C. J. (1999) A gentle vapor-diffusion technique for cross-linking of protein crystals for cryocrystallography. *J. Appl. Cryst.* **32,** 106–112.

12. Rodgers, D. W. (1997) Practical cryocrystallography, in *Macromolecular Crystallography Part A,* (Carter, C. and Sweet, R., eds.), Academic Press, New York, pp. 183–202.

13. Hamel, E. and Lin, C. M. (1981) Glutamate-induced polymerization of tubulin: characteristics of the reaction and application to the large-scale purification of tubulin. *Arch. Biochem. Biophys.* **209,** 29–40.

14. Vandecandelaere, A., Martin, S. R., Schilstra, M. J., and Bayley, P. M. (1994) Effects of the tubulin-colchicine complex on microtubule dynamic instability. *Biochemistry* **33,** 2792–2801.

15. Andreu, J. M. and Timasheff, S. N. (1982) Conformational states of tubulin liganded to colchicine, tropolone methyl ether and podophyllotoxin. *Biochemistry* **21,** 6465–6476.

16. Stura, E. A. (1999) Seeding techniques, in *Crystallization of Nucleic Acids and Proteins,* 2nd ed. (Ducruix, A. and Giege, R., eds.), University Press, Oxford, UK, pp. 177–208.

# 17

## Characterizing Ligand-Microtubule Binding by Competition Methods

### José Fernando Díaz and Rubén Martínez Buey

#### Summary

The knowledge of the thermodynamics and kinetics of drug-microtubule interaction is essential to understand the structure/affinity relationship of a given ligand family. When a ligand does not show an appropriate signal change (absorbance or fluorescence) upon binding, the extensive direct characterization of its binding affinities and kinetic rate constants of association and dissociation becomes a complex task. In those cases it is possible to obtain these parameters by competition of the ligand with a reference one of the same binding site that shows such change. Nevertheless, although the experimental setup of the competition measurements is easier, the treatment of the data is complex because simultaneous equilibrium/kinetic equations have to be solved. In this chapter, the taxoid-binding site of the microtubules will be used as an example to describe experimental competition and data analysis methods to determine the binding constants and kinetic rates of association and dissociation of ligands for microtubules.

**Key Words:** Microtubules; paclitaxel; epothilones; stopped flow; fluorescence; anisotropy.

## 1. Introduction

The pharmacopoeia of tubulin has been increased in the last decade with the discovery of a relatively wide range of the so called microtubule stabilizing agents (MSA); those compounds are paclitaxel mimetics in the sense that they mimic the paclitaxel mechanism of action, stabilizing the assembled form of tubulin. Although up to now two MSA binding sites are known (the taxoid-binding site and the laulimalide-binding site), most of the MSA discovered and developed target the taxoid-binding site (epothilones, eleutherobin, discodermolide, sarcodictyins, some steroid analogs, ciclostreptin, dictyostatin, and paclitaxel itself [1–6]).

From: *Methods in Molecular Medicine, Vol. 137: Microtubule Protocols*
Edited by: Jun Zhou © Humana Press Inc., Totowa, NJ

The binding affinity of the epothilones has been shown to correlate well with their cytotoxicity *(7)*, which makes determination of the binding affinity a very powerful tool for quick evaluation of a newly discovered compound or a large series of related synthetic compounds *(7)*.

The study of the interaction between paclitaxel and its mimetics with their binding site is greatly hampered by their mechanism of binding *(8,9)*. Because they preferentially bind to the assembled form of tubulin, inducing microtubule assembly, the assembly and binding are linked reactions *(8)*. Although empty sites can be assembled in the absence of ligand at high tubulin concentrations over the critical concentration (microtubule assembly is a noncovalent nucleated condensation polymerization, characterized by a cooperative behavior and the presence of a critical concentration Cr, below which no significant formation of large polymers takes place *(10)*, the high affinity of most of the MSAs makes it impossible to find conditions in which the reaction is not completely displaced towards the end state, thus making it difficult to obtain an exact measurement of the binding affinity of these compounds.

A procedure to stabilize assembled microtubules has made it possible to obtain taxoid-binding sites, which can be diluted to concentrations low enough to directly measure the binding affinity of two fluorescent derivatives of paclitaxel, Flutax-1 and Flutax-2 *(11)*, which could be used as reference ligands for a competition method that has been developed to measure the binding affinity of compounds that bind to the paclitaxel site *(7,12–15)*; and to distinguish them from compounds with similar effects, but which do not bind to the paclitaxel binding site, such as laulimalide *(13)* and peloruside A *(16)*.

The binding site of paclitaxel has been mapped in the β-tubulin subunit using photolabeling *(17–19)*. The labeled aminoacid residues are in agreement with the 3.5-Å resolution electron crystallographic structure of tubulin in paclitaxel-stabilized zinc-induced two-dimensional crystals *(20,21)*. Tubulin zinc-sheets, whose assembly is not GTP dependent *(22)*, consist of protofilaments similar to those that form the microtubules but organized in an antiparallel array. The docking of these protofilaments into electron microscopy density maps of paclitaxel-containing 14 and 15 protofilament microtubules *(23,24)* results in an atomic model of microtubules in which the binding site of paclitaxel is located on the microtubule inner surface *(24,25)*. Such luminal location will in principle make the binding site difficult to access for paclitaxel site ligands in assembled microtubules. However, it had previously been shown that paclitaxel modifies the flexibility of microtubules in a few seconds *(26)* and that the reversible binding of paclitaxel and its side chain analog docetaxel to an accessible site of microtubules changes the number of their protofilaments within a time range of 1 min *(27)*; so in principle, the binding site of paclitaxel seems to be accessible in preformed microtubules. The fluorescent derivatives of paclitaxel Flutax-1, Flutax-2, 7-Hexaflutax, and the stabilized binding sites have been employed to study the problem of the accessibility of the taxoid-binding site in preformed microtubules. The kinetics constants of binding and dissociation of both Flutax-1 and Flutax-2 for the paclitaxel site have been measured using standard kinetic methods and the determined constants have been employed to determine those of paclitaxel itself to its site using a competition method *(11,15,28)*.

In this chapter, we will describe the methodology for the preparation of stabilized crosslinked microtubules and its use for the determination of the equilibrium binding constants and kinetic rate constants of paclitaxel binding site ligands. In principle, the methods are of general application to any ligand with a good signal change (absorbance, fluorescence intensity, fluorescence anisotropy) upon binding to its site. Nevertheless, it has to be pointed out that competition methods require the constants to be measured to be lower or not much higher than those of the reference ligand. Because the analysis is based on the stoichiometric displacement or perturbation of the binding of the reference ligand, any ligand over a certain affinity (1000 times larger than the reference) or association rate (100 times larger) will behave essentially identically because competition tests are based on the amount of reference ligand free and bound to the site. If we mix 100 n$M$ of a problem ligand with 100 n$M$ of a reference ligand with $10^7/M$ binding constant and 100 n$M$ sites, the percentage of free reference ligand will depend on the binding constant of the problem ligand. If the constant is $10^7/M$, the percentage of free ligand will be 76.6%, if it is $10^8/M$, the free ligand will rise up to 81.9%, for $10^9/M$, the figure will be 92.8%, 97.5% for $10^{10}/M$, and 99.2 for $10^{11}/M$. So, with this reference ligand, it would be easy to determine with precision the binding constants of ligands in the interval between $10^7/M$ and $10^9/M$ but it would be very difficult to distinguish between two ligands of constants in the order of $10^{10}/M$. Because it is always possible to displace a ligand of high affinity with a large excess of one with lower affinity, reference ligands with the highest affinity possible are desired.

## 2. Materials

### 2.1. Buffers and Proteins

1. GAB, glycerol assembly buffer: 3.4 $M$ glycerol, 10 m$M$ sodium phosphate, 1 m$M$ EGTA, 6 m$M$ MgCl$_2$, 0.1 m$M$ GTP, pH 6.5. Prepare the same day of use and keep in the cold.
2. Tubulin, purify it by the modified Weisenberg procedure (*[29–31] also see* Chapter 2) and store it in liquid nitrogen.

### 2.2. Measuring the Binding Affinities and Kinetic Binding Constants Using a Competition Assay

1. Crosslinked stabilized microtubules, prepare as described next and keep them frozen in liquid nitrogen. Once defrozen, the concentration of sites slowly decay at 4°C with a half-life of approx 50 d (average of four batches).

2. Flutax-1 (Calbiochem).
3. Docetaxel (Sigma).

## 2.3. Software for Data Analysis

1. Equigra v5.0 (Díaz et al. unpublished) for binding competition data analysis. Available from the authors upon request.
2. Software for kinetic data simulation and fitting. One of the most powerful ones is FITSIM *(32)*, available at Prof. Carl Frieden's web page: http://biochem. wustl.edu/cflab/.

## 3. Methods

## 3.1. Preparation of Stabilized Taxoid-Binding Sites

1. Load 20 mg of tubulin in a 25 × 0.9-cm column of Sephadex G-25 (GE Healthcare Bioscience, Upsala, Sweden) equilibrated with two volumes of buffer 10 m$M$ phosphate, 1 m$M$ EGTA, 0.1 m$M$ GTP, 3.4 $M$ glycerol, pH 6.8 (prepare 100 mL of the buffer and keep at 4°C).
2. Measure the absorbance at 295 nm of the fractions containing protein, pool those whose absorbance is higher than 1.0.
3. Clarify the solution by centrifugation at 50,000$g$, 4°C, for 10 min using TL100.2 or TL100.4 rotors in a Beckman Optima TLX centrifuge or similar.
4. Make a 1:20 dilution of the solution to measure the concentration of tubulin spectrophotometrically (extinction coefficient 107,000/$M$/cm at 275 nm) *(31)* in 10 m$M$ sodium phosphate buffer containing 1% SDS pH 7.0.
5. Adjust the tubulin concentration to 50 µ$M$ and add 6 m$M$ MgCl$_2$ and up to 1 m$M$ GTP to the solution, final pH 6.5.
6. Assembly the tubulin by incubating it 37°C for 30 min. The solution should become turbid and viscous.
7. Add 4 µL of 50% glutaraldehyde (EMscope, microscopy grade) per milliliter of tubulin (final concentration 20 m$M$) to the solution and mix it with a 1-mL pipet, incubate the solution at 37°C for another 10 min.
8. Place in a 250-mL beaker 60 µL of a 1 $M$ NaBH$_4$ solution (Fluka) per milliliter of tubulin solution. Pour the assembled tubulin solution over it. Let the foam repose for 10 min degas it and recover the solution.
9. Dialyze the solution overnight using Slide-A-Lyzer 10K cassettes (Pierce) against GAB and drop freeze it in liquid nitrogen.

## 3.2. Measuring the Binding Sites Concentrations

1. Make a 1:20 dilution of the solution of crosslinked microtubules to measure the concentration of tubulin spectrophotometrically (extinction coefficient 107,000/$M$/cm at 275 nm, *(31)* in 10 m$M$ sodium phosphate buffer containing 1% SDS pH 7.0.
2. Make the calibration curves. Employing the previously determined concentration, prepare 0.2, 0.5, 1, 2, and 5 µ$M$ solutions (tubulin concentration) of crosslinked microtubules in 10 m$M$ sodium phosphate buffer containing 1% SDS

pH 7.0 and in a mixture 4:1 of 10 m$M$ sodium phosphate buffer containing 1% SDS pH 7.0 and GAB and measure their fluorescence intensity ($\lambda_{exc}$ 280 nm $\lambda_{ems}$ 323 nm) to get a fluorometric tubulin concentration calibration curve.

3. Prepare 0.2, 0.5, 1, 2, and 5 µ$M$ solutions of Flutax-1 in 10 m$M$ sodium phosphate buffer containing 1% SDS pH 7.0 and in a mixture 4:1 of 10 m$M$ sodium phosphate buffer containing 1% SDS pH 7.0 and GAB and measure their fluorescence intensity ($\lambda_{exc}$ 495 nm $\lambda_{ems}$ 520 nm) to get a fluorometric Flutax-1 concentration calibration curve.

4. Centrifuge two-hundred microliters of the following samples in GAB in a Beckman TL100 rotor in Beckman TLX Optima ultracentrifuge at 50,000 rpm for 20 min at 25°C:

   a. 20, 10, 5, and 2 µ$M$ crosslinked microtubules.
   b. 2 µ$M$ Crosslinked microtubules plus 5 µ$M$ Flutax-1.
   c. 2 µ$M$ Crosslinked microtubules plus 5 µ$M$ Flutax-1 plus 100 µ$M$ docetaxel.

5. Collect supernatants and dilute them 1:4 in 10 m$M$ sodium phosphate buffer containing 1% SDS pH 7.0. Add 200 µL of 10 m$M$ sodium phosphate buffer containing 1% SDS pH 7.0 to the pellets to resuspend them and dilute them 1:4 in 10 m$M$ sodium phosphate buffer containing 1% SDS pH 7.0.

6. Take the dilutions of supernatants and pellets of the tubes containing only crosslinked microtubules (**step 4a**) and measure the tubulin concentrations fluorometrically employing the appropriate calibration curves to know the percentage of tubulin polymerized, which should be more than 80%.

7. Take the dilutions of supernatants and pellets of the tubes containing crosslinked microtubules and Flutax-1 (**step 4b,4c**) and measure the Flutax-1 concentrations fluorometrically employing the appropriate calibration curves. The difference between the Flutax-1 concentration in tube b and c is the concentration of binding sites.

## 3.3. Measuring of the Binding Constant of the Problem Ligand

1. Take a 96-well plate suitable for florescence measurements (**Scheme 1**). Don't use the border wells: columns 1 and 12 and rows A and H. In the first five rows (B to F) dispense 200 µL of the mixture of 50 n$M$ Flutax-1 and 50 n$M$ binding sites. In row G dispense 200 µL of 50 n$M$ binding sites in GAB as blank in wells 2 and 3 and 200 µL of 50 n$M$ Flutax-1 in GAB in wells 4 and 5 as anisotropy standard.

2. In wells 2–11 of rows B–F, add the following problem ligand concentrations: 0 n$M$, 5 n$M$, 10 n$M$, 20 n$M$, 50 n$M$, 100 n$M$, 500 n$M$, 1 µ$M$, 2 µ$M$, and 5 µ$M$.

3. Set the plate at the desired temperature (note that the binding constant of the reference ligand should be known at this temperature [**Table 1**]) and measure $F_{VV}$ (fluorescence intensity with excitation 490 nm vertically polarized, emission 520 nm vertically polarized) and $F_{VH}$ (fluorescence intensity with excitation at 490 nm vertically polarized, emission at 520 nm horizontally polarized) of the wells, subtract the blank and calculate the G factor (the correction for the effi-

## Scheme 1
## Distribution of the 96-Well Plate for the Binding Constant Determination Assay

| | | | | | | | | | |
|---|---|---|---|---|---|---|---|---|---|
| Ligand A n$M$ | 0 | 5 | 10 | 20 | 50 | 100 | 500 | 1000 | 2000 | 5000 |
| Ligand B n$M$ | 0 | 5 | 10 | 20 | 50 | 100 | 500 | 1000 | 2000 | 5000 |
| Ligand C n$M$ | 0 | 5 | 10 | 20 | 50 | 100 | 500 | 1000 | 2000 | 5000 |
| Ligand D n$M$ | 0 | 5 | 10 | 20 | 50 | 100 | 500 | 1000 | 2000 | 5000 |
| Ligand E n$M$ | 0 | 5 | 10 | 20 | 50 | 100 | 500 | 1000 | 2000 | 5000 |
| | Blank | Blank | Standard | Standard | Standard | | | | | |

1. All the border wells are empty and all the wells containing problem ligands contain 200 µL of 50 n$M$ Flutax-1 and 50 n$M$ sites in GAB.

2. The blank wells contain 200 µL 50 n$M$ sites in GAB and the standard wells contain 200 µL 50 n$M$ Flutax-1 in GAB.

250

## Table 1
## Equilibrium Constants of Flutax-1 Binding to Microtubules

|                        | 27°C        | 32°C        | 37°C        | 40°C        | 42°C        |
|------------------------|-------------|-------------|-------------|-------------|-------------|
| $Ka \times 10^7/M$     | $6.0 \pm 0.2$ | $4.3 \pm 0.4$ | $2.9 \pm 0.3$ | $2.1 \pm 0.2$ | $1.5 \pm 0.1$ |
| $v_o$                  | 0.57        | 0.51        | 0.45        | 0.39        | 0.33        |

Data from **ref. *11***.

$v_o$ (Flutax-1)$_{bound}$/(Flutax-1)$_{total}$ is calculated from the concentration of Flutax-1, binding sites and the binding constant of Flutax-1 at the different temperatures for 50 n$M$ binding sites and 50 n$M$ Flutax-1.

ciency of each optical channel) *(33)* directly if your plate reader can measure $F_{HV}$ (fluorescence intensity with excitation at 490 nm horizontally polarized, emission at 520 nm vertically polarized) and $F_{HH}$ (fluorescence intensity with excitation at 490 nm horizontally polarized, emission at 520 nm horizontally polarized); otherwise calculate it from the standard (Flutax-1 free in GAB $r = 0.05$) and use it to calculate the anisotropy of each well using **Eq. 1**:

$$ r_x = \frac{\dfrac{Fvv}{Fvh * G} - 1}{\dfrac{Fvv}{Fvh * G} + 2} \tag{1} $$

4. Check whether the anisotropy value has reached the minimum value (that of the free Flutax-1) at the maximum problem ligand concentration; otherwise repeat the measurement with increased problem ligand concentrations in **step 2**, avoiding going over the solubility limit of the problem ligand, because precipitated ligand may cause scattering, which is polarized and will interfere with the measurement.

5. Transform the anisotropy values into fractional saturation values. The fractional saturation of Flutax-1 (reference ligand) ($v_x$=|Flutax-1|$_{bound}$/|Sites|) in the presence of the competitor ligand is calculated from the fluorescence anisotropy measurements as follows:

Because the anisotropy is an additive property, the anisotropy of a given mixture is the sum of the anisotropy of their components. The anisotropies of Flutax-1 in the two reference states (the bound $r_o = 0.24 \pm 0.02$ and the fully displaced $r_{min} = 0.05 \pm 0.02$ *(11)*, and the fractional binding of the reference ligand Flutax-1 in the absence of competitor ($v_o$, **Table 1**) are known, so the measured anisotropy values $r_x$ can transformed into fractional saturation values $v_x$ employing **Eq. 2**.

$$ v_x = \frac{vo * (rx - r\,min)}{(ro - r\,min)} \tag{2} $$

The binding constant of the problem ligand K(l) can be determined from the known values of the binding constant of the reference ligand K(r) and the total concentrations of binding sites, Flutax-1 and problem ligand by solving the simultaneous mass action equations (**Eqs. 3–7**).

$$K(l) = [Ligand]_{bound}/[Sites]_{free}*[Ligand]_{free} \qquad (3)$$

$$K(r) = [Flutax\text{-}1]_{bound}/[Sites]*[Flutax\text{-}1]_{free} \qquad (4)$$

$$[Flutax\text{-}1]_{free} = [Flutax\text{-}1]_{total} - [Flutax\text{-}1]_{bound} \qquad (5)$$

$$[Ligand]_{free} = [Ligand]_{total} - [Ligand]_{bound} \qquad (6)$$

$$[Sites]_{free} = [Sites]_{free} - [Ligand]_{bound} - [Flutax\text{-}1]_{bound} \qquad (7)$$

In **Fig. 1** the fractional saturation values of Flutax-1 at different competitor concentrations are best fitted (assuming unitary stoichiometry) to the equilibrium binding constant of different competitors (Epothilone B, paclitaxel, and an steroid analog of paclitaxel) and the binding constant of the reference ligand (Flutax-1) with a program developed *ad hoc* (Equigra v5) (*see* **Note 1**).

6. If the binding constant of the ligand at a different temperature is required, change the temperature setting of the plate reader and incubate the plate until it reaches equilibrium (usually 20 min is enough) and go back to **step 3**, (check Flutax-1 photobleaching after three to four measurements). If the binding constant of a ligand is measured at several temperatures, apparent thermodynamic parameters can be calculated using the van't Hoff and Gibbs' equations.

## 3.4. Measuring the Association Rate Constant of the Problem Ligand

1. Place in one of the syringes of the stop-flow device a 1.2-$\mu M$ solution of taxoid-binding sites in GAB.
2. Prepare 1 $\mu M$ solutions of Flutax-1 in GAB containing 0 and 1 $\mu M$ of the problem ligand.
3. In the second syringe, load the solution containing Flutax-1 and no competitor and measure the kinetic association curve (final concentrations 600 nM sites and 500 nM Flutax-1). Be sure that the reaction has reached equilibrium. Repeat the measurement with the solution containing 1 $\mu M$ of the problem ligand.
4. Compare the kinetic association curves and check if there is a significant decrease of the amplitude and velocity in the presence of the problem drug; otherwise increase the competitor concentration until the amplitude of the curve decreases by around 75%.
5. Prepare another four 1 $\mu M$ solutions of Flutax-1 in GAB containing 2, 1/2, 1/4, 1/8, and 1/16 of the problem ligand concentration determined in **step 4**. Measure the kinetic association curves of Flutax-1 to the taxoid sites in the presence of increasing concentrations of the problem ligand.
6. Scale the first curve to 1 at the equilibrium and to 0 in the origin (at the concentrations of Flutax-1 and sites employed, the half-life of the reaction is 2–3 s, so

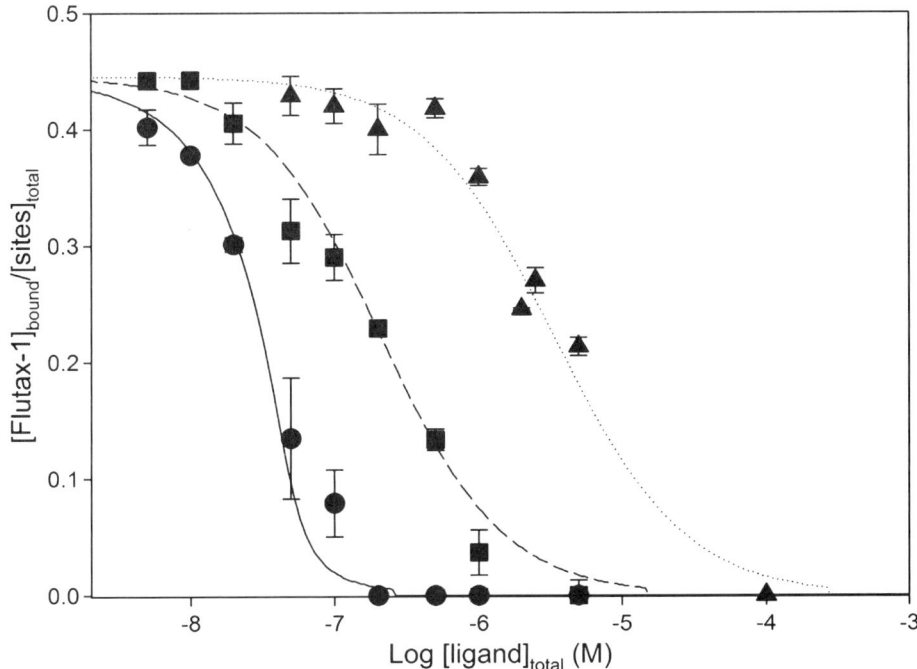

Fig. 1. Displacement of the fluorescent taxoid Flutax-1 (50 n*M*) from microtubule binding sites (50 n*M*) by taxoid-binding site ligands at 37°C. The points are data and the lines were generated with the best fit value of the binding equilibrium constant of each competitor, assuming a one to one binding to the same site. Ligands: Epothilone B (circles, solid line), Paclitaxel (squares, dashed line), 2-ethoxyestradiol analog (3,17β d-diacetoxy-2-ethoxy-6-oxo-B-homo-estra-1,3,5[10]-triene; [4]) (triangles, dotted line). The determined binding constants were $8 \times 10^8 M^{-1}$ for Epothilone B, $2 \times 10^7 M^{-1}$ for paclitaxel and $8.3 \times 10^5 M^{-1}$ for 2-ethoxyestradiol analog.

no significant part of the reaction should be observed within the typical dead time of the stopped flow instrument) and scale the other curves accordingly. Check that the proportion between the curves at equilibrium roughly corresponds with the equilibrium binding constant previously determined.

7. Fit the kinetic curve in the absence of competitor, employing FITSIM *(32)* *(see* **Note 2***)* and compare the kinetic rate of association obtained with the published one *(11)*.

8. Fit all kinetic curves simultaneously employing FITSIM *(32)* *(see* **Note 2***)*. **Figure 2A** shows the fitting of the association curves of Flutax-1 to the taxoid-binding site of the microtubules in the presence of different concentrations of Epothilone A.

9. If the association rate constant of the ligand at a different temperature is required, change the temperature setting of the stopped flow device and go back to **step 1**.

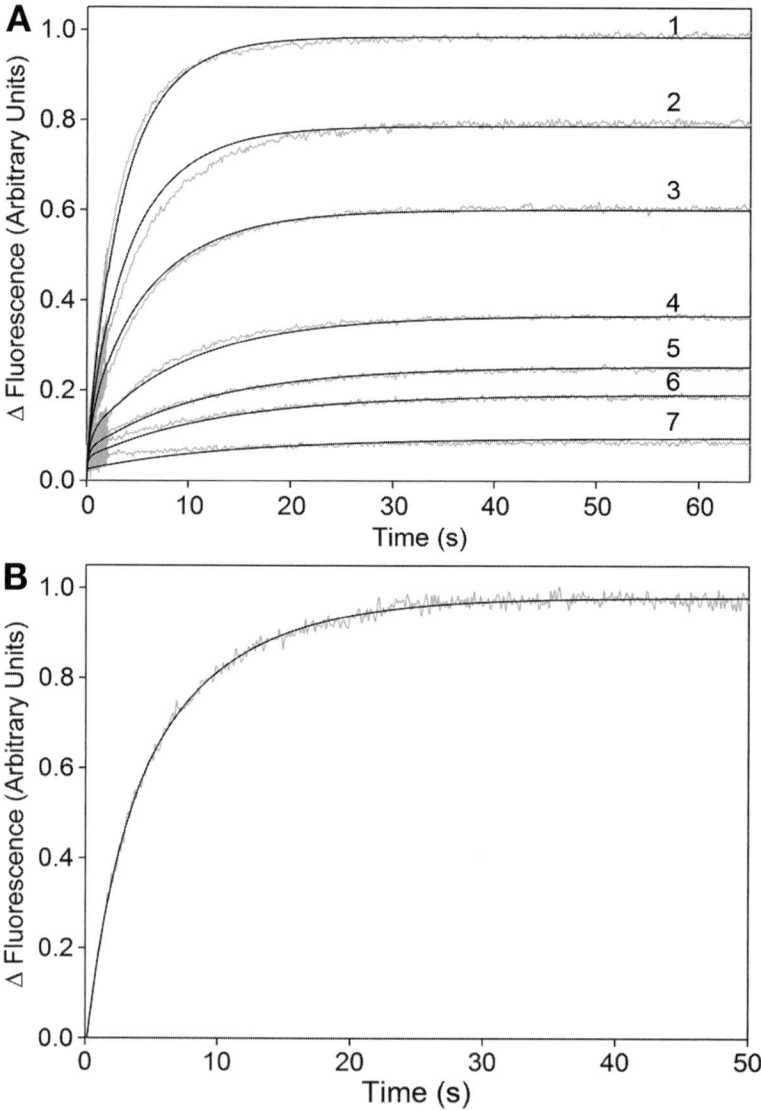

Fig. 2. (**A**) Kinetics of association of Epothilone A to microtubules at 37°C. In the stopped flow device, a solution of crosslinked microtubules containing 1.2 µ*M* sites was mixed 1:1 with a 1 µ*M* solution of Flutax-1, containing different concentrations of Epothilone A (curve 1) 0 µM, (curve 2) 0.5 µM, (curve 3) 1.0 µM, (curve 4) 2.0 µM, (curve 5) 3.0 µ*M*, (curve 6) 4.0 µM, (curve 7) 8.0 µ*M*. The gray lines are the experimental data and the solid lines are the result of the simultaneous fitting of all curves to a kinetic model of a single step association of Epothilone A to the paclitaxel binding site. The kinetic association rate constant of Epothilone A to the taxoid-binding site of

### 3.5. Measuring the Dissociation Rate Constant of the Problem Ligand

1. First, determine the binding constant of the problem ligand. If it is more than 20–30 times that of the reference ligand at the temperature of the measurement, it will not be possible to displace it with a 10 times excess making it not possible to measure the dissociation rate constant.
2. Place in one of the syringes of the stop-flow device a 1-$\mu M$ solution of taxoid-binding sites in GAB plus 1.2 $\mu M$ of the problem ligand. Using the association-binding constant, calculate that a significant percentage of the binding sites are filled with the problem ligand. Otherwise increase the concentration accordingly.
3. In the other syringe, place a 12 $\mu M$ solution of Flutax-1 in GAB (final concentrations 500 n$M$ taxoid-binding sites, 600 n$M$ problem ligand, 6 $\mu M$ solution of Flutax-1; if the concentration of binding sites and problem ligand have been increased and the Flutax-1/sites ratio is under 5:1 increase the Flutax-1 concentration accordingly). Measure the dissociation constant from the small change of fluorescence intensity upon Flutax-1 binding to the binding sites that are left empty by the problem ligand.
4. Analyze the kinetic curve as a sum of exponentials. If all the binding sites are filled with the problem ligand, only the kinetics of dissociation of the problem ligand should be observed. If a significant fraction of binding sites are empty, a fast initial phase corresponding to the filling of these sites by Flutax-1 will be observed. The dissociation of Epothilone A from the taxoid-binding site of the microtubules is shown in **Fig. 2B**.
5. If the dissociation rate constant of the ligand at a different temperature is required, change the temperature setting of the stop flow device and go back to **step 1**.

## 4. Notes

1. Equigra v5 may work both by introducing the data manually and by reading them from a file. If you want to introduce the data manually, the program will prompt you for:
   a. Ligand name.
   b. Total sites concentration (which should be equal for all data sets).
   c. Binding affinity of the reference ligand.

---

Fig. 2. (*continued*) microtubules at 37°C was found to be $3.30 \pm 0.03 \times 10^6$ M$^{-1}$s$^{-1}$. (**B**) Kinetics of dissociation of Epothilone A from microtubules at 37°C. At time 0 s, 10 $\mu M$ Flutax-1 was added to 1-$\mu M$ binding sites in crosslinked microtubules that contained 1.2 $\mu M$ Epothilone A. The reaction was followed by the change in fluorescence intensity (average of nine curves). The data were fitted to a double exponential (solid line) and the dissociation of Epothilone A from its site in the microtubules was found to be biphasic indicating the existence of two different species with kinetic constants of dissociation of $0.138 \pm 0.37$ s$^{-1}$ and $0.463 \pm 0.32$ s$^{-1}$, respectively.

d. The number of data sets (with different reference ligand concentrations).
e. The number of problem ligand concentrations for each data set (which should be equal for all data sets).
f. The lower and upper limits for the search of the binding constant.
g. The concentrations of the reference ligand for each data set.
h. The concentrations of the problem ligand for each data set.
i. The fractional saturation of reference ligard for each pair or reference ligand/ problem ligand concentration.

If you wish to prepare an input data file (which is much more useful) you should create a file with a name less than eight characters long and with .dat extension. The file should contain the same data except the ligand name **(Note 1a)**, which will be taken from the name of the input file. The format of the data should be the following (delete the comments in parenthesis):

| | |
|---|---|
| 50e-9 | (Concentration of binding sites) |
| 3.0e7 | (Binding constant of the reference ligand) |
| 1 | (One data set, multiple data sets with different reference ligand concentration can be employed) |
| 10 | (Ten problem ligand concentrations for each data set) |
| 1e7 | (Lower limit for the problem ligand binding affinity search) |
| 1e10 | (Upper limit for the problem ligand binding affinity search) |
| 50e-9 | (Reference ligand concentration) |
| | 0.0000 5e-9 10e-9 20e-9 50e-9 100e-9 200e-9 500e-9 1000e-9 5000e-9 (Problem ligand concentrations, M) |
| 0.4510 | (Fractional saturation for problem ligand concentration 0.0000) |
| 0.3806 | (Fractional saturation for problem ligand concentration 5e-9) |
| 0.3140 | (Fractional saturation for problem ligand concentration 10e-9) |
| 0.2112 | (Fractional saturation for problem ligand concentration 20e-9) |
| 0.1731 | (Fractional saturation for problem ligand concentration 50e-9) |
| 0.1379 | (Fractional saturation for problem ligand concentration 100e-9) |
| 0.0352 | (Fractional saturation for problem ligand concentration 200e-9) |
| 0.0376 | (Fractional saturation for problem ligand concentration 500e-9) |
| 0.0417 | (Fractional saturation for problem ligand concentration 1000e-9) |
| 0.0347 | (Fractional saturation for problem ligand concentration 5000e-9) |

Run the program and input the filename (without the .dat extension). The program will iteratively calculate the binding constant of the problem ligand, stopping after 100 iterations showing the quality of the fitness $D^2/N$ (the square of the deviations divided by the number of points) and the calculated binding constants and will prompt you to do another 100 iterations; type "y" until the quality of the fitness remains constant. When no more iterations are needed, type "n" and a plot of the fitting will appear. Press "e" to exit the program.

The program will generate two output files .out and .tab. The .out file contains the result of the fitting for the program to represent the result of a previous calculation. The .tab file is an ASCII file with the calculated competition curve.

2. This note does not aim to be a user guide for FITSIM *(32)* but just a small walkthrough of how to process competition kinetic data using this powerful tool. We can not give support on the tool and any question to the authors concerning the use of FITSIM will be unattended.

FITSIM is a program based on iterative solving of the differential equations that govern the kinetics of chemical reactions. In this way no pseudofirst order conditions are required for the experiments and a wide range of experimental setups can be analyzed.

To use FITSIM to analyze your data, the first thing you need is to set up a mechanism. This can be done with a text editor such as Windows notepad.

For the case of the competition of a ligand with Flutax-1 for the taxoid-binding site your mechanism will be as follow:

$Simple competition mechanism
A + T == AT
B + T == BT
*OUTPUT
F*AT

where A is Flutax-1, B the problem ligand, AT and BT the complexes between the ligands and the taxoid site and F is a factor to convert concentration of AT to the observed fluorescence change. Although Flutax-1 binding to microtubules follows a two-step mechanism *(11)*, only the first step is visible by fluorescence intensity change and useful for our purposes. Save the file with the extension .mec. Open KINSIM40 and compile the mechanism (Option O), then you will get a .sim file that will be required for the simulation.

Second, the data produced by the stopped flow instrument should be converted to the FITSIM format. To do this, the data should be in a two-column ASCII format: time and fluorescence change value (scaled from 0 to 1). Once the data are in ASCII format, they should be converted to FITSIM format using option A of KINSIM40, save the files with the extension .rdf.

Third, the parameter file of each data file has to be created. Use the C, F, and T options of the KINSIM40 menu to input the concentration of sites and ligands (C), the output factors (F) and the time factors for the calculation (T), and the K option to input any rate constant that wants to be fixed during the simulation.

The output factor is a conversion factor between the Flutax-1 concentration and the fluorescence signal. Because at the concentration of 500 n$M$ employed, the reaction is more than 99% displaced towards the bound state, it can be considered that, at the equilibrium in the absence of competitor all Flutax-1 is in the bound state (AT). Because the value of the association binding curve in the absence of competitor has been scaled to 1 at the equilibrium, the F factor for AT should be approx 1/500e-9, i.e. 2e6; note that the curve has been scaled to 0 at origin, so only the change of fluorescence is considered and therefore the F factor for A should be 0. In any case the F factor is a parameter that will be fitted later.

The time factors: (delta time, integrations [default = 1], run time, Ymax, flux tolerance [default = 0.02] and integral tolerance [default = 1E-6]) are parameters for the simulations that are used to calculate the constants. The user should introduce the run time of the curve in run time, Y max is always 1 and you should let the other parameters in the default values. Delta time determines the number of points in the simulation, i.e., if run time is 1 s and delta time 0.01 s there will be 100 points, but you can set it to a relatively large value (like 1) and the program will search for the smallest delta time that falls within an integral tolerance range. Save the data set as a .sav file.

Once you have the .sim file for the mechanism, a .rdf file and a .sav file for each file you may start the fitting.

Run FITSIM40 and choose 1 to make a .fdt file, which will contain the parameters for the fitting. Input the mechanism filename.sim (be sure of typing it with the .sim extension) and the program will prompt for the constants to be fitted (those that were selected to be fixed should be specified in the .sav file of the parameter file previously saved). Then it will prompt for the data sets to be employed, if you are not familiar with FITSIM it is recommended to just fit one measurement to learn the structure of the .fdt files, which are used for fitting. Afterwards, the program will ask if all the default parameters will be employed; answer "n" and include the output factor as a fitted parameter, let the other parameters in the default values unless you are familiar with them. Once the .fdt file is ready the system will return to the main menu, type 2 to fit the data curves and 3 to see the results of the fitting.

## References

1. Bollag, D. M., McQueney, P. A., Zhu, J., et al. (1995) Epothilones, a new class of microtubule-stabilizing agents with a taxol-like mechanism of action. *Cancer Res.* **55**, 2325–2333.
2. Hamel, E., Sackett, D. L., Vourloumis, D., and Nicolaou, K. C. (1999) The coral-derived natural products eleutherobin and sarcodictyins A and B: effects on the assembly of purified tubulin with and without microtubule-associated proteins and binding at the polymer taxoid site. *Biochemistry* **38**, 5490–5498.
3. Kowalski, R. J., Giannakakou, P., Gunasekera, S. P., Longley, R. E., Day, B. W., and Hamel, E. (1997) The microtubule-stabilizing agent discodermolide competitively inhibits the binding of paclitaxel (Taxol) to tubulin polymers, enhances tubulin nucleation reactions more potently than paclitaxel, and inhibits the growth of paclitaxel-resistant cells. *Mol. Pharmacol.* **52**, 613–622.
4. Verdier-Pinard, P., Wang, Z., Mohanakrishnan, A. K., Cushman, M., and Hamel, E. (1999) A steroid derivative with paclitaxel-like effects on tubulin polymerization. *Mol. Pharmacol.* **57**, 568–575.
5. Edler, M. C., Buey, R. M., Marcus, A. I., et al. (2005) Cyclostreptin (FR182877), a cytotoxic tubulin-polymerizing agent deficient in hypernucleating tubulin assembly. *Biochemistry* **44**, 11,525–11,538.

6. Paterson, I., Britton, R., Delgado, O., Meyer, A., and Poullennec, K. G. (2004) Total synthesis and configurational assignment of (-)-dictyostatin, a microtubule-stabilizing macrolide of marine sponge origin. *Angew. Chem. Int. Ed. Engl.* **43,** 4629–4633.

7. Buey, R. M., Díaz, J. F., Andreu, J. M., et al. (2004) Interaction of epothilone analogs with the paclitaxel binding site: relationship between binding affinity, microtubule stabilization, and cytotoxicity. *Chem. Biol.* **11,** 225–236.

8. Díaz, J. F., Menéndez, M., and Andreu, J. M. (1993) Thermodynamic of ligand-induced assembly of tubulin. *Biochemistry* **32,** 10,067–10,077.

9. Evangelio, J. A., Abal, M., Barasoain, I., et al. (1998) Fluorescent taxoids as probes of the microtubule cytoskeleton. *Cell Motil. Cytoskeleton* **39,** 73–90.

10. Oosawa, F. and Asakura, S. (1975) *Thermodynamics of the Polymerization of Protein.* Academic Press, London.

11. Díaz, J. F., Strobe, R., Engelborghs, Y., Souto, A. A., and Andreu, J. M. (2000) Molecular recognition of taxol by microtubules. Kinetics and thermodynamics of binding of fluorescent taxol derivatives to an exposed site. *J. Biol. Chem.* **275,** 26,265–26,276.

12. Andreu, J. M. and Barasoain, I. (2001) The interaction of baccatin III with the taxol binding site of microtubules determined by a homogeneous assay with fluorescent taxoid. *Biochemistry* **40,** 11,975–11,984.

13. Pryor, D. E., O'Brate, A., Bilcer, G., et al. (2002) The microtubule stabilizing agent laulimalide does not bind in the taxoid site, kills cells resistant to paclitaxel and epothilones, and may not require its epoxide moiety for activity. *Biochemistry* **41,** 9109–9115.

14. Nicolaou, K. C., Ritzén, A., Namoto, K., et al. (2002) Chemical synthesis and biological evaluation of novel epothilone B and trans-12,13-cyclopropyl epothilone B analogues. *Tetrahedron* **58,** 6413–6432.

15. Díaz, J. F., Barasoain, I., and Andreu, J. M. (2003) Fast kinetics of taxol binding to microtubules. Effects of solution variables and microtubule-associated proteins. *J. Biol. Chem.* **278,** 8407–8419.

16. Gaitanos, T. N., Buey, R. M., Díaz, J. F., et al. (2004) Peloruside A does not bind to the taxoid site on beta-tubulin and retains its activity in multidrug-resistant cell lines. *Cancer Res.* **64,** 5063–5067.

17. Rao, S., Krauss, N. E., Heerding, J. M., et al. (1994) 3'-(p-azidobenzamido)taxol photolabels the N-terminal 31 amino acids of beta-tubulin *J. Biol. Chem.* **269,** 3132–3134.

18. Rao, S., Orr, G. A., Chaudhary, A. G., Kingston, D. G., and Horwitz, S. B. (1995) Characterization of the taxol binding site on the microtubule. 2-(m-Azidobenzoyl)taxol photolabels a peptide (amino acids 217-231) of beta-tubulin. *J. Biol. Chem.* **270,** 20,235–20,238.

19. Rao, S., He, L., Chatkravarty, S., Ojima, I., Orr, G. A., and Horwitz, S. B. (1999) Characterization of the Taxol binding site on the microtubule. Identification of

Arg(282) in beta-tubulin as the site of photoincorporation of a 7-benzophenone analogue of Taxol. *J. Biol. Chem.* **274,** 37,990–37,994.

20. Löwe, J., Li, H., Downing, K. H., and Nogales, E. (2001) Refined structure of alpha beta-tubulin at 3.5 Å resolution. *J. Mol. Biol.* **313,** 1045–1057.

21. Nogales, E., Wolf, S. G., and Downing, K. (1998) Structure of the alpha-beta tubulin dimer by electron crystallography *Nature* **391,** 199–203.

22. Melki, R. and Carlier, M. F. (1993) Thermodynamics of tubulin polymerization into zinc sheets: assembly is not regulated by GTP hydrolysis *Biochemistry* **32,** 3405–3413.

23. Sosa, H., Dias, D. P., Hoenger, A., et al. (1997) A model for the microtubule-Ncd motor protein complex obtained by cryo-electron microscopy and image analysis *Cell* **90,** 217–224.

24. Meurer-Grob, P., Kasparian, J. and Wade, R. H. (2001) Microtubule structure at improved resolution *Biochemistry* **40,** 8000–8008.

25. Nogales, E., Whittaker, M., Milligan, R. A., and Downing, K. H. (1999) High-resolution model of the microtubule *Cell* **96,** 79–88.

26. Dye, R. B., Fink, S. P., and Williams, R. C. (1993) Taxol-induced flexibility of microtubules and its reversal by MAP-2 and Tau. *J. Biol. Chem.* **268,** 6847–6850.

27. Díaz, J. F., Valpuesta, J. M., Chac—n, P., Diakun, G., and Andreu, J. M. (1998) Changes in microtubule protofilament number induced by Taxol binding to an easily accessible site. Internal microtubule dynamics. *J. Biol. Chem.* **273,** 33,803–33,810.

28. Díaz, J. F., Barasoain, I., Souto, A. A., Amat-Gerri, F., and Andreu, J. M. (2005) Macromolecular accessibility of fluorescent taxoids bound at a paclitaxel binding site in the microtubule surface. *J. Biol. Chem.* **280,** 3928–3937.

29. Weisenberg, R. C., Borisy, G. G., and Taylor, E. W. (1968) The colchicine-binding protein of mammalian brain and its relation to microtubules. *Biochemistry* **7,** 4466–4479.

30. Lee, J. C., Frigon, R. P., and Timasheff, S. N. (1973) The chemical characterization of calf brain microtubule protein subunits. *J. Biol. Chem.* **248,** 7253–7262.

31. Andreu, J. M., Gorbunoff, M. J., Lee, J. C., and Timasheff, S. N. (1984) Interaction of tubulin with bifunctional colchicine analogues: an equilibrium study. *Biochemistry* **23,** 1742–1752.

32. Zimmerle, C. T. and Frieden, C. (1989) Analysis of progress curves by simulations generated by numerical integration. *Biochem. J.* **258,** 381–387.

33. Lackowicz, J. R. (1999) *Principles of Fluorescence Spectroscopy*. Kluwer Academic/Plenum Publishers, New York, pp. 291–318.

# 18

## Methods for Studying Vinca Alkaloid Interactions With Tubulin

### Sharon Lobert and John J. Correia

#### Summary

Vinca alkaloids play a vital role in chemotherapy protocols for a wide range of hematological and solid tumors. Studies of drug interactions with the drug target, tubulin or microtubules, have helped us to understand the cytotoxic and toxic effects. We present here in vivo and in vitro methods for studying vinca alkaloid interactions with tubulin. In vivo methods for examining drug effects on cell proliferation and intracellular tubulin or microtubules and direct visualization of drug effects by fluorescence microscopy are presented. In vitro methods for measuring drug affinity for tubulin by analytical ultracentrifugation, kinetics of drug binding by light scattering and drug effects on microtubules by turbidity are also presented.

**Key Words:** Tubulin; microtubules; vinca alkaloids; isoelectric focusing; analytical ultracentrifugation; tubular kinetics.

## 1. Introduction

For more than 40 yr vinca alkaloids have played a vital role in chemotherapy protocols for a wide range of hematological and solid tumors. Numerous studies of their mechanisms of action have helped us to understand the cytotoxic and toxic effects of vinca alkaloids. In addition to the original vinca alkaloids, vinblastine and vincristine *(1,2)*, a new generation of clinically useful agents, vinorelbine and vinflunine, with improved toxicity profiles have been developed *(3,4)*. The target for this class of drugs is tubulin, the major protein in microtubules and in mitotic spindles. These drugs alter the dynamics of microtubules, induce tubulin to form nonmicrotubule oligomers, and cause mitotic arrest and cell death *(5,6)*. The interaction of vinca alkaloids with tubulin has

From: *Methods in Molecular Medicine, Vol. 137: Microtubule Protocols*
Edited by: Jun Zhou © Humana Press Inc., Totowa, NJ

been extensively studied by intracellular methods and by solution methods with phosphocellulose purified (PC) tubulin or microtubule protein, tubulin that includes microtubule-associated proteins (MAPs).

Numerous studies suggest that an increase in the fraction of polymerized tubulin is associated with resistance to vinca alkaloids (reviewed in **ref. 7**). One can measure the drug concentration at which cell proliferation is reduced by 50% ($IC_{50}$) by exposing cell cultures to serial dilutions of drug stocks. The presence of drug sensitivity or resistance is established by comparing $IC_{50}$ measurements from parent cell lines or some standard set of cell lines like the National Institutes of Health (NIH) 60 cell line library (for an example *see* **ref. 8**). Intact microtubule cytoskeletons can then be extracted from these cell lines to evaluate the amount of polymerized vs free tubulin at physiological drug concentrations *(9–11)*. This discussion outlines the framework of our interest and experience in the field, and thus this chapter will focus on three in vivo methods; (1) $IC_{50}$ measurements, (2) extraction of intact microtubules and free tubulin pools from cells, and (3) direct visualization of fluorescently labeled vinca alkaloids interacting with cellular targets by fluorescence microscopy. Beyond the focus of this chapter are studies of microtubule dynamics using differential interference contrast (DIC) microscopy. DIC measurements have defined parameters underlying drug mechanisms of action for all vinca alkaloids currently in clinical use *(12–15)*. We will also not discuss intracellular drug accumulation studies utilizing radioactive drugs. Investigations into drug uptake and retention have differentiated the more potent agents, vincristine and vinblastine *(16,17)* and, although important, these issues are not within our experimental experience.

In vitro methods for studying vinca alkaloid binding to purified tubulin include radioactive drug binding *(18–20)* and gel filtration methods *(21–23)*. Fluorescence studies of tubulin quenching or binding of fluorescently labeled vinca alkaloids have been done to obtain quantitative comparisons of drug binding *(24–26)*. However, for the newer generation of agents that demonstrate weaker binding to tubulin, vinorelbine and vinflunine, these methods have limited usefulness. Because vinca alkaloid binding to tubulin is linked to tubulin self-association, analytical ultracentrifugation is a sensitive method for measuring binding constants for even the most weakly binding vinca alkaloids *(27)*. The kinetics of self-association of tubulin induced by the presence of drug can also be measured quantitatively by stopped flow light scattering techniques *(27,28)*. Turbidity experiments using microtubule absorbance at 350 nm can be used to compare inhibitory effects of single drugs and combinations of drugs *(29,30)*. Last, microscopy techniques and X-ray diffraction of crystals *(31)* and solution scattering have been successfully used to evaluate vinca alkaloid binding to tubulin *(32)*. This chapter will focus on three methods for in vitro studies ot tubulin-vinca alkaloid interactions that we have the most experience with: (1) analytical ultracentrifugation, (2) turbidity, and (3) stopped flow kinetics measurements. Microtubule dynamics, radioactive drug-binding, microscopy, and X-ray diffraction techniques are beyond the scope of this chapter.

## 2. Materials

### 2.1. Cell Culture

1. Phosphate buffered saline (PBS): 150 m$M$ NaCl, 2.7 m$M$ KCl, 20 m$M$ Na$_2$HPO$_4$, 1.8 m$M$ KH$_2$PO$_4$, pH 7.4.
2. 10X Trypsin (Mediatech, Inc. Herndon, VA).
3. 15-mL Polypropylene or polystyrene conical tubes (Becton Dickinson Labware, Franklin Lakes, NJ).
4. Plate cells in 75-cm polystyrene flasks (Corning Inc., Corning, NY) and grow to 80% confluence at 37°C and 5% CO$_2$ in medium as recommended by the supplier.

### 2.2. IC$_{50}$ Measurements

1. 12-Well polystyrene plates and 96-well polystyrene plates for cell culture from Fisher Scientific Co, LLC (Palatine, IL).
2. Trypan blue (Sigma-Aldrich, St. Louis, MO); filter with 0.2-μm syringe filter (surfactant free, cellulose acetate, Nalge Co., Rochester, NY).
3. CellTiter 96® AQ$_{ueous}$ One Solution (Promega, Madison, WI) diluted 1:5 with medium. This solution utilizes MTS a 3-(4,5-dimethylthiazol-2-yl)-5-(3-carboxymethoxyphenyl)-2-(4-sulfophenyl)-2H-tetrazolium, inner salt, in a colorometric reaction that is a direct measure of mitochondrial respiration.

### 2.3. Isolation of Intact Microtubules

1. Paclitaxel microtubule stabilization buffer: 20 m$M$ Tris-HCl, pH 6.8, 0.14 $M$ NaCl, 0.5% NP-40, 1 m$M$ MgCl$_2$, 2 m$M$ EGTA, 4 μg/mL paclitaxel, fresh protease inhibitor cocktail (Complete Protease Inhibitor Cocktail Tablets, Roche, Indianapolis, IN). This buffer is prepared and then frozen at –20°C in 1-mL aliquots.
2. Glycerol microtubule stabilization buffer: 100 m$M$ PIPES, pH 6.9, 1 m$M$ EGTA, 1 m$M$ MgSO$_4$, 30% glycerol, 5% DMSO, 5 m$M$ GTP, 1 m$M$ DTT, 0.02% sodium azide, 0.125% NP-40, protease inhibitor cocktail (Complete Protease Inhibitor Cocktail Tablets, Roche). The buffer is stored at –20ºC in 1.5-mL aliquots.
3. SDS Sample buffer: 0.624 $M$ TrisHCl pH 6.8, 4% SDS (sodium dodecyl sulfate), 2.73 $M$ glycerol, 5% β-mercaptoethanol. This buffer may be kept indefinitely at room temperature.
4. Bradford type assay for protein concentration: Advanced Protein Assay Reagent 5X Concentrate (Cytoskeleton Inc., Denver, CO).
5. Equipment: low speed centrifuge. For all low speed spins we use a tabletop 21000R Marathon MP centrifuge (Fisher Scientific Co., Houston, TX) with a swinging bucket rotor and appropriate adapters for 15-mL conical tubes.

## 2.4. Isolation of Total Intracellular Tubulin for Two-Dimensional-Isoelectric Focusing

1. Stock solutions: a.) 1 $M$ [$N$-morpholino]-ethane sulfonic acid (MES), pH 6.9, b.) 100 m$M$ EGTA, c.) 50 m$M$ MgSO$_4$, d.) 50 m$M$ GTP, e.) 0.1% acid violet 17.
2. Cell lysis buffer: 1 m$M$ MgSO$_4$ and 1 m$M$ EGTA, pH 6.9.
3. Microtubule isolation buffer: 0.1 m$M$ MES, pH 6.9, 1 m$M$ EGTA, 1 m$M$ MgSO$_4$, 1 m$M$ GTP buffer containing 20 μ$M$ paclitaxel.
4. MAP removal buffer: 0.1 m$M$ MES, pH 6.9, 1 m$M$ EGTA, 1 m$M$ MgSO$_4$,1 m$M$ GTP, 20 μ$M$ paclitaxel, and 0.35 $M$ NaCl.
5. Microtubule pellet solubilization buffer for two-dimensional-isoelectric focusing (2D-IEF): 7 $M$ urea, 2 $M$ thiourea, 2% 3-[(3-cholamidopropyl)dimethylammonio]-1-propanesulfonate (CHAPS), 0.5% Triton X-100, 2% 2-mercaptoethanol. Buffer may be made up in advance in 10-mL aliquots. It should be remade weekly.

## 2.5. Fluorescence Microscopy

1. Fluorescent vinblastine: vinblastine Bodipy Fl from Invitrogen-Molecular Probes, Carlsbad, CA.; vinblastine-coumarin from Dr. Susan Bane, SUNY Binghamton, New York.
2. Primary tubulin antibodies: monoclonal mouse anti-β class I SAP4G5 or anti-α tubulin DM1A (Sigma-Aldrich).
3. Secondary antibody: rhodamine-labeled goat antimouse IgG (Sigma-Aldrich).
4. Buffers for primary and secondary antibody dilutions: 1% BSA (bovine serum albumin) in PBS or 3% NFD (nonfat dry) milk/PBS.
5. Mounting media: SlowFade® Antifade (Molecular Probes/Invitrogen).

## 2.6. Phoshphocelluose-Purified Tubulin Prep

For in vitro studies of vinca alkaloid interactions with tubulin, phosphocelluose-purified tubulin (PC-tubulin) free of MAPs can be obtained by warm/cold polymerization/depolymerization with the addition of a final phosphocellulose chromatography step to separate tubulin from MAPs (for detailed procedures *see* **refs. 35** and **36**). Microtubule protein (MTP) is obtained by eliminating the phosphocellulose chromatography step. This material may be useful if you are interested in how MAPs affect drug-binding to tubulin. We store the protein obtained from these preps at –80°C by dropwise freezing into liquid N$_2$. Protein frozen by this rapid freezing method retains activity for months.

## 2.7. Other Materials

1. Determine tubulin concentrations using the extinction coefficient $ε_{278}$ = 1.2 L/g cm *(37)*.
2. Determine MTP concentrations using a Bradford type assay (Advanced Protein Assay Reagent 5X Concentrate, Cytoskeleton).
3. Drugs: vinblastine, vincristine (Sigma-Aldrich), vinorelbine (Glaxo Wellcome, Research Triangle Park, NC).

4. Spun columns: Sephadex G50 (Amersham Biosciences), 5-mL syringe barrel with frit. We cut 1-cm frits from sheets of porous polyethylene (Bel-Art Products, Pequannock, NJ).

5. 15-mL Polypropylene conical tubes (Becton Dickinson Labware).

6. Microtubule polymerization buffer: for PC-tubulin (1–3.5 mg/mL) in 100 m$M$ PIPES, pH 6.9, 1 m$M$ MgSO$_4$, 2 m$M$ EGTA, 1 m$M$ GTP, and 2 $M$ glycerol. Polymerize MTP (0.6–3 mg/mL) in the same buffer without glycerol.

7. Analytical ultracentrifugation buffer: 10 m$M$ PIPES, pH 6.9, 1 m$M$ MgSO$_4$, 2 m$M$ EGTA, 50 $\mu M$ GTP or GDP. This is the ultracentrifuge buffer used for many previous studies, but any typical tubulin buffer works as long as the [GXP] is kept <100 $\mu M$, and DTT ~0.1 m$M$ to avoid optical cutoff and baseline drifting. TCEP may be substituted for DTT to avoid these problems.

8. Stopped-flow buffer: equilibrate PC-tubulin (1–4 $\mu M$), using spun columns, into 10 m$M$ PIPES (pH 6.9), 2 m$M$ EGTA, 1 m$M$ MgSO$_4$, and 50 $\mu M$ GTP and up to 50 $\mu M$ vinorelbine, vinblastine, or vincristine.

## 3. Methods

We determine IC$_{50}$ values (concentration at which cell proliferation is reduced by 50%) for each cell line using two independent methods: cell counting and an assay for mitochondrial respiration.

### 3.1. IC$_{50}$ Measurement: Cell Counting

1. Prior to setting up IC$_{50}$ experiments, trypsinize adherent cells in a 75-cm$^2$ flask. To do this, wash two times with PBS, and then overlay cells with 1 mL 1X trypsin. When cells release from the plate, stop the enzymatic activity with 1 mL of media. Nonadherent cells do not require this trypsin release step.

2. Collect cells by pipetting and transfer to a 15-mL polystyrene conical tube to pellet them at low speed (2000 rpm for 5 min) at room temperature.

3. Resuspend cell pellets in 2–4 mL of medium and count using a hemocytometer.

4. Set up duplicate 12-well plates (5000 cells/well) for each cell counting experiment. Two to four independent experiments are done for each drug and cell line. Incubate cells for 24 h at 37°C and 5% CO$_2$.

5. Expose cells to twofold serial dilutions (one log unit above and below the predicted IC$_{50}$ value) for 48 h. For control wells, add an appropriate amount of drug diluent (e.g., DMSO, water, PBS, and so on).

6. Wash adherent cells twice with 1 mL PBS per well and release using 100 $\mu$L 1X trypsin. Stop the enzymatic reaction with 100 $\mu$L medium. For nonadherent cells add 100 $\mu$L of PBS to each well.

7. Add 100 $\mu$L of filtered trypan blue, mix well with a glass pipet and immediately count in a hemocytometer, excluding nonviable cells that stain with Trypan blue.

8. Plot data as cell number vs drug concentration (molar). Fit data to first-order exponential decay, and calculate the IC$_{50}$ values (**Fig. 1**). For exponential decay:

$$y = yo + A \exp(-x/t1)$$

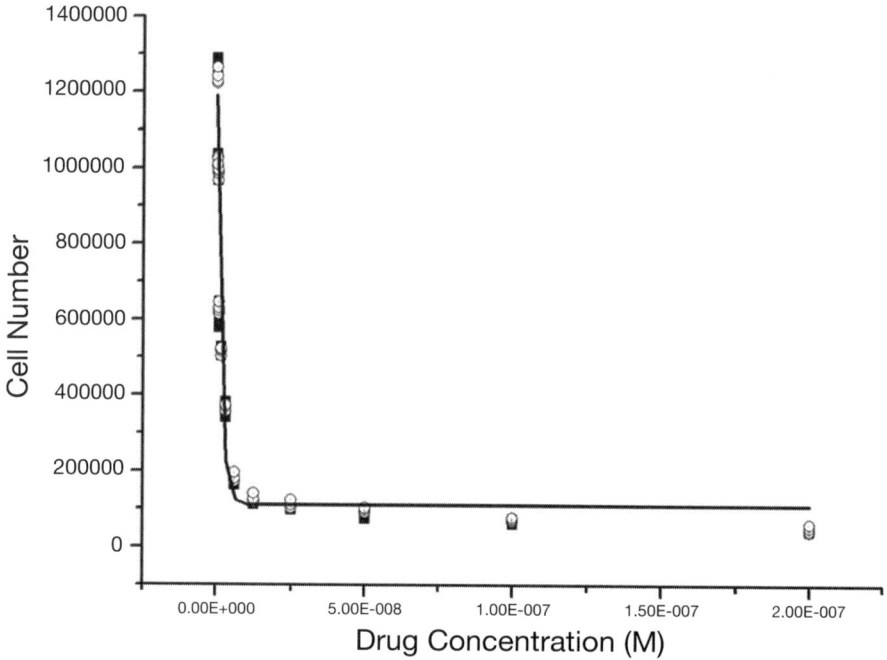

Fig. 1. $IC_{50}$ measurements for vinblastine treatment of colon cancer cells, COLO 205: cell number is plotted vs drug concentration. The data are from two separate experiments. The line represents the first order exponential decay fit of the data.

The $IC_{50}$ value in molar units is obtained directly by solving the exponential decay equation:

$$IC_{50} = -\tau_1 \ln \frac{1}{2}$$

## 3.2. $IC_{50}$ Measurement: Mitochondrial Respiration Assay (Used for Adherent Cells Only)

1. Follow **steps 1–3** for cell counting assay.
2. Plate adherent cells in 96-well microtiter plates at 8000–10,000 cells/well. The initial number of cells is determined empirically and is dependent upon the cell doubling time. Duplicate plates are set up for each experiment; and two to four independent experiments are done for each drug and cell line.
3. Incubate cells for 24 h, and then replace the medium with fresh medium containing appropriate concentrations of drug or solvent (as done in the cell counting assay no. 4).
4. After 72 h, remove the medium from the wells and add 100 µL of CellTiter 96® AQ$_{ueous}$ One Solution (Promega) diluted 1:5 with medium.

5. Incubate the plates for 2 h at 37°C and 5% $CO_2$, and then read the absorbance at 490 nm using a microplate reader. Absorbance values are corrected by subtracting baseline values obtained from wells with no cells.
6. Plot the baseline-corrected absorbance vs drug concentration and determine $IC_{50}$ values as described for the cell counting assay (no. 8).

### 3.3. Isolation of Intact Microtubules From Cells

When cell lines are determined to be sensitive or resistant to vinca alkaloids by $IC_{50}$ measurements, it is useful to compare the relative amounts of tubulin in the intracellular free and polymer pools. An increased fraction of tubulin in the polymer pool could be the cause of drug resistance. Extensive work has been done by Cabral and coworkers, examining tubulin extraction buffers for quantifying microtubule polymer levels in drug-sensitive and -resistant CHO cells *(11)*. They found that the amount of polymer is lysis buffer-dependent and that a paclitaxel-containing buffer at 4°C maintains the cytoskeleton intact, without inducing additional polymerization of free tubulin subunits. Thrower, Jordan and Wilson *(10)*, however, recommend a buffer that was not tested by Cabral and coworkers, containing glycerol, DMSO and GTP to stabilize microtubules at 37°C. Both extraction buffers result in about 40% polymerized tubulin in cell extracts (CHO cells for Cabral and coworkers; HeLa cells for Thrower, Jordan, and Wilson). Both methods for extraction of intact microtubules and free tubulin pools are presented here. We also present a procedure for extraction of total cellular microtubule populations for 2D IEF and SDS-PAGE analysis of isotype distributions. The role of tubulin isotypes in drug resistance continues to be controversial, but nonetheless important *(6,8)*.

### 3.3.1. Protocol Using Paclitaxel Buffer *(11)*

1. Wash cell samples ($2–4 \times 10^7$ cells/sample) with PBS and centrifuge at 2000 rpm for 5 min in a low speed centrifuge at 4°C.
2. Resuspend cell pellets in 1 mL paclitaxel microtubule stabilization buffer and homogenize on ice in a 1-mL dounce. Examine cells by light microscopy for complete cell lysis. Complete lysis may take up to 10 min.
3. Determine the total protein concentration using a Bradford type assay. We use the Advanced Protein Assay Reagent (Cytoskeleton) according to the manufacturer's instructions.
4. Centrifuge the whole cell lysate in a 1-mL tube for 10 min at 18,500 rpm ($12,000g$) and 4°C in a Beckmann TL100 tabletop centrifuge.
5. Remove the supernatant from the tube carefully without disturbing the pellet and add 1:3 hot SDS sample buffer. We use sample buffer warmed in a heating block at 100°C. Boil the diluted supernatant for 2 min to denature the proteins.
6. Resuspend the pellet in 400 µL hot SDS sample buffer and dounce to break up the pellet and boil for 2 min.

7. We use standard methods for SDS-PAGE and stain gels with Coomassie brilliant blue (R250). We load known amounts of pig brain tubulin in four wells to create a standard curve to permit densitometric quantification of unknown samples. For the standard curve, plot known amount of tubulin (milligram loaded) vs OD (determined by densitometry) and fit the data by linear regression. The unknowns are calculated from the fit of the data. Note that it is necessary to correct the supernatant and the pellet for the respective volumes to compare the total tubulin amounts.

### 3.3.2. Preparing Glycerol-Stabilized Microtubules (10)

1. Wash cell samples ($2–4 \times 10^7$ cells/sample) with PBS and centrifuge for 5 min at 37°C in a low speed centrifuge.
2. Resuspend cell pellets in 1 mL of glycerol containing microtubule stabilization buffer and homogenize in a prewarmed 1 mL dounce in a 37°C water bath. We do this in a test tube rack set directly into a water bath so that the temperature of the microtubule-containing solution does not vary. Cell lysis may take up to 10 min and can be monitored by light microscopy.
3. Determine total protein concentration by Bradford type assay as described for the paclitaxel-stabilized microtubules in above.
4. Centrifuge the whole cell lysate in a 1-mL tube for 5 min at 420,000$g$ and 37°C in a Beckman TL100 tabletop centrifuge.
5. Mix the cell pellets and supernatants with hot SDS sample buffer as in the paclitaxel stabilization buffer extraction (**Subheading 3.3.1., steps 5** and **6**) and quantify by SDS PAGE (as in **Subheading 3.3.1., step 7**).

### 3.3.3. Extracting Total Cellular Tubulin for 2D IEF and SDS-PAGE.

2D-IEF can be used to examine changes in tubulin isotypes and modifications of tubulin isotypes that may be present in vinca alkaloid- sensitive or -resistant cells. We describe here a protocol for extracting total tubulin from cells and resolving the tubulin isoforms using 2D-IEF.

1. Wash cell samples ($2–10 \times 10^7$ cells/sample) with PBS and centrifuge for 5 min at 37°C in a low speed centrifuge.
2. Dounce cells on ice in cell lysis solution (1 m$M$ MgSO$_4$ and 1 m$M$ EGTA, pH 6.9) and monitor by light microscopy to obtain complete cell lysis.
3. Add buffer to the sample to obtain a final concentration of 0.1 $M$ MES, pH 6.9, 1 m$M$ EGTA, 1 m$M$ MgSO$_4$, and 1 m$M$ GTP with protease inhibitor cocktail.
4. Centrifuge at 55,000 rpm for 1 h at 4°C to remove DNA and cell debris. All centrifugations are done in 1-mL tubes in a tabletop Beckman Coulter (Fullerton, CA) TL100 centrifuge.
5. Remove the supernatant from the pellet and bring the supernatant to 37°C for 1 h in 0.1 $M$ MES, pH 6.9, 1 m$M$ EGTA, 1 m$M$ MgSO$_4$ buffer containing 20 μ$M$ paclitaxel and 1 m$M$ GTP to isolate paclitaxel-stabilized microtubules.
6. Centrifuge the mixture at 48,000 rpm and 37°C through a 200 μL 10% sucrose cushion for 10 min to pellet microtubules. Free tubulin will remain in the supernatant.

7. Resuspend the microtubule pellet in MAP removal buffer with 20 μ*M* paclitaxel and 0.35 *M* NaCl and incubate at 37°C for 20 min to remove MAPs from the microtubules.

8. Centrifuge the mixture at 48,000 rpm and 37°C for 10 min to obtain purified microtubule pellets.

9. Rinse the final pellet with 0.1 m*M* MES, pH 6.9, 1 m*M* EGTA, 1 m*M* $MgSO_4$, 1 m*M* GTP, and 20 μ*M* paclitaxel.

10. Monitor the purification by SDS-PAGE and Western blotting using a pan-β-tubulin antibody (TU27, Sigma Chemical Co.). By comparing total tubulin in the initial supernatant and pellet with the amount of tubulin in the final pellet, we estimate that generally 90% or more of the total tubulin is extracted in the purification.

11. For 2D-IEF, resuspend the microtubule pellets in solubilization buffer (7 *M* urea, 2 *M* thiourea, 2% 3-[(3-cholamidopropyl)dimethylammonio]-1-propanesulfonate) (CHAPS), 0.5% Triton X-100, 2% 2-mercaptoethanol. Buffer may be made up in advance in 10-mL aliquots and should be remade weekly. A 30-s exposure to boiling water helps to solubilize the pellet, but may also cause deamidation and thus should be used cautiously.

12. Perform IEF in the first dimension using an IPGphor IEF system (Bio-Rad, Hercules, CA) and 17 cm IEF strips pH 4.7–5.9. Prior to focusing, hydrate strips at room temperature under humid conditions for 24 h. We wet the strip in a rehydration tray without oil, covering the tray with the cover, and place in a plastic zip lock bag with wet paper towels to prevent evaporation. Transfer the strips to the Protean IEF focusing tray and cover with mineral oil. Run the IEF strips according to the manufacturer's instructions. Immediately following IEF, stain strips with acid violet 17 (modified from **ref. *33***) or prepare for 2D SDS-PAGE.

13. For acid violet 17 staining, place the strips in a small plastic box or a glass tray. Wash about 20 s in acetone to remove the mineral oil. Next fix for 30 min in 50% methanol, 20% TCA. Then rinse three times for 1 min with 3% phosphoric acid. Next stain for 10 min with 0.1% acid violet 17 (100 mg/100 mL), 10% phosphoric acid. (The acid violet 17 dye [100 mg] should be dissolved in 50 mL of water and allowed to mix for a few hours, i.e., the time of the IEF run. Then just before staining mix with 50 mL of 20% phosphoric acid. A fresh solution every day is required to get good solubility without colloidal precipitation. It is often necessary to stain strips separately to keep track of different conditions or cell types so more than 100 mL of stain may be required.) Destain 1–2 h in 3% phosphoric acid. Wash with water three times for 1 min. Then impregnate with 2% glycerol for 20 min and air-dry.

14. For 2D SDS-PAGE, transfer the strip to a small plastic or glass tray and equilibrate by shaking in SDS sample buffer for 15–30 min. Typical mini gels are only 8–9 cm wide, so using a stained strip as a guide, cut the IPG strip at both ends (scissors or razor blade). Slide the cut strip between the glass plates onto a prepoured separation gel and embed it in melted 1% low temperature melt agarose made up in SDS stacking buffer. Run the gel using standard SDS-PAGE procedures, adding dye to the upper reservoir to track the front. Stain with either Coomassie brilliant blue or transfer to PVDF for a Western blot.

### 3.4. Protocol for Fluorescently Labeled Vinblastine Interaction With Tubulin by Fluorescence Microscopy

The direct interaction of fluorescently labeled vinca alkaloids with tubulin can be visualized in cells by fluorescent microscopy. Experiments with vinblastine-bodipy showed interaction with tubulin and with secondary sites such as mitochondria *(34)*. Microtubules and other organelles can be labeled with antibodies to identify sites of colocalization. We present here protocols for treatment of adherent cells with fluorescently labeled vinblastine and for labeling of tubulin with isotype-specific antibodies.

1. Plate cells on cover slips at $5 \times 10^5$ cells/mL and incubate overnight (5% $CO_2$ and 37°C) (e.g., six cover slips in a Petri dish with 5 mL medium).
2. Replace medium with medium plus fluorescently labeled drug at the $IC_{50}$ concentration and incubate for various times, 2–24 h.
3. Wash with filtered PBS at room temperature. We replace the medium with PBS and let stand at room temperature for 5 min. After removing the PBS with a pipet, the excess PBS is dried by capillary action at the plate edges with filter paper.
4. Fix and permeabilize cells with 100% methanol, –20°C on ice for 4 min.
5. Wash cells with PBS. This wash is done for 5 min as in **step 3**. Then the excess PBS is removed by capillary action with filter paper.
6. For tubulin antibodies dilute 1–3 mg/mL stocks 10-fold in PBS for a final antibody concentration of 0.1–0.3 mg/mL. Use 2 µL of this dilute (0.1–0.3 mg/mL) antibody solution in 2 mL 1% BSA/PBS or 3% NFD milk/PBS. Place 400-µL drops of primary antibody on cover slips and leave 30 min at RT.
7. Wash cells three times with PBS and dry with filter paper by capillary action at the edge of the cover slip.
8. Use 6 µL Rhodamine second antibody in 1200 µL 1%BSA/PBS or 3% NFD milk/PBS. Place 400-µL drops on cover slips for 30 min at RT.
9. Wash cells three times with PBS and dry with filter paper by capillary action at the edge of the cover slip.
10. Spread nail varnish along the edges of the cover slip. Apply 5 µL mounting media to a slide. Place the cover slip on top and seal. The slides are visualized immediately or kept for weeks at 4°C without loss of the fluorescence signal.

### 3.5. Spun Column Procedure to Equilibrate PC-Tubulin or MTP (38)

1. Equilibrate a 5-mL column of Sephadex G-50 with appropriate buffer. The G-50 is preswelled in water and stored cold with 0.01% NaAzide. We use a 5-mL syringe barrel with a frit in a 15-mL polypropylene conical tube and three 5-mL buffer exchanges by gravity on ice. It can also be done in a 1-mL syringe barrel with a smaller frit if sample volume is an issue.
2. Purified PC tubulin (~1–10 mg/mL) is typically frozen dropwise into liquid nitrogen. At this stage a few pellets of frozen tubulin are weighed (~0.1–0.2 g) in a 10-mL glass beaker. The exact amount depends upon the concentration of

the tubulin and the desired concentration for the experiment. The beaker is placed in a small Petri dish filled with room temperature water to facilitate thawing, and 300–400 µL of buffer are added to further speed up the melting rate and to prepare the right sample volume for adding to the spun column.

3. While tubulin is thawing, centrifuge the spun column in a 15-mL polypropylene conical tube in a tabletop clinical centrifuge at speed 5 (max speed 7) for 2 min (4°C).
4. Discard the flowthrough and place the column syringe in a dry 15-mL polypropylene conical tube.
5. Slowly dropwise load the PC-tubulin or MTP sample (450–600 µL) onto the dried column, being sure to not squirt material down the sides before the gel reswells, and centrifuge at speed 5 for 2 min. The column flow-through tubulin or MTP will be equilibrated in the chosen buffer. If the buffer contains high concentrations of glycerol, it may be necessary to repeat this step to achieve true equilibration.
6. Mix the flowthrough tubulin well and dilute an aliquot 10-fold (typically 60/600 µL) and measure the OD 278 nm blanking with the appropriate buffer. The tubulin concentration is 10*OD/1.2 in mg/mL or 10 times larger in micromolars (tubulin is a 100,000 Da heterodimer). Application of this technique to smaller tubulin binding proteins may require using a different resin (G-10) to achieve proper equilibration.

## 3.6. Turbidity: Critical Concentrations

We have used turbidity measurements to study vinca alkaloid interactions with tubulin. Single or combination drug effects can be measured. The critical concentration ($C_c$) is a measure of $K_p$ (polymer propagation constant) where $1/K_p = C_c$ *(39,40)*. This information, when collected as a function of temperature, can also be used to extract thermodynamic profiles including $\Delta H$, $\Delta S$, $\Delta C_p$. $C_c$ data can be used to calculate the free energy involved in drug interaction with tubulin. This type of analysis permits determination of whether drug combinations are additive, synergistic or inhibitory *(29)*. (It is worth noting that Cytoskeleton sells 96-well plates, CytoDYNAMIX Screen, useful for performing high throughput turbidity assays.)

1. Polymerize PC-tubulin (1–3.5 mg/mL) equilibrated in 100 m$M$ PIPES, pH 6.9, 1 m$M$ MgSO$_4$, 2 mM EGTA, 1 m$M$ GTP, and 2 $M$ glycerol. Polymerize MTP (0.6–3 mg/mL) in the same buffer without glycerol. Prior to polymerization degas samples on ice for 30 min. This can be done using a side-arm flask connected to a vacuum pump or a water-seal vacuum system.
2. Monitor microtubule formation using a spectrophotometer equipped with temperature control. We use a Gilson II UV-VIS scanning spectrophotometer equipped with a cooling Peltier cell holder and thin walled cuvets that hold 450 µL. Collect baseline data at 4°C and 350 nm.

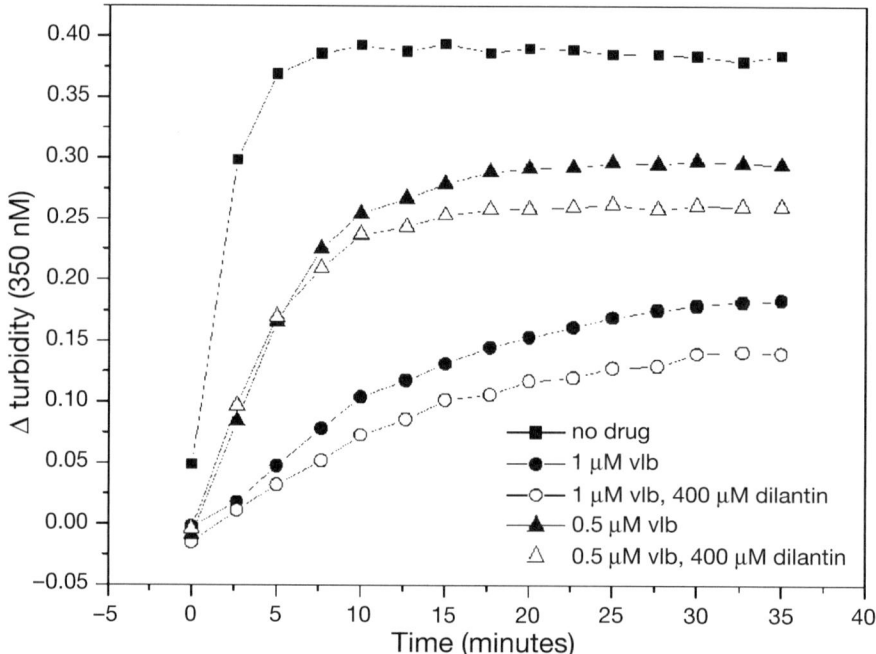

Fig. 2. Microtubule polymerization in the presence of dilantin and vinblastine (vlb). PC-tubulin was polymerized at 37°C and the change in turbidity was monitored at 350 nm. (Reprinted with permission from **ref. 29**.)

3. Increase the temperature to 37°C (or the desired temperature if you want to do a van't Hoft plot and thermodynamic analysis) and monitor the solutions at 350 nm for 45 min (**Fig. 2**).
4. Cool solutions to 0°C and record a second baseline.
5. Plot the change in optical density vs tubulin concentration after subtracting the second baseline from the plateau optical density at 45 min. The time at temperature to reach plateau may be longer at lower temperatures. The critical concentration, $C_c$, is determined from the x-intercept of the linear regression fits of the data. Typical data is linear but certain conditions may favor overshoot owing to sheet formation *(41)*.Critical concentrations are determined by combining the data from two or more independent experiments (**Table 1**).
6. Determine the effect of drugs on the critical concentrations by carrying out experiments in the presence or absence of drug over one to two log units (e.g., vinblastine (0.5–50 µM; *see* **ref. 42** for extinction coefficients of vinca alkaloids). (Vinca alkaloids are very water soluble and we make up stocks at typically 25–50 mM concentrations.) Control samples with the drug diluent (e.g., water or DMSO) are also monitored.

**Table 1**
**Critical Concentrations**

| PC tubulin | mg/mL |
|---|---|
| no drug | 0.310 (± 0.078) |
| 400 $\mu M$ dilantin | 0.375 (± 0.075) |
| 0.5 $\mu M$ vlb | 0.634 (± 0.151) |
| 0.5 $\mu M$ vlb, 400 $\mu M$ dilantin | 0.794 (± 0.103) |
| 1.0 $\mu M$ vlb | 0.914 (± 0.230) |
| 1.0 $\mu M$ vlb, 400 $\mu M$ dilantin | 1.150 (± 0.140) |
| 1.5 $\mu M$ vlb | 1.064 (± 0.311) |
| 1.5 $\mu M$ vlb, 400 $\mu M$ dilantin | 1.226 (± 0.225) |

Reprinted from **ref. 29**.

7. Calculate the free energy for microtubule polymer formation from the $K_p$. Complete thermodynamic information can be obtained as shown in **Table 2** for Dilantin and vinblastine (**29**). The additivity of the combined drug effects is apparent from the free energy comparisons.

### 3.7. Analytical Ultracentrifugation

Vinca alkaloids induce tubulin spiral formation and drug cytotoxicity correlates with spiral size (**42**). The energetics of spiral formation and the kinetics of tubulin self-association in the presence of drug can be measured by analytical ultracentrifugation (reviewed in **ref. 5**) and light scattering (**27,28**), respectively.

1. A typical titration experiment is done using from 0, 0.1, 1 $\mu M$ through 50, 60, 70 $\mu M$ drug. Because the XLA rotor holds three samples, the number of points in the titration is typically 15–18 samples. Twenty-five milliliters of buffer at each drug concentration is made by addition of drug from the vinca stock (typically 50 m$M$ in water). Each 5-mL spun column is equilibrated with approx 15 mL of buffer as previously described for preparing spun columns.
2. Because the concentration of free drug is a critical value in the analysis, it is important to dilute the protein prior to the spun column equilibration step to approximately achieve the desired concentration of drug and protein in the final centrifuge cell. The critical feature is that the sample must have an OD of 0.1 to 1 OD at the wavelength of choice, typically 278 nm with tubulin. Dilution of drug–tubulin complex releases free drug by mass action. The typical 1.2-cm cell holds 450 μL of sample, so the spun column sample should be 500–600 μL in volume.

**Table 2**
**Free Energy for Microtubule Polymerization (PC Tubulin, 37°C)[a]**

| No dilantin | $\Delta\Delta G$ | Dilantin |
|---|---|---|
| No drug<br>$\Delta G = -7.81 \times 10^3$ cal | $\Delta\Delta G = 110$ cal | 400 µ$M$ dilantin<br>$\Delta G = -7.70 \times 10^3$ cal |
| $\Delta\Delta G = 440$ cal | | $\Delta\Delta G = 470$ cal |
| 0.5 µ$M$ vinblastine<br>$\Delta G = -7.37 \times 10^3$ cal | $\Delta\Delta G = 139$ cal | 0.5 µ$M$ vinblastine, 400 µ$M$ dilantin<br>$\Delta G = -7.23 \times 10^3$ cal |
| $\Delta\Delta G = 224$ cal | | $\Delta\Delta G = 224$ cal |
| 1.0 µ$M$ vinblastine<br>$\Delta G = -7.15 \times 10^3$ cal | $\Delta\Delta G = 139$ cal | 1.0 µ$M$ vinblastine, 400 µ$M$ dilantin<br>$\Delta G = -7.01 \times 10^3$ cal |
| $\Delta G = 100$ cal | | $\Delta G = 40$ cal |
| 1.5 µ$M$ vinblastine<br>$\Delta G = -7.05 \times 10^3$ cal | $\Delta\Delta G = 80$ cal | 1.5 µ$M$ vinblastine, 400 µ$M$ dilantin<br>$\Delta G = -6.97 \times 10^3$ cal |

[a]The free energy was calculated from the $K_p$, equilibrium constant for microtubule polymerization, determined from turbidity experiments. $\Delta G$ is the free energy for the addition of a tubulin heterodimer to a microtubule and $\Delta\Delta G$ is the change in of free energy for microtubule polymerization due to the presence of dilantin, vinblastine or both. (Reprinted from **ref. *29*.**)

3. Prepare three samples at approximately the same tubulin concentration (2–10 µ$M$) and fill the cells using the appropriate buffer as a reference. (The absorbance of the drug contributes to the total signal and may preclude pseudo absorbance as a data collection method.) After sealing, align the cells in the rotor, place the rotor in the chamber, and allow the rotor to come to temperature equilibration before starting the run. Besides convective disturbances, temperature changes perturb the solution density, which complicates the assignment of solution density, viscosity, species buoyancy, and sedimentation behavior. It may also perturb the equilibrium state of the tubulin–drug system.

4. Data are collected using one flash of the lamp at a spacing of 0.002 cm. The speed is chosen to allow collection of at least 50–75 scans prior to pelleting of the boundary. Thus, samples are run in blocks of similar drug concentrations (e.g., 0, 0.1, 1 µ$M$; 2, 3, 5 µ$M$; 7.5, 10, 12.5 µ$M$; 15, 20, 25 µ$M$; 30, 35, 40 µ$M$; 50, 60, 70 µ$M$) that induce similar tubulin association and thus boundaries that sediment at comparable rates.

5. Analyze the data to determine the weight average sedimentation or Sw of the boundary. Using either DCDT to generate a g(s) distribution, or Sedfit to generate c(s), the distributions are integrated to produce Sw and C, or the concentration in the boundary expressed as absorbance at 278-nm units. Plot the data as Sw

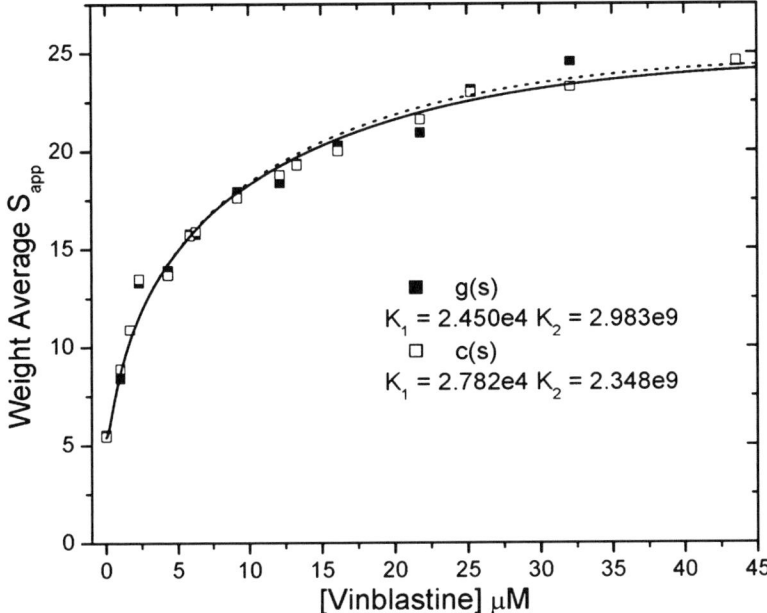

Fig. 3. Weight average Sw vs Vinblastine concentration. The points are derived from DCDT or g(s) and Sedfit or c(s) analysis of sedimentation velocity experiments done as various drug concentrations and fixed tubulin concentrations (2 μ*M*). Both sets of data fit equally well to a hydrodynamic 42-bead spiral model for drug-induced assembly of tubulin (*see* ref. *44* for details). The derived parameters correspond to drug binding to tubulin ($K_1$) and liganded tubulin assembly into a spiral ($K_2$).

vs C in mg/mL using extinction × length = 1.2 × 1.2 = 1.44 mL/mg × cm. Weight average values are fit to a ligand linked indefinite spiral assembly mechanism. We use an algorithm written in Fitall (MTR Software, Toronto, Canada) (**Fig. 3**) *(43)*.

6. More advanced analysis involves direct fitting the shape of the boundary to various hydrodynamic spiral models with SEDANAL *(44,45)*.

## 3.8. Stopped Flow Light Scattering

To investigate the kinetics of drug-induced tubulin association we use rapid-mixing stopped-flow light scattering *(27,28)*.

1. Equilibrate PC-tubulin (1–4 μ*M*), using spun columns, into buffer (10 m*M* PIPES, pH 6.9), 2 m*M* EGTA, 1 m*M* MgSO$_4$, and 50 μ*M* GTP) plus drug.
2. Degas samples for 1 h at room temperature. This can be done using a side arm flask connected to a vacuum pump or a water-sealed vacuum system.
3. Initiate stopped-flow rapid-mixing experiments by diluting tubulin samples 1:1 with the same buffer without drug. We use a manual stopped-flow apparatus (Hi-

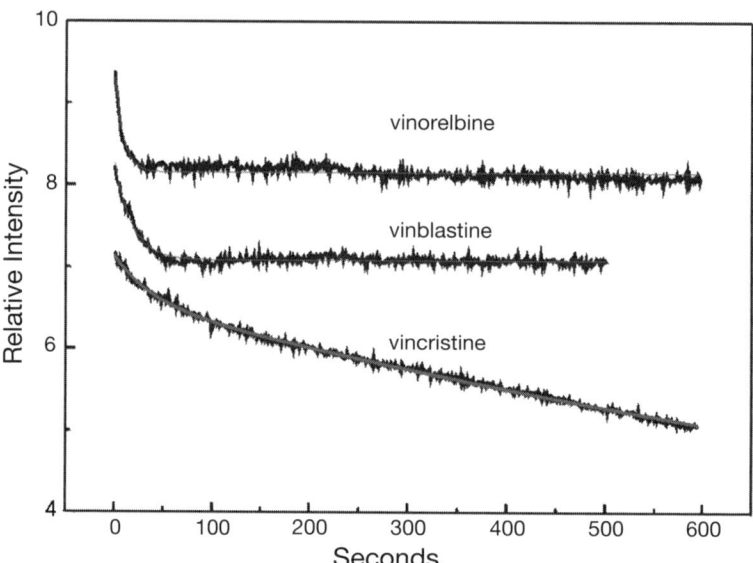

Fig.4. Stopped-flow light scattering experiments. Relative intensity is plotted vs time (seconds). Following dilution to 25 μ*M* drug, scattering was monitored over 10 min at 1 s intervals: (**A**) vinorelbine, (**B**) vinblastine, (**C**) vincristine. (Reprinted with permission from **ref. 28**.)

Tech Wiltshire, England). The final drug and tubulin concentrations are 25 and 0.5–2 μ*M*, respectively.

4. Monitor relaxation using a fluorimeter at 350 nm (90° scattering) over 10 min at 10-s intervals for vinflunine, vinorelbine, and vinblastine and over 30–45 min for vincristine at 1-s intervals (**Fig. 4**).

5. Fit data by exponential decay to obtain relaxation times, $\tau$, using the following equation, $x = a_0 \exp(-1/\tau)$, where $a_0$ is the change in amplitude associated with $\tau$ over time, t (seconds) *(46)* (**Table 3**). Use chi-square to select the best fits of the data.

6. Alternatively, two relaxation times, $\tau_1$ and $\tau_2$, are determined by fitting data with the summation $x = a_1 \exp(-1/\tau_1) + a_2 \exp(-1/\tau_2)$, where $a_1$ and $a_2$ are the amplitude changes associated with $\tau_1$ and $\tau_2$.

7. To calculate the on rate, $k_a$, and the off rate, $k_d$, for vinblastine, vinorelbine, and vinflunine, plot $1/\tau^2$ against the final tubulin concentration. Fit the data by linear regression and use the slope and y-intercept to determine $k_a$ and $k_d$, respectively *(47)*. The vincristine kinetic data include a significant contribution from large spiral condensates or aggregates that dominate the light scattering and slow the relaxation process, and therefore, $k_a$ and $k_d$ can not be estimated in this way (**Table 4**).

**Table 3**
**Stopped-Flow Light Scattering With a Drug Dilution of 50 to 25 μM**

| Drug | [tubulin] μM | A1 | $\tau$ (s) |
|---|---|---|---|
| Vinorelbine | 0.5 | 0.66 +/- 0.10 | 12.77 +/- 3.44 |
| | 1.0 | 0.99 +/- 0.18 | 9.33 +/- 1.04 |
| | 1.5 | 1.25 +/- 0.11 | 7.56 +/- 1.29 |
| Vinblastine | 0.5 | 0.55 +/- 0.07 | 34.60 +/- 2.29 |
| | 1.0 | 0.96 +/- 0.05 | 26.18 +/- 0.33 |
| | 1.5 | 1.25 +/- 0.11 | 19.97 +/- 4.98 |

| Drug | [tubulin] μM | A1 | $\tau_1$ (s) | A2 | $\tau_2$ (s) |
|---|---|---|---|---|---|
| Vincristine | 0.5 | 1.15 +/- 0.12 | 145.72 +/- 34.54 | 0.86 +/- 0.28 | 561.47 +/- 56.92 |
| | 1.0 | 0.37 +/- 0.05 | 51.29 +/- 33.25 | 3.05 +/- 0.17 | 538.73 +/- 122.53 |
| | 1.5 | 0.51 +/- 0.11 | 54.80 +/- 18.71 | 2.54 +/- 0.03 | 417.53 +/- 1.20 |
| | mean | | 92.76 +/- 56.17 | | 513.85 +/- 89.53 |

Reprinted from **ref. 28**.

**Table 4**
**Kinetic and Drug Binding Parameters From Stopped-Flow Measurements**

|  | $k_a$ ($M^{-1}s^{-1}$) | $k_d$ ($s^{-1}$) | $K$ ($M^{-1}$) | $K_2$ ($M^{-1}$) |
|---|---|---|---|---|
| Vinorelbine | $1.1 \times 10^5$ | 0.0210 | $5.4 \times 10^6$ | $1.5 \times 10^6$ |
| Vinblastine | $4.2 \times 10^5$ | 0.0082 | $5.2 \times 10^6$ | $5.1 \times 10^6$ |

Reprinted from **ref. 28**.

## Acknowledgments

We thank Dr. Laree Hiser, Ashish Aggarwal, and Valeria Cucchiarelli for helpful discussions during the preparation of this chapter.

## References

1. Noble, R. L, Beer, C. T., and Cutts, J. H. (1958) Role of chance observation in chemotherapy: vinca rosea. *Ann. N.Y. Acad. Sci.* **76,** 882–894.
2. Johnson, I. S., Wright, H. F., and Svoboda, G. H. (1959) Experimental basis for clinical evaluation of antitumor principles derived from *Vinca rosea* Linn. *J. Lab. Clin. Med.* **54,** 830.
3. Hill, B. T. (2001) Vinflunine, a second generation novel Vinca Alkaloid with a distinctive pharmacological profile, now in clinical development and prospects for future mitotic blockers. *Curr. Pharm. Des.* **7,** 1199–1212.
4. Kruczynski, A. and Hill, B. T. (2001) Vinflunine, the latest Vinca alkaloid in clinical development. A review of its preclinical anticancer properties. *Crit. Rev. Oncol. Hematol.* **40,** 159–173.
5. Lobert, S. and Correia, J. J. (2000) Energetics of vinca alkaloid interactions with tubulin. *Meth. Enzymol. Energetics of Macromolecules* **323,** 77.
6. Correia, J. J. and Lobert, S. (2001) Physiochemical aspects of tubulin-interacting, antimitotic drugs. *Curr. Pharm. Design* **7,** 1213–1228.
7. Cabral, F. (2006) Mechanisms of resistance to drugs that interfere with microtubule assembly, in *Microtubules in Health and Disease*, (Fojo, T., ed.), Humana Press, Totowa, NJ, in press.
8. Hiser, A., Aggarwal, R., Young, A., et al. (2006) Correlation of beta-tubulin mRNA and protein levels in twelve human cancer cell lines. *Cell Motil. Cytoskel.* **63,** 41.
9. Kavallaris, M., Tait, A. S., Walsh, B. J., et al. (2001) Multiple microtubule alterations are associated with vinca alkaloid resistance in human leukemia cells. *Cancer Res.* **61,** 5903–5809.
10. Thrower, D., Jordan, M. A., and Wilson, L. (1993) A quantitative solid-phase binding assay for tubulin. *Methods Cell Biol.* **37,** 129–145.
11. Minotti, A. M., Barlow, S. B., and Cabral, F. (1991) Resistance to antimitotic drugs in Chinese hamster ovary cells correlates with changes in the level of polymerized tubulin. J. Biol. Chem. **266,** 3987–3994.

12. Dhamodharan, R., Jordan, M. A., Thrower, D., Wilson, L., and Wadsworth, P. (1995) Vinblastine suppresses dynamics of individual microtubules in living interphase cells. *Mol. Biol. Cell* **6**, 1215–1229.
13. Jordan, M. A. and Wilson, L. (1998) Use of drugs to study role of microtubule Assembly dynamics in living cells. *Methods Enzymol.* **298**, 252–276.
14. Jordan, M. A. (2002) Mechanisms of action of antitumor drugs that interact with microtubules and tubulin. *Curr. Med. Chem. Anti-Cancer Agents* **2**, 1–17.
15. Wilson, L., Panda, D., and Jordan, M. S. (1999) Modulation of microtubule dynamics by drugs: a paradigm for the actions of cellular regulators. *Cell Struct. Funct.* **24**, 329–335.
16. Beck, W. T., Cirtain, M. C., and Lefko, J. L. (1983) Energy-dependent educed drug binding as a mechanism of vinca alkaloid resistance in human leukemic lymphoblasts. *Mol. Pharmacol.* **24**, 485–492.
17. Ferguson, P. J., Phillips, J. R., Selner, M., and Cass, C. E. (1984) Differential activity of vincristine and vinblastine against cultured cells. *Cancer Res.* **44**, 3307–3312.
18. Bai, R., Swartz, R. E., Kepler, J. A., Pettit, G. R., and Hamel, E. (1996) Characterization of the interaction of cryptophycin 1 with tubulin: binding in the Vinca domain competitive inhibition of dolastatin 10 binding, and unusual aggregation reaction. *Cancer Res.* **56**, 4398–4406.
19. Wilson, L., Jordan, M. S., Morse, A., and Margolis, R. L. (1982) Interaction of vinblastine with steady-state microtubule in vitro. *J. Mol., Biol.* **159**, 125–149.
20. Jordan, M. A. and Wilson, L. (1990) Kinetic analysis of tubulin exchange at microtubule ends at low vinblastine concentrations. *Biochemistry* **29**, 2730–2739.
21. Wilson, L., Creswell, K. M., and Chin, D. (1975) The mechanism of action of vinblastine. Binding of [*acetyl*-$^3$H] vinblastine to embryonic chick brain tubulin and tubulin from sea urchin sperm tail outer doublet microtubules. *Biochemistry* **14**, 5586–5592.
22. Lee, J. C., Harrison, D., and Timasheff, S. N. (1975) Interaction of vinblastine with calf brain microtubule protein. *J. Biol. Chem.* **250**, 9276–9282.
23. Singer, W. D., Hersh, R. T., and Himes, R. H. (1988) Effect of solution variables on the binding of vinblastine to tubulin. *Biochem. Pharmacol.* **37**, 2691–2696.
24. Prakash, V. and Timasheff, S. N. (1991) Mechanism of interaction of vinca alkaloids with tubulin: catharanthine and vindoline. *Biochemistry* **30**, 873–880.
25. Sackett, D. L. (1995) Vinca site agents induce structural changes in tubulin different from and antagonistic to changes induced by colchicines site agents. *Biochemistry* **34**, 7010–7019.
26. Rai, S. S. and Wolff, J. (1987) Localization of the vinblastine-binding site on β-tubulin. *J. Biol. Chem.* **271**, 14,707–14,711.
27. Lobert, S., Ingram, J. W., Hill, B. T., and Correia, J. J. (1998) A comparison of thermodynamic parameters for vinorelbine- and vinflunine-induced tubulin self association by sedimentation velocity. *Mol. Pharmacol.* **53**, 908–915.
28. Lobert, S., Vulevic, B., and Correia, J. J. (1996) Interaction of vinca alkaloids with tubulin: a comparison of vinblastine, vincristine and vinorelbine. *Biochemistry* **35**, 6806–6814.

29. Lobert, S., Ingram, J. W., and Correia, J. J. (1999) Additivity of dilantin and vinblastine inhibitory effects on microtubule assembly. *Cancer Res.* **59,** 4816–4822.
30. Verdier-Pinard, P., Gares, M., and Wright, M. (1999) Differential in vitro association of vinca alkaloid-induced tubuline spiral filaments into aggregated spirals. *Biochem. Pharmacol.* **58,** 959–971.
31. Gigant, B., Wang, C., Ravelli, R. B. G., et al. (2005) Structural basis for the regulation of tubulin by binblastine. *Nature* **435,** 519–522.
32. Nogales, E., Medrano, F. J., Diakun, G. P., Mant, G. R., Towns-Andrews, E., and Bordas, J. (1995) The effect of temperature on the structure of vinblastine-induced polymers of purified tubulin: detection of a reversible conformational change. *J. Mol. Biol.* **254,** 416–430.
33. Williams, R. C., Shah, C., and Sackett, D. (1999) Separation of tubulin isoforms by isoelectric focusing in immobilized pH gradient gels. *Anal. Biochem.* **275,** 265–267.
34. Lobert, S. and Correia, J. J. (2001) Characterization of fluorescent vinca alkaloid (bobipy-vinblastine) to probe drug targets and apoptosis. *Mol. Biol. Cell* **12,** 431a.
35. Williams, R. C., Jr. and Lee, J. C. (1982) Preparation of tubulin from pig brain. *Methods Enzymol.* **85,** 376–408.
36. Correia, J. J., Baty, L. T., and Williams, R. C., Jr. (1987) $Mg^{2+}$ dependence of guanine nucleotide binding to tubulin. *J. Biol. Chem.* **262,** 17,278–17,284.
37. Detrich, H. W. and Williams, R. C., Jr. (1978) Reversible dissociation of the αβ dimer of tubulin from bovine brain. *Biochemistry* **34,** 3900–3907.
38. Penefsky, H. S. (1979) A centrifuged-column procedure for the measurement of ligand binding by beef heart F1. *Methods Enzymol.* **56,** 527–531.
39. Lee, J. C. and Timasheff, S. N. (1977) In vitro reconstitution of calf brain microtubules: effects of solution variables. *Biochemistry* **16,** 1754–1764.
40. Vulevic, B. and Correia, J. J. (1997) Thermodynamic and structural analysis ofmicrotubule assembly: the role of GTP hydrolysis. *Biophys. J.* **72,** 1357–1375.
41. Detrich, H. W., Jordan, M. A., Wilson, L., and Williams, R. C. (1985) Mechanism of microtubule assembly. Changes in polymer structure and organization during assembly of sea urchin egg tubulin. *J. Biol. Chem.* **260,** 9479–9490.
42. Lobert, S., Fahy, J., Hill, B. T., Duflos, A., Entievant, C., and Correia, J. J. (2000) Vinca alkaloid-induced tubulin spiral formation correlates with cytotoxicity in the leukemic L1210 cell line. *Biochemistry* **39,** 12,053–12,062.
43. Correia, J. J. (2000) The analysis of weight average sedimentation data. *Methods Enzymol.* **321,** 81–100.
44. Sontag, C. A., Stafford, W. F., and Correia, J. J. (2004) A comparison of weight average and direct boundary fitting of sedimentation velocity data for indefinite polymerizing systems. *Biophys. Chem.* **108,** 215–230.
45. Correia, J. J., Sontag, C. A., Stafford, W. F., and Sherwood, P. J. (2005) Models for direct boundary fitting of indefinite ligand-linked self-association, in *Analytical Ultracentrifugation: Techniques and Methods,* (Scott, D., Harding, S., and Rowe, A., eds.). Royal Society of Chemistry, Cambrige, UK, pp. 51–63.
46. Bernasconi, C. F. (1976) *Relaxation Kinetics.* Academic Press, New York.
47. Thusius, D., Dessen, P., and Jallon, J. M. (1975) Mechanism of bovine liver glutamate dehydrogenase self-association: I. Kinetic evidence for a random association of polymer chains. *J. Mol. Biol.* **92,** 412–432.

# 19

## High-Throughput Screening of Microtubule-Interacting Drugs

### Susan L. Bane, Rudravajhala Ravindra, and Anna A. Zaydman

#### Summary

Drugs that affect microtubule dynamics are among the most effective anticancer agents in routine clinical use. The standard assay for antimicrotubule agents observes the ability of a particular substance to affect in vitro microtubule assembly. We have modified these procedures so that they can be performed in 96-well plates using a standard fluorescence plate reader. Two different protocols are provided in this chapter. One of these protocols is for ligands that inhibit microtubule polymerization, such as colchicine and related molecules. The second is for ligands that promote in vitro microtubule assembly, such as Taxol.

**Key Words:** Microtubules; tubulin polymerization; antitumor agents; antimicrotubule drugs; high-throughput screening; drug discovery; colchicine; taxol; 4',6-diamidino-2-phenylindole.

## 1. Introduction

Because of the success of tubulin-binding agents in cancer chemotherapy *(1)*, there is continuing interest in discovering new substances that interfere with microtubule-mediated processes. One of the simplest assays for antimicrotubule agents involves monitoring microtubule assembly in vitro by light scattering, which can be performed on an absorption spectrophotometer *(2,3)*. Alternatively, it was observed that the nucleic acid stain 4',6-diamidino-2-phenylindole (DAPI) undergoes fluorescence changes upon tubulin binding and upon tubulin assembly, and that the change in emission intensity can be used as a signal to monitor polymerization of the protein *(4,5)*.

The turbidity method has been adapted to a 96-well plate format by a number of groups *(6–10)*. A drawback to monitoring apparent absorption in microtiter plates is that absorption intensity is a function of the path length of

From: *Methods in Molecular Medicine, Vol. 137: Microtubule Protocols*
Edited by: Jun Zhou © Humana Press Inc., Totowa, NJ

the solution, and therefore miniaturizing the sample decreases the signal strength. Fluorescence can be measured from the surface of a solution, and therefore miniaturization of the sample will not necessarily affect the intensity of the signal. We adapted the fluorescence method first described by Bonne et al. to a robust 96-well plate assay that can be used to evaluate both inhibitors and promoters of in vitro microtubule assembly using a standard fluorescence plate reader *(11)*.

The procedures described here are for identification of potential active molecules from a library of compounds. Quantitative analysis of the active compounds (to determine an $IC_{50}$) can also be performed in a high throughput format. Protocols for these quantitative assays are described elsewhere *(11,12)*.

Two different protocols are provided. One of these protocols is for ligands that inhibit in vitro microtubule assembly, such as colchicines and related molecules. This assay uses microtubule protein (MTP, tubulin plus the microtubule associated proteins [MAPs] that copurify with tubulin during cycles of assembly and disassembly). The second is for ligands that promote in vitro microtubule assembly, such as Taxol. Pure tubulin is used in this assay.

## 2. Materials

### 2.1. Inhibition of Tubulin Polymerization

1. Microtubule protein, purified from bovine (or other ruminant) brain (*see* **Note 1**).
2. PME buffer: 100 m*M* PIPES, 1 m*M* MgSO$_4$, 2 m*M* EGTA (*see* **Note 2**).
3. Desalting column: G-25 or G-50 1-mL syringe columns (*see* **Note 3**).
4. DMSO, spectroscopic grade.
5. DAPI in H$_2$O (1 m*M*). Freeze in 250-µL aliquots. Thaw necessary amount and discard what is not used.
6. Drugs to be tested in DMSO.
7. Colchicine in DMSO: 1 m*M* solution. Keep this solution in the dark when not in use. Make fresh weekly (*see* **Note 4**).
8. Stock solution of GTP in H$_2$O (50 m*M*).
9. 96-Well plate (for example, NUNC, cat no. 137101).
10. Fluorescence plate reader, capable of producing excitation light in the vicinity of 360 nm and detect emission in the vicinity of 450 nm (*see* **Note 5**).

### 2.2. Promotion of Tubulin Assembly

1. Purified tubulin, purified from bovine (or other ruminant) brain (*see* **Note 6**).
2. Buffer: PME (*see* **Note 2**).
3. Desalting column: G-25 or G-50 1-mL syringe columns (*see* **Note 3**).
4. DMSO, spectroscopic grade.
5. Drugs to be tested: stock solutions of 1 m*M* in DMSO.
6. Taxol in DMSO: 1 m*M* solution (*see* **Note 7**).
7. Stock solution of GTP in H$_2$O (50 m*M*).
8. 96-Well plate (such as NUNC, cat no. 137101).

9. DAPI in $H_2O$ (1 m$M$). Freeze in 250 µL aliquots. Thaw necessary amount and discard what is not used.

10. Fluorescence plate reader, capable of producing excitation light in the vicinity of 360 nm and detect emission in the vicinity of 450 nm.

## 3. Methods

### 3.1. Inhibition of Tubulin Polymerization

1. Prepare solutions of compounds to be tested in DMSO, at a known concentration if possible. The DMSO concentration in the standard assay is 3%, so a 1-m$M$ solution of ligand will yield a final concentration of 30 µ$M$.

2. Thaw MTP by placing the pellets in the bottom of a small beaker. The warmth of the hand is usually sufficient to melt the pellets. Place the beaker on ice once the melting is complete.

3. Desalt the protein into PME buffer using either a G-25 (medium) chromatography column or 1 mL G-50 (fine) syringe columns. Measure the protein concentration (*see* **Note 8**).

4. Determine the final volume of solution required. For 96-well plates, 200 µL per well is sufficient. Add 10–15% to this value to allow for losses during pipetting.

5. Dilute the protein with PME buffer and with DAPI in $H_2O$ to a final concentration of 1 mg/mL. Add the PME buffer first, mix gently, then add the DAPI solution to yield a final concentration of 20 µ$M$ (1/100th volume). Keep this solution on ice until use.

6. Using a multipipettor, add 189 µL of the protein solution into each well, excluding those wells that will be part of the control. This can be done on the lab bench at room temperature.

7. Prepare the control wells. Do each of these in duplicate. All control wells will contain 189 µL PME with tubulin and DAPI. Background: 6 µL of 1 m$M$ colchicine in DMSO, 5 µL of $H_2O$ (no GTP). No inhibition: 6 µL DMSO, 5 µL of GTP (1 m$M$ final concentration). Complete inhibition: 6 µL of 1 m$M$ colchicine in DMSO, 5 µL of GTP (1 m$M$ final concentration).

8. Prepare the test wells. Add 6 µL of the ligand to be tested to each well. Mix well using the multipipettor (*see* **Note 9**).

9. Preincubate at room temperature for 40 min (*see* **Note 10**).

10. Add 5 µL of 1 m$M$ GTP to each well (to initiate tubulin polymerization) and mix well.

11. Incubate plate at 37°C for 45 min (*see* **Note 11**).

12. Measure fluorescence using plate reader (*see* **Note 12**).

13. Data analysis: determine percent inhibition of polymerization by

$\Delta F(sample)/\Delta F(control) \times 100$

$\Delta F\ control = F(no\ inhibition) - F(complete\ inhibition)$

$\Delta F\ sample = F(sample) - F(complete\ inhibition)$

The fluorescence of the background wells should be comparable to that of the fluorescence of the complete inhibition wells (*see* **Note 13**).

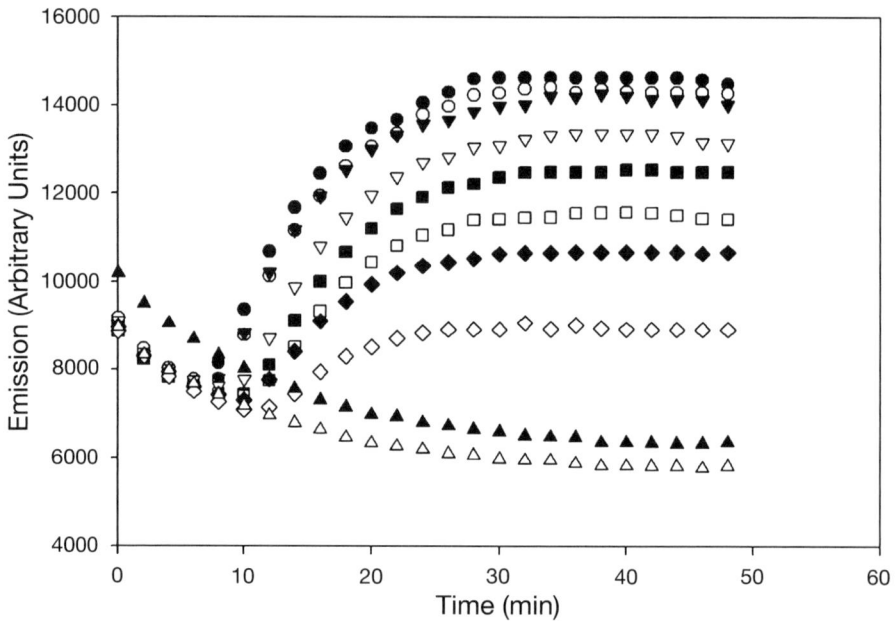

Fig. 1. Inhibition of MTP polymerization by colchicine. The procedure for protocol 1 was followed, except that the incubation in step 11 was performed in the instrument rather than prior to data collection. Note the decrease in fluorescence intensity during the first 10 min of the incubation. This fluorescence change is due to the increase in temperature (room temperature to 37°C). Note also that a steady state is reached by 40 min, and the fluorescence signal remains constant for at least 20 min after steady state is achieved. Data were collected using a Cytofluor® Series 4000 plate reader (settings: excitation 360/40, emission 460/40; the reaction was monitored for 60 min with acquisition of data points at every one min interval). Symbols: filled circles, control, no colchicine; solid triangles, control, complete inhibition (5 μ*M* colchicine); open triangles, background, no GTP. The other curves are MTP containing various concentrations of colchicine (0.01–2.5 μ*M*).

**Figure 1** provides an illustration of data obtained for colchicine using this assay.

### 3.2. Promotion of Tubulin Assembly

1. Prepare solutions of compounds to be tested in DMSO, at a known concentration if possible. The DMSO concentration in the standard assay is 3%, so a 1-m*M* solution of ligand will yield a final concentration of 30 μ*M*.
2. Thaw tubulin by placing the pellets in the bottom of a small beaker. The warmth of the hand is usually sufficient to melt the pellets. Place the beaker on ice once the melting is complete.

3. Desalt the protein into PME buffer using either a G-25 (medium) chromatography column or 1-mL G-50 (fine) syringe columns (*see* **Note 3**). Measure the protein concentration (*see* **Note 8**).
4. Determine the final volume of solution required. For 96-well plates, 200 µL per well is sufficient. Add 10–15% to this value to allow for losses during pipetting.
5. Add 1 m$M$ GTP to PME buffer and dilute the protein with this buffer to obtain a final concentration of 5 µ$M$ of tubulin. Add tubulin to PME buffer first, mix gently, and then add DAPI solution (10 µ$M$, final concentration). Keep this solution on ice until use (*see* **Note 14**).
6. Using a multipipettor, add 194 µL of the protein solution into each well, excluding those wells that will be part of the control. This can be done on the lab bench at room temperature.
7. Prepare the control wells. Do each of these in duplicate. All control wells will contain 194 µL PME with tubulin and DAPI. No assembly: 6 µL DMSO. Complete assembly: 6 µL of 1 m$M$ Taxol in DMSO.
8. Prepare the test wells. Add 6 µL of the ligand to be tested to each well. Mix well (*see* **Note 9**).
9. Incubate plate at 37°C for 40 min (*see* **Note 11**).
10. Measure fluorescence using plate reader. Collect data (*see* **Note 12**).
11. Data analysis: determine percent assembly promotion by

$\Delta$F(sample)/ $\Delta$ F(control) × 100

$\Delta$F control = F(complete assembly) – F(no assembly)

$\Delta$F sample = F(sample) – F(no assembly)

The fluorescence of the background wells should be approximately equal to that of the fluorescence of the no assembly wells. This control is important because DMSO can promote tubulin assembly in the absence of Taxol.

## 4. Notes

1. Microtubule protein is prepared from bovine brain by cycles of assembly and disassembly. Two cycles from the procedure of Williams and Lee *(13)* provides MTP of sufficient purity for these experiments. The MTP is diluted to about 10 mg/mL with PMEG buffer, mixed gently but well, and drop frozen into liquid nitrogen.
2. This buffer is made from stock solutions of each component. All buffer stock solutions are filtered through a 0.45-µm filter prior to storage. A 400-m$M$ stock of PIPES buffer is made from the free acid, which does not dissolve in water. The pH of the solution is raised using 5 $M$ NaOH until pH 6.9 is reached. PIPES stock solution is stored at 4°C. A 50-m$M$ stock solution of $MgSO_4$ is filtered and stored at room temperature. A 100-m$M$ stock solution of EGTA is prepared from the free acid. The pH is adjusted to pH 6.90 using 1 $M$ NaOH prior to filtering. This solution is stored at room temperature. A 50-m$M$ stock solution of GTP is prepared in water and is frozen in 1-mL aliquots. Unused portions of the aliquots may be quickly refrozen for later use. It is convenient to make up a buffer containing just the first three components (PME) and store in the refrigerator. If the

modification described in **Note 10** is performed, a portion of buffer required for the assay can be removed and the proper volume of 50 m*M* GTP added for the day's experiments.

3. A 1 × 20 cm G-25 column is usually sufficient for desalting up to 2 mL of thawed MTP. Up to 200 µL can be desalted using a 1-mL syringe column by the method of Penefsky *(14)*.

4. Colchicine is very soluble in aqueous solution and does not require DMSO. It is important, however, to have an equal concentration of DMSO in all wells containing tubulin, and it is convenient to do this by adding it with the colchicine. Colchicine is prone to photochemical decomposition, which decreases its activity, and solutions should be kept in the dark when not in use.

5. If an absorption plate reader is used, DAPI is omitted from the mixture. It is important to note that colchicine has an absorption maximum near 350 nm, and 350 nm is the optimum wavelength for detection of tubulin assembly. If absorption detection is used, then the intrinsic absorptivity of colchicine must be taken into account. Alternatively, podophyllotoxin rather than colchicine can be used as the control. Podophyllotoxin binds very rapidly to the colchicine site on tubulin, so preincubation is not required. Podophyllotoxin is more active (and more expensive) than colchicine, and its absorption maximum is less than 300 nm.

6. The microtubule protein prepared for procedure A is subjected to phosphocellulose chromatography as described in the Williams and Lee *(13)* procedure prior to drop freezing into liquid nitrogen.

7. Taxol is not soluble in aqueous solution. We use DMSO to prepare Taxol stock solutions. It is important to have an equal concentration of DMSO in all wells containing tubulin.

8. A Bradford assay or other simple assay for protein concentration is acceptable.

9. Care must be taken to avoid introducing bubbles into the wells. Bubbles will produce inaccurate data.

10. This incubation is to allow the drug–tubulin complex to form. Colchicine binds to tubulin at an unusually slow rate, and many colchicine analogs bind equally slowly. The procedure can be modified to omit the first incubation time if podophyllotoxin is used as the standard. In this case, GTP will be included in the buffer in step rather than added separately, and **steps 9** and **10** are omitted.

11. This incubation time is sufficient for steady state to be achieved in the polymerization system.

12. Ideally, the plate reader is temperature controlled and is at the same temperature as the incubation. If the plate reader is not temperature controlled, the plate should be cooled to room temperature (or the appropriate temperature for the plate reader) before measurements are made. We typically collect about 5 min of data. If the system is at a steady state, then each well will show a flat line in the data panel. If the intensity of the reading changes with time, then more data should be taken until a flat line is observed. The most common cause of a change in the intensity in a well as a function of time is temperature fluctuation. Temperature

affects the extent of microtubule assembly, and decreasing the temperature from the incubation will cause a decrease in the amount of assembled protein, but it will reach a steady state at the lower (room) temperature. Altering temperature also affects fluorescence intensity.

13. This control should be performed to ensure that the colchicine worked correctly. Colchicine is prone to photochemical decomposition, which decreases its activity.
14. If an absorption plate reader is used, DAPI is omitted from the mixture

## Acknowledgment

This work was supported by National Institutes of Health grant CA 69571.

## References

1. Jordan, M. A. and Wilson, L. (2004) Microtubules as a target for anticancer drugs. *Nat. Rev. Cancer* **4,** 253–265.
2. Gaskin, F., Cantor, C. R., and Shelanski, M. L. (1974) Turbidimetric studies of the in vitro assembly and dis-assembly porcine neurotubules. *J. Mol. Biol.* **89,** 737–758.
3. Gaskin, F. (1982) Techniques for the study of microtubule assembly in vitro. *Methods Enzymol.* 85, 433–439.
4. Bonne, D., Heusele, C., Simon, C., and Pantaloni, D. (1985) 4',6-Diamidino-2-phenylindole, a fluorescent probe for tubulin and microtubules. *J. Biol. Chem.* 260, 2819–2825.
5. Heusele, C., Bonne, D., and Carlier, M. F. (1987) Is microtubule assembly a biphasic process? A fluorimetric study using 4' 6-diamidino-2-phenylindole as a probe. *Eur. J. Biochem.* 165, 613–620.
6. Huang, Y. T., Huang, D. M., Guh, J. H., Chen, I. L., Tzeng, C. C., and Teng, C. C. (2005) CIL-102 interacts withmicrotubule polymerization and causes mitotic arrest following apoptosis in the human prostate cancer PC-3 cell line. *J. Biol. Chem.* **280,** 2771–2779
7. Lopes, N. M., Miller, H. P., Young, N. D., and Bhuyan, B. K. (1997) Assessment of microtubule stabilizers by semiautomated in vitro microtubule protein polymerization and mitotic block assays. *Cancer Chemother. Pharmacol.* **41,** 37–47.
8. Keminitzer, W., Drewe, J., Jiang, S., et al. (2004) Discovery of 4-aryl-4H-chromosomes as a new series of apoptosis inducers using a cell- and caspase-based high-throughput screening assay. 1. Structure-activity relationships of the 4-aryl group. *J. Med. Chem.* **47,** 6299–6310.
9. Dong, H., Li, Y. Z., and Hu, W. (2004) Analysis of purified tubulin in high concentration of glutamate for application in high throughput screening for microtubule-stabilizing agents. *Assay Drug Dev. Technol.* **2,** 621–628.
10. Hu, W., Dong, H., Li, Y. Z., Han, G. J., and Qu, Y. B. (2004) A high throughput model for screening anti-tumor agnets capable of promoting polymerization of tubulin in vitro. *Acta Pharmacol. Sin.* **25,** 775–782.

11. Barron, D. M., Chatterjee, S. K., Ravindra, R., et al. (2003) A fluorescence-based high-throughput assay for antimicrotubule drugs. *Anal. Biochem.* **315,** 49–56.

12. Bane, S. L. (2006) Tubulin assembly induced by taxol and other microtubule assembly promoters, in *Cell Biology Protocols,* (Harris, R., Graham, J., and Rickwood, D., ed.), John Wiley & Sons, London, UK.

13. Williams, R. C. and Lee, J. C. (1982) Preparation of tubulin from brain. *Methods Enzymol* **85,** 376–385.

14. Penefsky, H. S. (1979) A centrifuged-column procedure for the measurement of ligand binding by beef heart F1. *Methods Enzymol.* **56,** 527–530.

# 20

## Strategies for the Development of Novel Taxol-Like Agents

Susan L. Mooberry

### Summary

Taxol, the first microtubule stabilizer identified, is one of the most important new anticancer drugs to be brought to the clinic in the past 20 yr. The clinical success of Taxol™ led to the development of a second-generation taxane, docetaxel (Taxotere™), and multiple third-generation taxane derivatives are under development. Non-taxane microtubule-stabilizers of diverse chemical structures, including the epothilones and discodermolide, show promising preclinical activities and several epothilones are progressing through clinical trials. One important advantage of the new stabilizers is their ability to circumvent drug resistance mechanisms. The clinical development of these new classes of agents suggests that microtubule stabilizers will continue to be important drugs for the treatment of cancer. This chapter provides a brief history of Taxol and the discovery and development status of other classes of microtubule stabilizers. Although all microtubule-stabilizers share similar mechanisms of action, interesting subtle differences among the stabilizers are being detected. This chapter also provides some strategies for identifying the differences among microtubule stabilizers that may help prioritize them for development and clinical use.

**Key Words:** Taxol; paclitaxel; docetaxel; Taxotere™; microtubule stabilizers; antimitotic agents; epothilones; discodermolide; laulimalides; drug resistance; drug synergism.

## 1. Introduction

Despite great advances in understanding the molecular causes of cancer and the advent of targeted therapy, a critical need for more effective therapies to treat many metastatic cancers remains. Drugs that target microtubules have been used for the treatment of cancer for decades, and they continue to play an important role in the curative and palliative therapy of adult and childhood malignancies (*1*). The first microtubule stabilizers, Taxol and docetaxel, are

From: *Methods in Molecular Medicine, Vol. 137: Microtubule Protocols*
Edited by: Jun Zhou © Humana Press Inc., Totowa, NJ

some of the most important drugs used today for the treatment of a wide range of tumors. Combinations of microtubule stabilizers with molecularly targeted therapies provide additional new therapeutic opportunities. Trastuzumab (Herceptin™), the first molecularly targeted cancer therapy, together with a taxane, is particularly effective against breast cancer *(2)*. Early clinical data suggests that new classes of microtubule stabilizers have remarkable activity against taxane-resistant tumors. The development of structurally diverse microtubule stabilizers with advantages over the taxanes will increase the therapeutic opportunities for cancer therapy. Evidence suggests that microtubule stabilizers will remain important for the treatment of cancer and that new stabilizers might have advantages over the taxanes.

## 2. Discovery and Development of Microtubule Stabilizing Agents
### 2.1. Taxol

The bark of the Pacific yew *Taxus breifolia* was collected in 1962 as part of a large-scale drug discovery effort at the National Cancer Institute (NCI). The extract was found to have cytotoxic activity and the active constituent was isolated in the late 1960s by Drs. Wani and Wall *(3)*. Taxol was evaluated by the NCI and it was found to have an excellent spectrum of activity in vitro and in human xenographs in vivo and in 1977 it was prioritized for preclinical development *(4)*. At this time there were significant concerns about the supply of the compound, its lack of aqueous solubility, and the fact that it was just another antimitotic *(4)*. A breakthrough in the development of taxol occurred in 1979 with the discovery of its unique mechanism of action by Dr. Susan Horwitz et al. *(5,6)*. Unlike the earlier tubulin targeting drugs, which at high concentrations cause microtubule depolymerization, taxol caused an increase in the density of cellular microtubules. This discovery stimulated the clinical development of taxol, as did rational adjustments in drug scheduling and addition of premedications early in clinical evaluations. The supply issue that was thought by some to be an insurmountable problem was solved by the semi-synthesis of Taxol from 10-deacetylbaccatin, the chemical backbone of the complex diterpenoid, which is obtained from cuttings of the abundant Eastern Yew. The name selected for this compound by its discoverers, Drs. Wall and Wani, was trademarked by Bristol Meyers Squibb and they provided the generic name paclitaxel. The clinical successes of Taxol led to the clinical development of a second-generation taxane, docetaxel (Taxotere™). Clinically, docetaxel may have some advantages over Taxol *(7)*. Third-generation taxanes with improved biological properties including better aqueous solubility and lower affinities for transport by P-glycoprotein (Pgp) are in clinical development *(8)*.

## 2.2. New Chemically Diverse Microtubule Stabilizing Agents

The clinical success of Taxol led to a search for other microtubule stabilizing compounds. The first non-taxane microtubule stabilizers identified were the epothilones *(9)*. Epothilones A and B were isolated from a myxobacterium and found to stimulate the polymerization of tubulin in vitro, cause the formation of abnormal interphase and mitotic microtubules, and displace [$^3$H]Taxol from tubulin *(9)*. These data suggest that the epothilones share a binding site with Taxol on mammalian tubulin. In contrast, the epothilones A and B can stabilize yeast tubulin and paclitaxel cannot, indicating some differences in their binding to yeast tubulin *(10)*. The epothilones have an advantage over Taxol in vitro in that they are effective in multidrug-resistant cells, including those that overexpress Pgp *(9,11)*. Epothilones that are or have been in clinical development including epothilone B (EPO 906, Patupilone), aza-epothilone B (BMS-247550, Ixabepilone), epothilone D (KOS862), a semisynthetic derivative of epothilone B (BMS-310705), and ZK310705 *(12)*. New epothilone derivatives with impressive antitumor activities in xenograft models are in preclinical development *(13)*. Early clinical trial data suggest that the epothilones have broad anticancer activity and, importantly, that they are effective against taxane-resistant tumors *(12,14,15)*. The results also show that the epothilones are effective in the same tumor types as the taxanes *(16)*. Although two of the epothilones in clinical trials, epothilone B and aza-epothilone B, differ in structure by one atom, they exhibit different limiting toxicities *(12,16)*. As with the vinca alkaloids, clinical activity or limiting toxicities cannot be predicted on the basis of chemical structure. Although there is great optimism for the eventual clinical use of the epothilones, some questions remain about their ultimate utility *(16)*.

The marine environment has provided the richest source of new microtubule stabilizers with the discovery of discodermolide *(17)*, eleutherobin and the structurally related sarcodictyins A and B *(18,19)*, the laulimalides *(20)*, peloruside A *(21)*, and dictyostatin 1 *(22)*. Discodermolide is a potent stabilizer of cellular microtubules and it initiates microtubule protein assembly under conditions in which Taxol is inactive *(17)*. Discodermolide competitively displaces [$^3$H]Taxol from tubulin, suggesting binding to the taxane-binding site *(23)*. Discodermolide inhibits microtubule dynamics, but in a unique way. It decreases the rate and degree of microtubule shortening, but it does not concomitantly increase the frequency of microtubule catastrophe, a term used to describe a shift between microtubule growth and shrinkage *(24)*. Interestingly, Taxol and discodermolide are synergistic cytotoxins *(25)* that synergistically inhibit microtubule dynamics *(26)*. These data suggest that Taxol and discodermolide have similar, but not identical, mechanisms of action.

Discodermolide was evaluated in phase I clinical trials *(27)* and discodermolide analogs are under preclinical evaluation by Kosan Inc.

The laulimalides were identified in a mechanism-based screening program. Laulimalide is a potent inhibitor of cellular proliferation and it can circumvent drug resistance mediated by overexpression of Pgp and by mutations in the Taxol-binding site *(20,28)*. The polymerization of bovine brain tubulin is promoted by laulimalide, suggesting a direct interaction with tubulin. Pryor et al. *(28)* showed that laulimalide does not displace [$^3$H]Taxol or Flutax-2, a fluorescent Taxol derivative, suggesting that laulimalide does not bind to the Taxol binding site. Synthetic laulimalide analogs that retain the mechanism of action of the natural compound and the ability to circumvent multidrug resistance have been identified *(29)*. Peloruside A and dictyostatin 1 are two other sponge-derived macrolides with potent microtubule stabilizing properties; they are also poor substrates for transport by Pgp *(21,30)*. Similar to laulimalide, peloruside A does not bind to β-tubulin within the taxane binding site *(30)*. In contrast, dictyostatin 1 binds within the taxane site binding site and its chemical macrocyclic structure is proposed to represent the model for the biological active form of discodermolide *(31)*.

Other new classes of microtubule stabilizers include the taccalonolides, the only plant-derived microtubule stabilizers identified since Taxol *(32)* and cyclostreptin (FR1828770, WS9885b), the second microtubule stabilizer isolated from a bacterial source *(33)*. The taccalonolides A and E cause Taxol-like reorganization of interphase and cellular microtubules, mitotic accumulation and initiation of apoptosis. They are effective inhibitors of proliferation and initiate cytotoxicity in cancer cell lines *(32)*. Unlike the other stabilizers identified, taccalonolides A and E do not stimulate the polymerization of bovine brain tubulin or microtubule protein, and the nature of their cellular binding site is under investigation *(34)*. Cyclostreptin causes classic Taxol-like microtubule bundling in cells and mitotic accumulation and it has a high affinity for the taxane-binding site on β-tubulin *(33,34)*. Interestingly, it is relatively inactive at promoting the polymerization of purified tubulin, suggesting subtle differences in its interaction with tubulin *(33)*.

## 3. Circumventing Drug Resistance

### 3.1. Drug Resistance Limits the Activity of the Taxanes

Drug resistance is a serious problem in the treatment of cancer. Some tumors are intrinsically resistant to chemotherapy and others develop resistance following therapy. Several different mechanisms of resistance have been identified for the taxanes including expression of ATP-dependent membrane transport proteins, target-based resistance arising from changes in β-tubulin isotype composition, mutations in β-tubulin that affect drug binding or intrinsic stability, or changes in levels of tubulin-binding proteins that impact microtubule stability (reviewed in **refs. *35*** and ***36***). Deregulation of proteins involved in apoptosis including Bcl-2 and survivin are also linked with taxane resistance *(37–39)*. One major goal for the discovery and clinical development of chemically diverse microtubule stabilizers was the expectation that they would circumvent some of the limitations of the taxanes, including drug resistance *(40)*.

A significant proportion of tumors are multidrug resistant (MDR) because of the aberrant expression of membrane proteins that act as drug efflux pumps, including the ABC (ATP-binding cassette) transport proteins. Overexpression of the *MDR-1* gene product, Pgp (ABCB1), leads to diminished intracellular drug accumulation and to attenuated cytotoxic effects in vitro and in vivo *(41)*. Clinically, the expression of Pgp in both hematologic and solid tumors is associated with poor treatment response to Taxol *(42–44)*. In numerous models and in clinical practice Pgp expression limits the utility of many currently available agents including Taxol and docetaxel. Microtubule-stabilizing agents that can circumvent Pgp-mediated mechanisms of resistance may provide significant advantages for the treatment of taxane-refractory tumors. The epothilones, discodermolide, the laulimalides, peloruside A and the taccalonolides are poor substrates for transport by Pgp in cancer cell lines. The epothilones are effective in Pgp-expressing tumors *(45–47)*. Clinically, the epothilones show activity in taxane-resistant tumors *(12,14,15,48)*, yet whether these differences in sensitivity are due to Pgp expression has not yet been evaluated. In addition to Pgp, other ABC transporters of the MRP family have been implicated in resistance to the taxanes including MRP7 (ABCC10), MDR2 (ABCB4), and BSEP (ABCB11) *(49–51)*. Whether expression of these transporters limits response to treatment is under investigation.

Other mechanisms of resistance linked to microtubule stabilizing agents include mutations in β-tubulin, and aberrant expression of β-tubulin isotypes and tubulin binding proteins *(35,36)*. Resistant cell lines derived from exposure of sensitive cell lines to Taxol or epothilone A and B have been developed and several were found to be resistant because of point mutations in the Taxol binding site on β-tubulin *(52,53)*. Other Taxol-resistant cells have mutations at sites on β-tubulin that are important for lateral protofilament association, and these mutations impair normal microtubule stability *(36,54,55)*. Several clinical studies have evaluated whether β-tubulin mutations lead to clinical taxane resistance. With the exception of one report, the data suggest that mutations in β-tubulin are not associated with either intrinsic or acquired taxane resistance in the clinical setting *(56–60)*. Changes in the expression of specific β-tubulin isotypes are implicated in a second tubulin-based mechanism of taxane resistance *(35)*. Mammals have seven β-tubulin genes and seven different β-tubulin protein isotypes. The tubulin isotypes are highly homologous, but differ primarily in the 10–15 amino acids of the COOH terminus *(61)*. In a variety of cell lines, overexpression of βIII tubulin is associated with resistance to Taxol and other tubulin-binding antimitotic agents *([62,63]*; and reviewed in **ref.** *35*). The cell line data are supported with several biochemical studies that were conducted using tubulin enriched in βIII or lacking βIII. These data show that microtubules composed of βIII are less sensitive to the effects of Taxol and removal of βIII isotype leads to enhanced Taxol-induced microtubule assembly *(64,65)*. Recent mechanistic data show that microtubules from cells with high levels of βIII exhibit normal dynamics but are less susceptible to Taxol-induced suppression of dynamics *(66)*. Clinically, expression of the βIII tubulin isotype in ovarian *(67)*, breast *(68)*, and non-small lung cancer *(69)* is linked with resistance to Taxol. Although many other tubulin-based mechanisms of taxane-resistance have been identified in cell lines, to date only expression of Pgp and the βIII tubulin isotype have been linked with clinical resistance. New microtubule stabilizers are able to circumvent Pgp-mediated resistance in preclinical models, and there are promising data that a third generation taxane, IDN5390, is able to circumvent βIII tubulin-mediated drug resistance in vitro *(70)*.

A clear need exists for new drugs that can circumvent the resistance mechanisms that reduce the effectiveness of the taxanes. Preclinical and emerging clinical data that structurally diverse microtubule stabilizers will be able to bypass some taxane resistance mechanisms has raised expectations. Unfortunately, the emerging clinical data also show that the epothilones have clinical activity in the same tumor types that are sensitive to the taxanes, particularly breast and prostate, but not in tumor types such as colon cancer that are traditionally insensitive to the taxanes *(16)*. Further studies are needed to define the complex nature of drug sensitivity and resistance. This information, together with understanding the susceptibilities of each microtubule stabilizer to the various resistance mechanisms, will allow for the design of rational drug combinations that avoid tumor resistance.

### 3.2. Strategies to Generate Drug-Resistant Cell Lines

Drug-resistant cell lines have proven to be extremely useful to begin to define the various mechanisms of drug resistance that diminish the utility of Taxol. Traditionally, two different methods have been used to create drug-resistant cell lines and the advantages and disadvantages of each method and the resultant resistance mechanisms have been described in detail in a recent review by Fernando Cabral *(35)*. Briefly, the two methods employ a single exposure strategy or multiple step-wise exposures to increasing concentrations ot the targeted drug. Very different results are obtained from the two exposure schemes and yet both are very useful for identifying the molecular sequence of events that occur during acquisition of drug resistance. The single exposure protocols result in cell lines that have a low level of resistance, two- to threefold, due to alterations specific to tubulin *(35,36)*. Multiple-step exposure schemes generated multiple cell lines with point mutations in the taxane-binding site *(52,53)*. Further evaluation of these cell lines led to the discovery that the earliest isolated drug-resistant clones were approx 10-fold resistant and that they had a point mutation in the Taxol binding site of the most prevalent tubulin isotype, βI *(71)*. Clones isolated after later step exposures exhibited 30- to 50-fold resistance and loss of heterozygosity of the second allele of the βI-tubulin gene *(71)*. Similar results have been obtained with other taxane-resistant cell lines, thus demonstrating that the multiple-step exposure protocols cause multiple mutational events and high levels of drug resistance *(36,72,73)*. These drug-resistant cell lines are very useful in identifying several different mechanisms of cellular resistance to Taxol and other microtubule stabilizers, but follow-on studies are critically important to identify which of the various resistance mechanisms are clinically relevant.

## 4. Strategies for Evaluating Subtle Differences in Mechanism of Action of Microtubule Stabilizers

### 4.1. Mechanisms of Action of Microtubule Stabilizers

Chemically diverse compounds can stabilize microtubules and, whereas they appear to share a common general mechanism of action, the mechanisms of action are not identical. The general consensus of the work of many groups over the last 20 yr is that at clinically relevant concentrations Taxol binds to β-tubulin in microtubules and inhibits microtubule dynamics. This interrupts mitotic progression during the transition from metaphase to anaphase, leading ultimately to initiation of apoptosis (reviewed in **refs.** *74* and *75*). Although Taxol at high concentrations dramatically shifts the equilibrium of cellular tubulin to the polymerized state, it is now believed that these changes in microtubule polymerization are not the basis of the antimitotic actions of Taxol. The work of Mary Ann Jordan and Leslie Wilson and their collaborators in a series of studies showed that the ability of Taxol to inhibit the dynamic instability of microtubules at low clinically relevant concentrations was likely the basis for its antiproliferative and cytotoxic effects *(76–78)*.

Early studies show that the actions of microtubule stabilizers are not necessarily identical. Discodermolide, although it binds within the same binding site as the taxanes, has a slightly different mechanism of inhibiting microtubule dynamics. Discodermolide suppresses microtubule dynamics by decreasing the rate and degree of microtubule shortening, and in these effects it is similar to Taxol *(24)*. However, discodermolide, unlike Taxol and epothilone B, does not concomitantly increase the frequency of microtubule catastrophe and it actually increased the frequency of catastrophe rescue *(24)*. These data suggest that paclitaxel and discodermolide have similar, but not identical, mechanisms of action. As we learn more about the signaling pathways initiated by microtubule stabilizers that lead to apoptosis and the subtle variations among the stabilizers and their effects on these pathways, this too will provide the opportunity to identify slight distinctions in their mechanisms of action.

Combinations of drugs with complementary actions are the mainstay of modern cancer chemotherapy. Identification of differences in cellular targets and mechanisms of action provides the scientific rationale for evaluating specific combinations of microtubule stabilizers for synergistic actions that might ultimately lead to rational combinations for more effective chemotherapy.

### 4.2. Synergism Studies to Detect Slight Differences in Mechanism

All microtubule stabilizers appear to share a common global mechanism of action but the detection of minor mechanistic variations is significant because it might lead to the identification of mechanistic differences. The discovery of the ability of discodermolide and Taxol to act as synergistic cytotoxins by Susan Horwitz's laboratory opened new horizons for understanding the importance of slight mechanic differences among microtubule stabilizers. Although the combination of discodermolide and Taxol caused synergistic effects, no synergism was observed when Taxol was used in combination with the epothilones. These pivotal synergism studies led to detailed mechanistic evaluations of the nature of the synergistic activities of these two drugs. The results showed that discodermolide and Taxol can synergistically inhibit microtubule dynamics because they have slightly different mechanisms of inhibiting dynamic instability *(26)*.

Initial studies to evaluate the ability of microtubule stabilizers to act synergistically can provide the opportunity to detect mechanistic differences. It will be very interesting to evaluate the effects of combinations of microtubule stabilizers that bind to tubulin at different binding sites. Initial studies show that laulimalide and Taxol synergistically stimulate the assembly of purified tubulin. Whether the combination of laulimalide and Taxol will yield synergistic cytotoxicity actions is also of interest. It is anticipated that the results of these investigations will initiate further studies to identify mechanistic differences among the various classes of microtubule stabilizers. Synergism evaluations are relatively straightforward and a number of different analyses are available to define synergism. Helpful reviews detailing the design and conduct of experiments for these types of studies are available *(79–82)*. Additionally, a commercially available program can be helpful in calculating combination index values. It is critically important that investigators understand the definition of synergism before embarking on these types of analyses. Synergism is defined as an effect greater than the additive effect anticipated by each compound when used alone, i.e., a superadditive effect *(83)*. Secondarily, it is important to note in the design of synergism experiments that calculation of combination indices requires the generation of dose–response curves for each agent alone and then again for the two agents in combination *(82)*. Synergism studies, in addition to their utility in identifying mechanistic differences among microtubule stabilizers, can also be used to identify drug combinations that should be investigated in vivo for antitumor effects as a prelude for potential application in the clinic.

## References

1. Rowinsky, E. K. and Tolcher, A. W. (2001) Antimicrotubule agents, in *Cancer Principles and Practice of Oncology, Vol. 1,* (DeVita, V. T. J., Hellman, S., and Rosenberg, S. A., eds.), Lippincott, Williams and Wilkins, Philadelphia, PA, pp. 431–447.
2. Pegram, M. D., Konecny, G. E., O'Callaghan, C., Beryt, M., Pietras, R., and Slamon, D. J. (2004) Rational combinations of trastuzumab with chemotherapeutic drugs used in the treatment of breast cancer. *J. Natl. Cancer Inst.* **96,** 739–749.
3. Wani, M. C., Taylor, H. L., Wall, M. E., Coggon, P., and McPhail, A. T. (1971) Plant antitumor agents. VI. The isolation and structure of taxol, a novel antileukemic and antitumor agent from *Taxus brevifolia. J. Am. Chem. Soc.* **93,** 2325–2327.
4. Wall, M. E. and Wani, M. C. (1995) Camptothecin and taxol: discovery to clinic— thirteenth Bruce F. Cain Memorial Award Lecture. *Cancer Res.* **55,** 753–760.
5. Schiff, P. B., Fant, J., and Horwitz, S. B. (1979) Promotion of microtubule assembly in vitro by taxol. *Nature* **277,** 665–667.
6. Schiff, P. B. and Horwitz, S. B. (1980) Taxol stabilizes microtubules in mouse fibroblast cells. *Proc. Natl. Acad. Sci. USA* **77,** 1561–1565.
7. Crown, J., O'Leary, M., and Ooi, W. S. (2004) Docetaxel and paclitaxel in the treatment of breast cancer: a review of clinical experience. *Oncologist* **9,** 24–32.
8. Camps, C., Felip, E., Sanchez, J. M., et al. (2005) Phase II trial of the novel taxane BMS-184476 as second-line in non-small-cell lung cancer. *Ann. Oncol.* **16,** 597–601.
9. Bollag, D. M., McQueney, P. A., Zhu, J., et al. (1995) Epothilones, a new class of microtubule-stabilizing agents with a taxol-like mechanism of action. *Cancer Res.* **55,** 2325–2333.
10. Bode, C. J., Gupta, M. L., Jr., Reiff, E. A., Suprenant, K. A., Georg, G. I., and Himes, R. H. (2002) Epothilone and paclitaxel: unexpected differences in promoting the assembly and stabilization of yeast microtubules. *Biochemistry* **41,** 3870–3874.

11. Kowalski, R. J., Giannakakou, P., and Hamel, E. (1997) Activities of the microtubule-stabilizing agents epothilones A and B with purified tubulin and in cells resistant to paclitaxel (Taxol™). *J. Biol. Chem.* **272,** 2534–2541.

12. Goodin, S., Kane, M. P., and Rubin, E. H. (2004) Epothilones: mechanism of action and biologic activity. *J. Clin. Oncol.* **22,** 2015–2025.

13. Chou, T. C., Dong, H., Zhang, X., Tong, W. P., and Danishefsky, S. J. (2005) Therapeutic cure against human tumor xenografts in nude mice by a microtubule stabilization agent, fludelone, via parenteral or oral route. *Cancer Res.* **65,** 9445–9454.

14. Low, J. A., Wedam, S. B., Lee, J. J., et al. (2005) Phase II clinical trial of ixabepilone (BMS-247550), an epothilone B analog, in metastatic and locally advanced breast cancer. *J. Clin. Oncol.* **23,** 2726–2734.

15. Galsky, M. D., Small, E. J., Oh, W. K., et al. (2005) Multi-institutional randomized phase II trial of the epothilone B analog ixabepilone (BMS-247550) with or without estramustine phosphate in patients with progressive castrate metastatic prostate cancer. *J. Clin. Oncol.* **23,** 1439–1446.

16. de Jonge, M. and Verweij, J. (2005) The epothilone dilemma. *J. Clin. Oncol.* **23,** 9048–9050.

17. ter Haar, E., Kowalski, R. J., Hamel, E., et al. (1996) Discodermolide, a cytotoxic marine agent that stabilizes microtubules more potently than taxol. *Biochemistry* **35,** 243–250.

18. Long, B., Carboni, J., Wasserman, A., et al. (1998) Eleutherobin, a novel cytotoxic agent that induces tubulin polymerization, is similar to paclitaxel (Taxol). *Cancer Res.* **58,** 1111–1115.

19. Hamel, E., Sackett, D. L., Vourloumis, D., and Nicolaou, K. C. (1999) The coral-derived natural products eleutherobin and sarcodictyins A and B: effects on the assembly of purified tubulin with and without microtubule-associated proteins and binding at the polymer taxoid site. *Biochemistry* **38,** 5490–5498.

20. Mooberry, S. L., Tien, G., Hernandez, A. H., Plubrukarn, A., and Davidson, B. S. (1999) Laulimalide and isolaulimalide, new paclitaxel-like microtubule-stabilizing agents. *Cancer Res.* **59,** 653–660.

21. Hood, K. A., West, L. M., Rouwe, B., et al. (2002) Peloruside A, a novel antimitotic agent with paclitaxel-like microtubule-stabilizing activity. *Cancer Res.* **62,** 3356–3360.

22. Isbrucker, R. A., Cummins, J., Pomponi, S. A., Longley, R. E., and Wright, A. E. (2003) Tubulin polymerizing activity of dictyostatin-1, a polyketide of marine sponge origin. *Biochem. Pharmacol.* **66,** 75–82.

23. Kowalski, R. J., Giannakakou, P., Gunasekera, S. P., Longley, R. E., Day, B. W., and Hamel, E. (1997) The microtubule-stabilizing agent discodermolide competitively inhibits the binding of paclitaxel (Taxol) to tubulin polymers, enhances tubulin nucleation reactions more potently than paclitaxel, and inhibits the growth of paclitaxel-resistant cells. *Mol. Pharmacol.* **52,** 613–622.

24. Honore, S., Kamath, K., Braguer, D., Wilson, L., Briand, C., and Jordan, M. A. (2003) Suppression of microtubule dynamics by discodermolide by a novel

mechanism is associated with mitotic arrest and inhibition of tumor cell proliferation. *Mol. Cancer Ther.* **2,** 1303–1311.

25. Martello, L. A., McDaid, H. M., Regl, D. L., et al. (2000) Taxol and discodermolide represent a synergistic drug combination in human carcinoma cell lines. *Clin. Cancer Res.* **6,** 1978–1987.

26. Honore, S., Kamath, K., Braguer, D., et al. (2004) Synergistic suppression of microtubule dynamics by discodermolide and paclitaxel in non-small cell lung carcinoma cells. *Cancer Res.* **64,** 4957–4964.

27. Mita, A., Lockhart, A., Chen, T. -L., et al. (2004) A phase I pharmacokinetic (PK) trial of XAA296A (discodermolide) administered every 3 wks to adult patients with advanced solid malignancies. *J. Clin. Oncol.* **22,** 2025.

28. Pryor, D. E., O'Brate, A., Bilcer, G., et al. (2002) The microtubule stabilizing agent laulimalide does not bind in the taxoid site, kills cells resistant to paclitaxel and epothilones, and may not require its epoxide moiety for activity. *Biochemistry* **41,** 9109–9115.

29. Mooberry, S. L., Randall-Hlubek, D. A., Leal, R. M., et al. (2004) Microtubule-stabilizing agents based on designed laulimalide analogues. *Proc. Natl. Acad. Sci. USA* **101,** 8803–8808.

30. Gaitanos, T. N., Buey, R. M., Diaz, J. F., et al. (2004) Peloruside A does not bind to the taxoid site on beta-tubulin and retains its activity in multidrug-resistant cell lines. *Cancer Res.* **64,** 5063–5067.

31. Madiraju, C., Edler, M. C., Hamel, E., et al. (2005) Tubulin assembly, taxoid site binding, and cellular effects of the microtubule-stabilizing agent dictyostatin. *Biochemistry* **44,** 15,053–15,063.

32. Tinley, T. L., Randall-Hlubek, D. A., Leal, R. M., et al. (2003) Taccalonolides E and A: plant-derived steroids with microtubule-stabilizing activity. *Cancer Res.* **63,** 3211–3220.

33. Edler, M. C., Buey, R. M., Gussio, R., et al. (2005) Cyclostreptin (FR182877), an antitumor tubulin-polymerizing agent deficient in enhancing tubulin assembly despite its high affinity for the taxoid site. *Biochemistry* **44,** 11,525–11,538.

34. Buey, R. M., Barasoain, I., Jackson, E., et al. (2005) Microtubule interactions with chemically diverse stabilizing agents: thermodynamics of binding to the Paclitaxel site predicts cytotoxicity. *Chem. Biol.* **12,** 1269–1279.

35. Cabral, F. (2001) Factors determining cellular mechanisms of resistance to antimitotic drugs. *Drug Resist. Updat.* **4,** 3–8.

36. Orr, G. A., Verdier-Pinard, P., McDaid, H., and Horwitz, S. B. (2003) Mechanisms of Taxol resistance related to microtubules. *Oncogene* **22,** 7280–7295.

37. Gazitt, Y., Rothenberg, M. L., Hilsenbeck, S. G., Fey, V., Thomas, C., and Montegomrey, W. (1998) Bcl-2 overexpression is associated with resistance to paclitaxel, but not gemcitabine, in multiple myeloma cells. *Int. J. Oncol.* **13,** 839–848.

38. Buchholz, T. A., Davis, D. W., McConkey, D. J., et al. (2003) Chemotherapy-induced apoptosis and Bcl-2 levels correlate with breast cancer response to chemotherapy. *Cancer J.* **9,** 33–41.

39. Zhou, J., O'Brate, A., Zelnak, A., and Giannakakou, P. (2004) Survivin deregulation in beta-tubulin mutant ovarian cancer cells underlies their compromised mitotic response to taxol. *Cancer Res.* **64,** 8708–8714.
40. Lee, J. J. and Swain, S. M. (2005) Development of novel chemotherapeutic agents to evade the mechanisms of multidrug resistance (MDR). *Semin. Oncol.* **32,** S22–S26.
41. Borst, P. and Elferink, R. O. (2002) Mammalian ABC transporters in health and disease. *Annu. Rev. Biochem.* **71,** 537–592.
42. Chiou, J. F., Liang, J. A., Hsu, W. H., Wang, J. J., Ho, S. T., and Kao, A. (2003) Comparing the relationship of Taxol-based chemotherapy response with P-glycoprotein and lung resistance-related protein expression in non-small cell lung cancer. *Lung* **181,** 267–273.
43. Penson, R. T., Oliva, E., Skates, S. J., et al. (2004) Expression of multidrug resistance-1 protein inversely correlates with paclitaxel response and survival in ovarian cancer patients: a study in serial samples. *Gynecol. Oncol.* **93,** 98–106.
44. Yeh, J. J., Hsu, W. H., Wang, J. J., Ho, S. T., and Kao, A. (2003) Predicting chemotherapy response to paclitaxel-based therapy in advanced non-small-cell lung cancer with P-glycoprotein expression. *Respiration* **70,** 32–35.
45. Chou, T. C., Zhang, X. G., Harris, C. R., et al. (1998) Desoxyepothilone B is curative against human tumor xenografts that are refractory to paclitaxel. *Proc. Natl. Acad. Sci. USA* **95,** 15,798–15,802.
46. Hofstetter, B., Vuong, V., Broggini-Tenzer, A., et al. (2005) Patupilone acts as radiosensitizing agent in multidrug-resistant cancer cells *in vitro* and *in vivo.* *Clin. Cancer Res.* **11,** 1588–1596.
47. Lee, F. Y., Borzilleri, R., Fairchild, C. R., et al. (2001) BMS-247550: a novel epothilone analog with a mode of action similar to paclitaxel but possessing superior antitumor efficacy. *Clin. Cancer Res.* **7,** 1429–1437.
48. Eng, C., Kindler, H. L., Nattam, S., et al. (2004) A phase II trial of the epothilone B analog, BMS-247550, in patients with previously treated advanced colorectal cancer. *Ann. Oncol.* **15,** 928–932.
49. Ross, D. D. and Doyle, L. A. (2004) Mining our ABCs: pharmacogenomic approach for evaluating transporter function in cancer drug resistance. *Cancer Cell* **6,** 105–107.
50. Fojo, T. and Bates, S. (2003) Strategies for reversing drug resistance. *Oncogene* **22,** 7512–7523.
51. Hopper-Borge, E., Chen, Z. S., Shchaveleva, I., Belinsky, M. G., and Kruh, G. D. (2004) Analysis of the drug resistance profile of multidrug resistance protein 7 (ABCC10): resistance to docetaxel. *Cancer Res.* **64,** 4927–4930.
52. Giannakakou, P., Sackett, D. L., Kang, Y. K., et al. (1997) Paclitaxel-resistant human ovarian cancer cells have mutant beta-tubulins that exhibit impaired paclitaxel-driven polymerization. *J. Biol. Chem.* **272,** 17,118–17,125.
53. Giannakakou, P., Gussio, R., Nogales, E., et al. (2000) A common pharmacophore for epothilone and taxanes: molecular basis for drug resistance conferred by tubulin mutations in human cancer cells. *Proc. Natl. Acad. Sci USA* **97,** 2904–2909.

54. Gonzalez-Garay, M. L., Chang, L., Blade, K., Menick, D. R., and Cabral, F. (1999) A beta-tubulin leucine cluster involved in microtubule assembly and paclitaxel resistance. *J. Biol. Chem.* **274**, 23,875–23,882.

55. Schibler, M. J. and Cabral, F. (1986) Taxol-dependent mutants of Chinese hamster ovary cells with alterations in alpha- and beta-tubulin. *J. Cell Biol.* **102**, 1522–1531.

56. Sale, S., Sung, R., Shen, P., et al. (2002) Conservation of the class I beta-tubulin gene in human populations and lack of mutations in lung cancers and paclitaxel-resistant ovarian cancers. *Mol. Cancer Ther.* **1**, 215–225.

57. Kelley, M. J., Li, S., and Harpole, D. H. (2001) Genetic analysis of the beta-tubulin gene, TUBB, in non-small-cell lung cancer. *J. Natl. Cancer Inst.* **93**, 1886–1888.

58. Kohonen-Corish, M. R., Qin, H., Daniel, J. J., et al. (2002) Lack of beta-tubulin gene mutations in early stage lung cancer. *Int. J. Cancer* **101**, 398–399.

59. Tsurutani, J., Komiya, T., Uejima, H., et al. (2002) Mutational analysis of the beta-tubulin gene in lung cancer. *Lung Cancer* **35**, 11–16.

60. Monzo, M., Rosell, R., Sanchez, J. J., et al. (1999) Paclitaxel resistance in non-small-cell lung cancer associated with beta-tubulin gene mutations. *J. Clin. Oncol.* **17**, 1786–1793.

61. Luduena, R. F. (1998) Multiple forms of tubulin: different gene products and covalent modifications. *Int. Rev. Cytol.* **178**, 207–275.

62. Kavallaris, M., Kuo, D. Y., Burkhart, C. A., et al. (1997) Taxol-resistant epithelial ovarian tumors are associated with altered expression of specific beta-tubulin isotypes. *J. Clin. Invest.* **100**, 1282–1293.

63. Hari, M., Yang, H., Zeng, C., Canizales, M., and Cabral, F. (2003) Expression of class III beta-tubulin reduces microtubule assembly and confers resistance to paclitaxel. *Cell Motil. Cytoskeleton* **56**, 45–56.

64. Lu, Q. and Luduena, R. F. (1993) Removal of beta III isotype enhances taxol induced microtubule assembly. *Cell Struct. Funct.* **18**, 173–182.

65. Derry, W. B., Wilson, L., Khan, I. A., Luduena, R. F., and Jordan, M. A. (1997) Taxol differentially modulates the dynamics of microtubules assembled from unfractionated and purified beta-tubulin isotypes. *Biochemistry* **36**, 3554–3562.

66. Kamath, K., Wilson, L., Cabral, F., and Jordan, M. A. (2005) Beta III-tubulin induces paclitaxel resistance in association with reduced effects on microtubule dynamic instability. *J. Biol. Chem.* **280**, 12,902–12,907.

67. Mozzetti, S., Ferlini, C., Concolino, P., et al. (2005) Class III beta-tubulin overexpression is a prominent mechanism of paclitaxel resistance in ovarian cancer patients. *Clin. Cancer Res.* **11**, 298–305.

68. Paradiso, A., Mangia, A., Chiriatti, A., et al. (2005) Biomarkers predictive for clinical efficacy of taxol-based chemotherapy in advanced breast cancer. *Ann. Oncol.* **16**, iv14–iv19.

69. Seve, P., Mackey, J., Isaac, S., et al. (2005) Class III beta-tubulin expression in tumor cells predicts response and outcome in patients with non-small cell lung cancer receiving paclitaxel. *Mol. Cancer Ther.* **4**, 2001–2007.

70. Ferlini, C., Raspaglio, G., Mozzetti, S., et al. (2005) The seco-taxane IDN5390 is able to target class III beta-tubulin and to overcome paclitaxel resistance. *Cancer Res.* **65,** 2397–2405.

71. Wang, Y., O'Brate, A., Zhou, W., and Giannakakou, P. (2005) Resistance to microtubule-stabilizing drugs involves two events: beta-tubulin mutation in one allele followed by loss of the second allele. *Cell Cycle* **4,** 1847–1853.

72. Martello, L. A., Verdier-Pinard, P., Shen, H. J., et al. (2003) Elevated levels of microtubule destabilizing factors in a Taxol-resistant/dependent A549 cell line with an alpha-tubulin mutation. *Cancer Res.* **63,** 1207–1213.

73. He, L., Yang, C. P., and Horwitz, S. B. (2001) Mutations in beta-tubulin map to domains involved in regulation of microtubule stability in epothilone-resistant cell lines. *Mol. Cancer Ther.* **1,** 3–10.

74. Jordan, M. A. and Wilson, L. (2004) Microtubules as a target for anticancer drugs. *Nat. Rev. Cancer* **4,** 253–265.

75. Zhou, J. and Giannakakou, P. (2005) Targeting microtubules for cancer chemotherapy. *Curr. Med. Chem. Anti-Canc. Agents* **5,** 65–71.

76. Jordan, M. A., Toso, R. J., Thrower, D., and Wilson, L. (1993) Mechanism of mitotic block and inhibition of cell proliferation by taxol at low concentrations. *Proc. Natl. Acad. Sci. USA* **90,** 9552–9556.

77. Jordan, M. A., Wendell, K., Gardiner, S., Derry, W. B., Copp, H., and Wilson, L. (1996) Mitotic block induced in HeLa cells by low concentrations of paclitaxel (Taxol) results in abnormal mitotic exit and apoptotic cell death. *Cancer Res.* **56,** 816–825.

78. Yvon, A. M., Wadsworth, P., and Jordan, M. A. (1999) Taxol suppresses dynamics of individual microtubules in living human tumor cells. *Mol. Biol. Cell* **10,** 947–959.

79. Tallarida, R. J. (2002) The interaction index: a measure of drug synergism. *Pain* **98,** 163–168.

80. Tallarida, R. J., Stone, D. J., Jr., and Raffa, R. B. (1997) Efficient designs for studying synergistic drug combinations. *Life Sci.* **61,** PL 417–425.

81. Grabovsky, Y. and Tallarida, R. J. (2004) Isobolographic analysis for combinations of a full and partial agonist: curved isoboles. *J. Pharmacol. Exp. Ther.* **310,** 981–986.

82. Chou, T. C. and Talalay, P. (1984) Quantitative analysis of dose-effect relationships: the combined effects of multiple drugs or enzyme inhibitors. *Adv. Enzyme Regul.* **22,** 27–55.

83. Tallarida, R. J. (2001) Drug synergism: its detection and applications. *J. Pharmacol. Exp. Ther.* **298,** 865–872.

# Index